MAKING SENSE OF CITIES

A GEOGRAPHICAL SURVEY

Blair Badcock

Policy Manager, National Office,
Housing New Zealand Corporation,
Wellington, New Zealand

HODDER
EDUCATION
AN HACHETTE UK COMPANY

First published in Great Britain in 2002 by
Hodder Education, an Hachette UK Company,
338 Euston Road, London NW1 3BH

http://www.hoddereducation.co.uk

British Library Cataloguing in Publication Data
A catalogue record for this book is available from the British
Library

Library of Congress Cataloging-in-Publication Data
A catalog record for this book is available from the Library
of Congress

ISBN 978 0 340 74224 2

Typeset in 9.5/11.5 pt Palatino by Macmillan Publishing Solutions
(www.macmillansolutions.com)

What do you think about this book? Or any other Hodder Education title?
Please send your comments to www.hoddereducation.co.uk

CONTENTS

PREFACE

Apart from the other formalities normally found in a preface, this one tells instructors and students what kind of book *Making Sense of Cities* is trying to be.

Making Sense of Cities has been developed to serve as a new-generation urban geography text for use in undergraduate programmes. It aims to fulfil the needs of students coming to the subject matter of urban geography for the first time, presumably after a first year course in human geography. Most of the contents are drawn from the second and third year courses currently taught in Australian universities. The text will also be useful in related disciplines where there is a need to build upon a systematic and up-to-date working knowledge of urbanisation and cities. It offers a solid foundation for later coursework in disciplines like urban and regional planning, sociology, environmental studies, urban design and architecture, and maybe political studies and economics.

On completion of *Unfairly Structured Cities* in 1984, I promised myself that I would never take on the task of writing another textbook! So why did I finally give in to a persistent commissioning editor way back in June 1997, knowing full well what would be involved? Especially when urban geography, having given ground to 'new cultural geography' in the 1990s, seems to have lost its way a bit in geography curricula on both sides of the Atlantic. With urban geography currently suffering somewhat from a crisis of confidence of sorts, it is fair to ask then, 'What is urban geography these days?' At a time when cities have never been more significant in shaping the material and environmental conditions that now govern the lives of the majority of humankind, this surely must strike the outsider as a peculiar state of affairs within geography.

First, the opportunity to write a new-generation urban geography text coincides with the advent of what some commentators anticipate will be the 'urban millennium'. The forces of urbanisation and burgeoning mega-cities will dominate the organisation of human affairs as never before. But, despite the unbridled faith that some futurists place in technological solutions, the massing of population in cities globally is bound to become increasingly unsustainable, in both human and environmental terms. While counter-urbanisation might be helping to relieve the pressure on some cities in affluent societies, the transport, housing and environmental systems that are crucial to the efficient functioning of a city are typically overloaded and close to breaking point in parts of Asia, Africa and Latin America.

In view of this there is a need for a confidently argued urban text that demonstrates just how pivotal the study of cities remains for a discipline like geography. This to me is the real intellectual challenge ahead of urban geographers, for it requires no less than a reconciliation of the debates around the relative importance of economics, politics and culture for a global population in the twenty-first century that will be overwhelmingly urban. How can human-geography teachers with the responsibility of educating coming generations of students afford to neglect the urban? If they do, it will be at the risk of increasing irrelevance in policy circles, whether it be at the level where the big global issues are played out or at the level where local communities struggle to work out their own affairs. It follows that an understanding of urbanisation processes and how cities develop, their place in the economy, their impact on the environment, and how they sustain social life, is going to be vital to managing the future growth of cities and improving living conditions in them. Quite simply, then, this is to assert the value of a problem-centred approach to the study of cities.

Educationally speaking, what does the particular approach to urban geography adopted in *Making Sense of Cities* have to offer students? Geographers accept that the way cities are organised spatially can tell us a great deal about the economic, political and cultural systems of the societies in which they have evolved. Cities exhibit, often with stunning clarity, the outcome of broad social processes. For example, the effects of economic restructuring, together with the politics of public policy, social conflict and changing gender relations, are etched and imprinted in the geography of cities. This is a geographical perspective that boldly claims, 'It all comes together in cities.' Not only does this hark back in a way to geographers' time-honoured mission as 'grand synthesisers'; this places them in the company of environmentalists, who are also conscious of the systemic interrelatedness of all aspects of human endeavour and nature.

Taking this approach ultimately involves making an educational judgement, which at the same time is epistemologically 'loaded', in relation to how urbanisation and the geography of cities are to be represented in this text. Does one set out to develop the unifying themes and to identify those aspects that cities share in common? Or does one pursue the themes of distinctiveness and difference, and forgo the search for generalisation? In opting for a cross-cultural comparative analysis of cities, this textbook presents evidence of both convergent and divergent forms of urbanisation and urbanism, and highlights both commonalities and differences in outcomes with respect to patterns and processes in cities. I aim to show that an integrative approach to the study of cities can sensibly go hand in hand with acknowledging 'diversity and difference'. But the lasting impression I want students to take away from *Making Sense of Cities* is of the patterns and processes that generally apply to urbanisation and cities.

Secondly, now that we have just rounded the corner to a new millennium, this is a fascinating time to be interpreting events on the world stage and how they are impacting upon cities and urban communities. Part of the great appeal of studying urban change and the spatial organisation of human activity in cities is that it raises deeper questions about the shaping of society that simply cannot be avoided. Any new text, even if confined largely to the needs of English-speaking students, has to reflect the realities of a world that is growing increasingly interdependent. While geography texts have traditionally been dominated by Anglo-American experience, in a global context the Asia-Pacific region is becoming more and more pivotal – economically and culturally. Environmental stresses are going to press these cities as never before. This requires a shift in orientation without overlooking the impact of recent developments in an expanded European region. UK cities, for example, are in the throes of changes that can be traced to the direct and indirect effects of integration into the European Union.

Yet for us to make sense of the urban transformations taking place, they also have to be measured against the geographies of cities past and present. However, despite the overwhelming tendency to eurocentrism in texts written by English speakers, this writer has nevertheless erred on the side of selecting case studies that students on both sides of the Atlantic will mostly be able to relate to. Within these practical constraints, I have tried to select materials that are vibrant, relevant, and inclusive of the experience of other settings and cultures. Extensive use is made of material sourced from current periodicals like the *Guardian Weekly*, The *Economist* and several daily newspapers. No one can do justice to urban studies these days without keeping up to date with current affairs around the globe.

Thirdly, in calling this textbook *Making Sense of Cities* I want to signal an intellectual commitment to the Enlightenment project without getting mired in an epistemological morass. Students ought to be alive to the fact that the sense we make of the natural and social world around us is the product of our powers of reasoning. Yet the knowledge acquired through this process always remains open to question. Naturally our ideas are going to be contested – and so they should be! Indeed, the history of ideas suggests that the only certainty is that eventually a mode of thought or an idea will be overtaken. Nevertheless, there are significant ideas and content within urban geography from earlier generations that are still relevant and ought to be carried forward into a new text. On this basis I am convinced that, whatever the intellectual fashions of the day, there needs to be a clear structure and sure hand steering students through an introductory text so long, as teachers, we remember the cautionary wisdom of Zen: 'Use your finger to point to the moon, but don't mistake your finger for the moon.'

This is a good time to be developing a new urban text because, by now, the relative importance of globalisation, economics, politics and culture has at least had a good airing in the social sciences, and we also have the benefit of a variety of post-structuralist and feminist critiques of what has gone before. Gender issues need to be integrated into all discourse, and not simply because a majority of students are likely to be women. As well as this, after the economic fundamentalism of the 1980s and 1990s, governments and their advisers in countries like the UK, Canada, Australia and New Zealand have come to realise that the limit has been reached in the pursuit of market solutions. Although decidedly problematic, the search is on for a 'Third Way' to manage better economic and social policy, urban and regional development, and the affairs of communities within cities. The UK's Prime Minister, Tony Blair, and perhaps the strongest advocate of a 'Third Way', has even gained a measure of support for his government's reforms from some of Europe's most powerful social democrats. This all has implications for the future management of cities and the well-being of their citizens. One is now able to be much more confident about the kind of balance to strike in relation to these debates since some time has elapsed since they first began.

Fourthly, there is one other development in the realm of ideas that helped to spark my enthusiasm for this task. When I was first approached to write this book, there were signs that the suspension of ethical judgement that had hitherto blighted so much writing in a post-modernist vein was under challenge. I must say I wouldn't be able to muster much enthusiasm for an assignment that didn't hold the spotlight on issues of social and environmental justice in cities. The point, ultimately, of trying to make sense of cities is not for its own sake, but to advance ideas that can help to contribute to an improvement in the human condition. In my experience students appreciate critical analysis that is followed up by canvassing options for bringing about progressive change. Students should be aware of, and encouraged by, the fact that there are cities where urban managers and communities are successfully tackling some of the same seemingly intractable problems that beset their metro area or neighbourhood.

This book was conceived and written during a topsy-turvy time. It would have been impossible to carry on at work and rebuild my life after Carole's death without the understanding of very caring departmental colleagues and lots of treasured friends. Simon and Jonathan were instrumental in urging me to press ahead with the writing. Leaving Adelaide after 30 years to return to New Zealand opens up new possibilities for me, but not without some cost to those left behind. With the decision to move from university life to a public service position, I also owe a special debt of thanks to the generation of students who have elected to take my courses over the years. Their own career moves will continue to be watched with a kind of paternal pride. And, for me, the time has come to show that I've also got what it takes to 'walk the talk' …

On the production side, I have been helped by two very professional members of the Department of Geographical and Environmental Studies at Adelaide University, Mrs Joanna Rillo (tables) and Mrs Chris Crothers (figures). I have worked very closely with Mrs Marian Browett on a series of research projects over the years and she must take the credit for some of the quite complex computations sitting behind Figure 6.7. I am grateful to the Institute of Geography at Victoria University in Wellington for providing a Visiting Fellowship and other logistical support whilst writing. Dr Phillip Morrison at 'Vic' has been especially supportive and Chapter 5 is much the better for Phil's generous offer of access to a manuscript that has subsequently appeared in *Urban Studies*. Discussions held with Professor Chris Hamnett and Dr Loretta Lees at University College London in April 2000 reassured me that I was getting at least some things right! The three commissioning editors associated with this book contract – Laura McKelvie, Luciana O'Flaherty and Liz Gooster – have shown extraordinary patience and forbearance during its gestation. In the latter stages, Liz Gooster and her team at Arnold have been a constant source of advice and practical support.

Finally, this book is dedicated to two special women who were at school together in the 1950s. Without their love and encouragement, this book would never have been written. Carole's will to live inspired me to write it; Rosemary's healing hand made it possible.

INTRODUCTION

Our experience asks to be understood.

(from James K. Baxter's poem *Autumn Testament*)

What matters is that we try to understand what is going on ...

(Chris Hamnett 1999a, 255)

Urban life muddles through the pace of history. When this pace accelerates, cities – and their people – become confused, spaces turn threatening, and meaning escapes from experience. In such disconcerting, yet magnificent times, knowledge becomes the only source to restore meaning, and thus meaningful action.

(Manuel Castells 1994, 20)

This introductory section of *Making Sense of Cities* takes the form of a scoping exercise. First and foremost, this is a geographical survey of urbanisation and cities that, none the less, represents just one of several possible ways a geographer may approach the task. This text places the emphasis on urbanisation processes and how cities are spatially organised. However, after looking at 'urbanisation patterns and processes' and 'urban systems and the growth of cities' in the first two chapters, the remainder of the textbook is devoted to the geography of the city. Cities are among the most complex organisational and institutional forms created by human endeavour. A city is an arena for economic development, political action, social life, cultural enrichment and environmental stress.

There is a sense, then, in which 'it all comes together in cities', but because of the complexity of cities an introductory textbook like this one can only ever hope to be a kind of 'hitch-hiker's guide to the galaxy'. In setting out to explore the city, in just the same way as Arthur Dent has to in the Douglas Adams text of the same name, when confronted with the vastness of our galaxy, we have to decide on our destination. We begin by asking some basic questions along the lines of 'Why study cities?' and 'What do geographers bring to the study of cities?' We should also be clear about some of the working assumptions that influence questions of a conceptual nature about cities, and focus our attention upon particular aspects of city growth and urban existence at the expense of others.

Our accumulated wisdom about urbanisation and cities is the product of an intellectual process whereby key thinkers in urban studies, including urban geographers, have developed theories and interpreted events in an attempt to make sense of that part of the observed world – nature and society – that comes together in cities. In this way our knowledge about cities is 'constructed' and remains open to question. This discussion of how our knowledge has been constructed may be illustrated by examining the top ten books on 'urban' matters that I think have been among the most influential in shaping our understanding of cities. This constitutes a potted history of urban studies and the evolution of the ideas that we are able to draw upon in trying to make sense of cities as geographers.

However, although you can expect to gain a healthy regard from these pages of the contestability of knowledge and competing theories available to geographers trying to reach an understanding of cities, in its orientation the book leans towards 'post-Marxist' urban political economy. What distinguishes it from some of the other geographical approaches that, by contrast, draw heavily on cultural or feminist studies is the stress it still places on structural explanations of urban space and material differences in the quality of life in cities. The extent to which these so-called post-modern approaches in human geography have lost sight of the growing inequalities that have accompanied globalisation at every geographical scale – internationally, regionally and *within cities* – is a matter of real concern to some geographers.

One reason for favouring a political economy of urban space is that I want students to get a real taste of the potential relevance of urban geography to the problems arising in cities around the globe. In the process of stimulating a series of searching questions about the subject matter of urban geography, a problem-centred approach gives added meaning and purpose to the task of learning.

Of course, the social and environmental problems found in cities vary from place to place, in kind, and by degrees. Yet this doesn't negate the fact that, in all cities, regardless of their differences, administrators and planners are attempting to solve essentially the same range of 'urban' problems. They are 'urban' problems to the extent that social and environmental processes can take a different form in cities, and are invariably more complicated as a result. At the same time, ordinary citizens are actively involved in community affairs in an attempt to strengthen the democratic base of cities. With their expertise and skills, graduates in the social sciences, including urban geographers, are well placed to participate in these forms of political action. Indeed, being well informed is an indispensable condition of responsible citizenship. This is a text for good global citizens.

In my view, such an approach to understanding cities offers much more to people developing public policy that will shape cities than a post-modern interpretation. And many geography graduates do eventually get jobs, if not as city planners or managers, in other areas of public policy that indirectly play a part in shaping urban development and city environments. Others contribute at the local level by joining a resident activists' group, an environmental lobby or a social movement working for better cities.

Therefore, while none of the core concepts and ideas from other bodies of urban theory is overlooked, we will draw quite heavily on the conceptual frameworks developed by Michael Dear, an urban geographer, and Frank Stilwell, an urban and regional political economist. Their ways of seeing through the complexity of the economic, political and cultural processes that shape geographies of the city are outlined below. This leads on to the last part of this Introduction, where I map out the structure of the text and say something about the selection of themes, conceptual signposts and case studies.

A SCOPING EXERCISE

Why study cities?

When the history books are written I strongly believe that the next century will be known as the 'urban century' – where most of the world's critical issues were urban issues and where the leadership to address these challenges came from local democracies.

This confident prediction was made in a speech to a gathering of New Zealand mayors in the run-up to the new millennium (August 1999) by the Mayor of Honolulu, Jeremy Harris. In 2000, for the first time, a majority of the world's population was living in cities. According to United Nations estimates (1996), this trend shows no sign of slowing over the next 25 years or so. In highly urbanised countries like the UK, the Netherlands or Australia, over 80 per cent of the population live in urban areas. And, even though counter-urbanisation processes are slowing the growth of major cities in many western countries, the concentration of people in cities will continue to dominate settlement patterns.

Often staggering rates of urbanisation are occurring in a broad group of countries that are popularly known as the 'Third World' or the 'South'. However, although widely used, there are some specialists who object to these terms. In-migration and natural increase are disproportionately feeding the growth of the very largest cities, which will grow to a size within a generation that is beyond the experience of 'First World' countries. In mega-cities like Shanghai, Mexico City, São Paulo, Lagos, Bangkok and Jakarta, the problems generated by housing shortages, transport congestion, poor sanitation and other forms of environmental stress can be expected to increase in severity.

To complicate matters, these unprecedented rates of urbanisation and urban development are taking place in the context of massive technological changes, which are remapping the world's economic geography (Dicken 1998). Globalisation is the financial, economic, social and political response to technological change. An important feature of the changing global order is the leadership role taken up by a growing number of economically powerful global cities. Not only are the traditional relationships between the governance and power of nation-states and cities breaking down, but the economic interests of cities like London, Paris, Frankfurt, New York and Tokyo are now factored into central government policy. Our shrinking world and increasingly borderless states are fundamentally altering the way in which scale-effects mediate city-forming processes. The forces of globalisation are creating a new world network of cities that are no longer so dependent upon national governments. In 1995, Singapore's Minister for Information predicted that 'In the next century, the most relevant unit of economic production, social organisation and knowledge generation will be the city' (Parker 1995). It behoves us as geographers, therefore, to keep abreast of the profound changes accompanying globalisation because they are transforming our cities economically, culturally and environmentally.

Globalisation is enhancing the role of select cities that are leading the way internationally because they are especially suited in terms of their factor endowments and assets to dominate as the powerhouse of the nation-state. This is shifting the balance of investment and market power between cities and regions in Europe, North America and Southeast Asia. Globally, people are moving between countries and places as never before, and radically modifying the cultural and social mosaic of cities, sometimes unsettling a pre-existing social order. Consequently, this new cultural and social heterogeneity in those cities open to the global movement of workers is becoming a defining trait of urban society and city living. In Third World cities, the environmental counterpart of 'Green politics' is the 'Brown Agenda'. In large measure, one is the product of affluence, the other the product of poverty. Although cities in affluent societies have their versions of housing, transport and servicing problems, they are not to be compared with the pollution and congestion, housing shortages, water quality and unhealthy living conditions that form the basis of the Brown Agenda in Third World mega-cities.

The changes taking place globally in conjunction with rapid urbanisation are so far-reaching that Chris Hamnett (1999a, 225), a British urban geographer, recommends, almost as a matter of necessity, 'We need to know how cities are changing, and what the implications for urban life are and will be in the future.' For these reasons, then, 'the study of the city, the spatiality of metropolitan life and the policy challenges of contemporary urbanism have come to hold a powerful position within the contemporary social and policy sciences' (Amin and Graham 1997, 412).

The 'so what?' question

Our quest, therefore, is for *useful* knowledge. This is an urban geography text in which attention is paid to the theory, concepts, methods and data (in the broadest sense) that are likely to be of greatest relevance in identifying, analysing and then solving the complex social and environmental problems that beset cities. Our theoretical body of knowledge, together with our system of values, provides the intellectual framework that largely determines how we perceive problems and issues arising in cities, and then the forms of action we might take in response to them. Significantly, these days almost all the strategic plans prepared for cities around the world, regardless of the specific urban problems they face, contain the same set of normative goals: economic efficiency; social justice; urban liveability; environmental sustainability (eco-cities). In practice, these constructs have become so overworked by planners that they have taken on the status of parenthood statements.

How these goals are reconciled in principle, let alone realised in practice, is always a matter of intense debate among politicians, treasury officials, urban managers and the planners. In government we find that public policy is routinely shaped by the political and economic theories that politicians and administrators are attracted to. Social scientists, including urban geographers, have a crucial role as researchers in developing the ideas and gathering the information that gets fed into this process. Naturally research staff in government agencies put these ideas and information before decision-makers as well. This is the body of knowledge that ultimately shapes urban policy and we must never forget that there is a crucial sense in which it is both 'constructed' and used selectively.

Urban geographers who have made notable contributions to the development of urban policy, either through their writing or by becoming actively involved as advisers to government, include some of the following people. Europeans like Peter Hall, Arie Shachar and Frans Dieleman, Canadians like Larry Bourne, and Australians like Mal Logan, have all acted as high-level advisers to central government in the areas of urban and regional development and metropolitan strategy. In the UK, Ray Forrest and Alan Murie have consistently commented on housing issues and social exclusion. Bill Clarke at UCLA was an expert witness before enquiries into racial segregation and bussing in the United States during the 1980s. The research of Doreen Massey and Ron Martin has informed discussions about manufacturing decline and labour markets in UK cities. Australians Kathy Gibson and Jenny Cameron are working with communities hit by restructuring in regional Victoria to develop 'capacity-building' skills and local economic development strategies. But these are really only a few of the numerous urban geographers putting their expertise and skills to good use.

The knowledge that becomes available to urban managers or urban citizens playing an active part in community affairs is not immutable. *Ideas are constructs.* They take shape in the mind and as such are contestable. It's crucial, then, that we have a sense of how our knowledge of cities is derived; that it is a source of power; and that, in informing and shaping public policy, ideas are used to achieve political ends. Hence ideas and data need to be approached critically. This questioning ought to include consideration of the motives behind a claim or argument. The values and ideology intruding into a debate need to be made explicit. Does the conceptual model assume forms of social action that are prone to domination or subordination, marginalisation or inclusion, conflict or consensus, collectivism or individualism? What meanings are invested in the forms of language that we choose to use? Who stands to win or lose from the way a particular slant on reality is being put to use?

THE CONTESTABILITY OF IDEAS AND KNOWLEDGE

Well over 20 years ago, a little book called *Urban Planning in Rich and Poor Countries* by Hugh Stretton

(1978, 15) began with a discussion of 'Rival theories and methods' in which he stated that 'every social science ought to harbour rival theories and methods'. The sense we make of the natural and social world around us is the product of our powers of reasoning. But the knowledge we acquire via this process always remains open to question. The debates that take place across the frontiers of knowledge between philosophers of science and within disciplines between the proponents of competing theories are positively healthy! Not only does this lead to new discoveries; this is how disciplines continue to be renewed and invigorated.

When formalised, this concern about the sources and framing of knowledge constitutes the underlying epistemology of a subject area or discipline. According to Leonie Sandercock, who teaches in urban planning at the University of Melbourne, the epistemological questions at the heart of urban studies are:

> What do I know? How do I know that? What are my sources of knowledge? How is knowledge produced … ? How and when do I know what I know? How secure am I in my knowledge? What level of uncertainty or ambiguity can I tolerate? What forms of knowledge offer me most security? How adequate is my knowledge for the purpose at hand? How can I improve the knowledge base of my (and others') actions? … What responsibilities do I assume for the application of what I claim to know? What is valid knowledge … ? Who decides that?
>
> (Sandercock 1998, 58)

These questions are really worth returning to from time to time. This applies as much to practitioners who think they know it all as it does to students embarking on the study of cities and urbanisation for the first time. Enormous social damage has been visited upon the populations of developed and less developed countries since the 1970s by economists working as advisers for organisations like the International Monetary Fund (IMF), the World Bank and the World Trade Organization (WTO), convinced of the immutability of their own knowledge and theory. One way of illustrating the implications of privileging a particular body of knowledge is to select some of the ideas that have had a lasting influence on urban studies. As well as changing the direction of thinking within urban disciplines, these ideas reflect the extent to which knowledge about cities has been contested and revised. Naturally this is just a personal assessment of the seminal ideas in urban studies. You might want to give more prominence to the writing of contemporaries like Doreen Massey, Delores Hayden or Linda McDowell, but this means leaving others out. Once you are more familiar with urban knowledge and ideas, you will be in a position to draw up your own equivalent to my suggested 'top ten'.

The other reason for starting with a brief, though admittedly patchy, history of urban ideas is that it helps to locate the development and portability of ideas in a scholarly and societal context. Urban geographers do not rely exclusively on geographical or spatial concepts and facts in making sense of cities. They frequently draw on contributions to our knowledge of cities developed by researchers in other disciplines like urban and regional planning, economics, sociology, history, politics and public administration.

My 'top ten' would start chronologically with Charles Booth's survey of working-class living conditions in the East End of London. As well as documenting the dire consequences of rapid urbanisation in Victorian England, it represents the forerunner of the social survey research that has become the mainstay of all social science disciplines. Next came the pioneering work in the 1920s by staff and students associated with the Department of Sociology at Chicago University and written up by Park, Burgess and McKenzie in *The City* (1925). In what is now regarded as a classic, they set out their general theory linking a city's spatial form and social organisation with processes that find a parallel in a natural ecosystem. Borrowing from Charles Darwin and Herbert Spencer, they decided that the processes shaping the Chicago of their day could be likened to those operating within natural ecosystems. Given the intense competition among successive waves of immigrants arriving in Chicago and seeking work from the 1860s onwards, certain neighbourhoods were bound to be 'naturally selected' by those groups that could make the best use of them. In due course, this process of invasion and succession would produce 'natural' areas that were distinguishable because of their social or ethnic homogeneity – the ghetto, the 'Gold Coast', the slum, Little Italy or Chinatown. Significantly, the Chicago ecologists were the first to appreciate that, while language, culture, religion and race provide the motivation for residential segregation in cities,

geographical barriers and physical distance along with improved mobility provide the means to practise it. They 'discovered' that the physical features and two-dimensional spaces of the city are used by different cultural groups to accomplish social distancing and residential segregation.

Alonso's *Location and Land Use: Toward a General Theory of Land Rent* (1964) represents a stage in the development of urban theory when economic modelling held sway. This approach to understanding the organisation of land uses within cities is characteristic of positive science because it sets its sights on generalisation and theory building, and employs scientific methods in the process. Alonso's general theory sets out to predict the location of firms and households within a city with a single centre. The model can be solved using calculus so long as a series of restrictive assumptions about economic conditions and the behaviour of firms and households apply. The general family of residential location models that emerged from the subsequent attempts to predict the spatial organisation of urban land use is typical of the search for universal generalisations that characterised the social sciences in the 1960s and 1970s. This was also an era when transportation engineers thought they could meet the future demand for cars in cities by building urban freeways (Fig. 0.1). Similarly, public architects were building high-rise council housing estates and project housing in Europe, North America and Australia, as part of comprehensive urban renewal programmes. The source of these ideas can be traced directly to the French architect Le Corbusier, who dreamed of building 'cities in the sky' (Fig. 0.2).

FIGURE 0.1 *The high point of modernist city-building in the 1950s: Chicago's Dan Ryan Expressway and Taylor homes*
Source: unknown

FIGURE 0.2 *Le Corbusier's prototype for 'cities in the sky', 1922–1925: the 'Voisin' Plan for Paris*
Source: Le Corbusier (1933, 206). Reproduced by permission of the Foundation Le Corbusier

Behind this lay a belief in the superiority of *modernist* design and construction techniques.

When the reaction to positive science and modernism began to set in, it took a variety of forms. At the beginning of the 1960s a journalist for the *New York Times*, Jane Jacobs, wrote a devastating critique of the excesses of modernist architecture and urban design called the *Death and Life of Great American Cities* (1961). In particular, Jacobs condemned both modern planning philosophy and government bureaucracies for so strictly segregating land uses and bulldozing traditional inner-city neighbourhoods to make way for public housing. On both counts, according to Jacobs, the assumptions behind these policies failed to deliver to citizens the benefits they promised. Her challenge to planning wisdom

helped to change the way cities and suburban housing are now designed. About the same time, Lewis Mumford was writing *The City in History* (1961). As a member of the Regional Planning Association of New York, he gave voice to the mounting concern about uncontrolled urban growth that was prevalent among urban and regional planners at the time. But *The City in History* is much more than that: it is a historical *tour de force* that links the flowering of western civilisation to urbanism and urban institutions. By the same token, the unprecedented physical growth of cities that accompanied mass car ownership in the post-war affluence of the 1950s was seen as a looming threat not only to environmental values, but perhaps to civilised life in the West itself. Both Jane Jacobs and Lewis Mumford were instrumental in starting debates that eventually led on to the framing of principles like urban sustainability and urban liveability.

The next important intellectual development in the field of urban ideas visualised the city as a resource-distributing mechanism in its own right. By the early 1970s a number of commentators in affluent capitalist societies were growing increasingly disillusioned at the persistence of poverty and inequality that the welfare state was meant to eradicate. In *Social Justice and the City*, David Harvey (1973, 68), an urban geographer, wrote: 'I ... regard the city as a gigantic resource system, most of which is man-made.' Harvey opened up the possibility that the welfare transfers targeted at individuals by government agencies are negated to some degree by countervailing effects. These can be traced to the way cities are structured (and not just spatially) in capitalist societies. For example, in big cities especially, the residential property market and housing provision often operate to increase, rather than dampen, inequalities. This idea of the city as a resource-distributing mechanism continued to dominate urban research beyond the 1970s. In as much as this is a 'man-made' system – while Harvey would avoid the gender bias these days, the responsibility still mainly lies with males – 'urban managers' and 'gatekeepers' working in human services agencies are in a position to control urban resources and regulate the life chances of many citizens.

The revival of Marxist theory in the 1970s introduced a completely new perspective into urban studies and urban geography. A fully fledged version of this structuralist perspective was developed by an urban sociologist based at the University of Paris, where he was exposed to the debates that were raging between Marxist political and social philosophers in Europe. But even though these ideas were in circulation by the late 1960s, the book by Manuel Castells (1977) that seeks to show that cities are the product of the structural forces that shape capitalist societies was not translated into English for another decade. Subsequently, in the skilful hands of Mike Davis, who teaches in an urban planning school at UCLA, we see how much Marxist political economy can add to our understanding of a city like Los Angeles. His book *City of Quartz* (1990) portrays Los Angeles, with all its pretensions to postmodernity, as the prototypical twenty-first-century city. One of the defining features, he suggests, is the emergence of 'Fortress LA' whereby public space is progressively alienated and the affluent withdraw into protected 'gated communities'.

William Cronon is an urban historian who, for the first time in a book called *Nature's Metropolis*, set out to show how much a rapidly growing city like nineteenth-century Chicago transforms the natural environment and the landscape of its hinterland as it develops. As he says, 'no city played a more important role in shaping the landscape and economy of the mid-continent during the second half of the nineteenth century than Chicago' (Cronon 1991, p. xv). The other side of this coin is the complex story of the 'elaborate and intimate linkages' between Chicago, the city, and the mid-west, its countryside. At the time, there was really no other model for a book like this one, which tries to tell this city/country story as a unified narrative.

Lastly, Leonie Sandercock's book *Towards Cosmopolis: Planning for Multicultural Cities* (1998) brings together and applies the best of the insights that have emerged from the recent debates about the shortcomings of structuralism. There are two main theoretical strands to this critique – post-structuralism and feminism – that run together in places. Post-structuralists dismiss the efforts of Marxists to construct all-embracing 'grand' theory. Part of this critique also involves rejecting unacceptable forms of essentialism and foundationalism. They argue that these precepts always smacked of intellectual 'imperialism' and are no longer tenable in a 'post-colonial' world of social and cultural diversity and difference. These criticisms of structural theory find a resonance in the thinking of a number of women who pioneered the development of feminist theory in urban studies. They include Dolores Hayden, Doreen Massey, Linda McDowell and Susan Hanson (to name a few). In particular, they took exception to the way Marxists objectify class relations and class structure – almost to the point where they smother consideration of gender relations and patriarchy.

Sandercock points out that the acceptance of foundational truths, universally held values and a common purpose helps to perpetuate the falsehood of an indivisible 'public interest' within the domain of planning. Traditionally, males have dominated the planning and engineering professions, and built cities to their own liking. But the rise of movements for civil rights, women and the environment in the 1960s and 1970s, together with the mass migration of people with vastly differing cultural backgrounds into cities around the world in the 1980s and 1990s, have created a plethora of interest groups based upon their separate identities. There are now 'multiple publics' where once there was only the 'public interest'. However, given the existing structure of power, the interests of minorities are easily subordinated in conflicts arising from major urban development. According to Sandercock, planning for multicultural cities now requires not only an awareness of these 'multiple publics' but their active participation in the planning process. She contends that without a politics of recognition it is very difficult for the powerless in cities to have their needs met.

Where does this leave us, then? Following a critical review of the contribution to social theory of post-structuralism and feminism, Mark Davis (1999b, 268) decides that, 'New knowledges … don't necessarily replace the old. But they do affect the way old knowledges are used.' There is a strict philosophical sense in which the innate superiority of a given approach to understanding can never be decided. But this is the slippery slope of relativism, which is the doctrine that knowledge is relative and not absolute. This is to imply, for example, that equal credence should be given to 'creation' theory and the theory of evolution. For most people this is plainly untenable.

In the everyday world, anyone with an enquiring mind and seeking to get to the bottom of a problem tends to opt for one or other forms of explanation or interpretation. Except for those who piece together differing parts from competing theories, the rest of us are generally drawn to the theory that provides the best 'fit' with our view of reality. In the case of the physical and natural sciences, the scientist is well satisfied if a model based on a theory of the processes involved actually works under test conditions. Maybe a philosopher will be more impressed by the cogency and elegance of a set of formal propositions. On the other hand, it is commonplace within literary or cultural studies to make use of representational devices like metaphor and poetic imagery to get a message across. Devotees contend that story-telling is an equally valid way of going about understanding the worlds we inhabit.

But in each case this involves *informed* judgement. Is the theory believable? Does a proposition 'ring true'? ('Truth' is philosophically problematic as well.) Is what is being proposed consistent with our personal experience? And even here, because individual identity can vary, our sense of shared experience will not always coincide. But if these conditions do hold, we credit our 'fabricated' ideas and the body of knowledge they help to form with a degree of authenticity and plausibility. Then, once accepted, they take on definite authority in the minds of people that are in agreement intellectually.

All these points about the construction and contestability of knowledge are vividly illustrated by the debate about the underlying causes of inner-city poverty in the United States. The poorest inner-city neighbourhoods of many of the older industrial cities of the north-east (New York, Newark, Philadelphia) and the mid-west (Chicago, Pittsburgh, Detroit) are home to black communities that are dreadfully poor, and prone to social dislocation and breakdown. (This is treated more extensively in Chapters 7 and 9.) Social conservatives largely attribute the problems that overwhelm many black neighbourhoods to the formation and acceptance of a ghetto subculture of welfare dependence and crime. Opposing them are liberal commentators who are accused of turning a 'blind eye' to the higher incidence of family breakdown and anti-social behaviour in the black ghetto. This has the effect of heightening the popular appeal of the conservative arguments set out in a book like Charles Murray's *Losing Ground* (1984).

A third set of essentially structuralist arguments is best represented in two monographs written by a black sociologist, William Julius Wilson (1987; 1996). They are based upon his own research carried out in the neighbourhoods of Chicago's 'black belt'. He argues that de-industrialisation and the disappearance of blue-collar jobs have created a sense of despair in these poor neighbourhoods, particularly among young urban blacks. They comprise the hard core of a 'new urban poor'. According to Wilson, *When Work Disappears* (1996), and in the face of systematic discrimination, *The Truly Disadvantaged* (1987) are tempted to direct their anger against society and turn to pushing drugs as a way out of poverty. While Wilson is not about to exonerate socially unacceptable behaviour, he does insist that 'the disappearance of work and the consequences of that disappearance for both social and cultural life are the central problems in the inner city ghetto' (Wilson 1996, p. xix).

These competing theories, setting out reasons for the appearance of 'new urban poverty' in advanced economies like the United States, are not just academic. The way the debate between the three main groups of theorists has been resolved is all the more important because it has had a crucial bearing on the direction of social welfare and labour market programmes in the United States. Support for the respective claims that welfare payments produce a culture of dependency and a refusal to work, and that the breakdown of family life is responsible for higher levels of promiscuity among young blacks, led in turn to 'the end of welfare as we used to know it' under President Clinton.

It is in this regard that voting citizens, as well as the more powerful people framing public policy, are definitely making judgements about the practical *adequacy* of the different theories and ideas on offer. At the same time it is equally important to realise that, in exercising our powers of reasoning, we are unavoidably prone to sources of subjectivity such as a personal belief system and our own individual life experience.

WHAT DO GEOGRAPHERS BRING TO THE STUDY OF CITIES?

What we want from urban geography today is not a consensus that any one approach to the study of urbanisation and cities is inherently superior to another, 'but rather an awareness of when and where a particular approach is appropriate, and a sensitivity to what it can and cannot achieve' (Smith 1999, 22). This textbook adopts an approach that is problem-centred and capable of informing both the management of cities and the participation of urban citizens in community affairs. In my opinion, a loose form of 'post-Marxist' urban political economy offers the best means of achieving this.

Alternatively, 'Getting to grips with the culture of the modern city is as much about film, video and novels as it is about census data, interviews and questionnaire surveys' (Hamnett 1999a, 255). Indeed, 'the culture' of the city has come under new scrutiny (Jacobs 1993). This so-called cultural turn is partly in response to material shifts accompanying the passage of late capitalism towards postmodernity, and partly in response to the emergence of other fresh, and sometimes thought-provoking, ways of viewing the city – e.g. post-structuralist and feminist critiques.

This has created tensions within human geography that are affecting the very questions we now ask about our theoretical object, the city, and how we, in turn, approach the study of cities: 'What is urban geography these days?' The main divide is between those who want to keep structural and material considerations to the fore in making sense of cities. Urban geographers like Allan Scott, Ed Soja, Doreen Massey, David Harvey, Neil Smith, Richard Walker, Michael Storper, Paul Knox, Jan van Wessep and Kathy Gibson are prominent within this group. This means paying greater attention to economic and political processes in their analysis. Then there are those writing about cities who, along with other human geographers, spend their time interpreting cultural processes. Leading figures include Peter Jackson, Sharon Zukin, Jane Jacobs, David Ley, Kay Anderson and Judith Burgess. With culture assuming the mantle of theoretical object, the focus shifts to the representation and interpretation of symbols and artefacts in the urban landscape. Divining meaning is a subjective exercise that pays heed to the different significance of cultural markers in the city according to location, generation, gender, ethnicity and class.

Yet, more and more, in trying to make sense of the transformations taking place in cities around the globe, we are belatedly coming to appreciate 'the extraordinary difficulty of separating out something called "the economic" from "the social" or "the cultural" or "the political" or "the sexual" or what have you' (Thrift and Olds 1996, 312). Urban scholars like Sharon Zukin (1997) have voiced their concerns about the damaging impact of the separate economic and cultural agendas in urban studies, and are now calling for a pooling of effort. One of the distortions produced by the 'cultural turn' in human geography and the overshadowing of urban geography has been the comparative neglect in the geography curriculum of what might be termed the urbanisation of injustice (Merrifield and Swyngedouw 1997).

In fact the problem is more disabling than this separation between urban political economy and the culture of cities (Zukin 1995). Amongst the disciplines studying urban processes, whether they relate to the political, the social, the cultural or the institutional assets of cities, there is an overwhelming tendency to concentrate exclusively on one element of urban life and city development. This could be culture, social polarisation, housing, industrial districts, transport, governance, property development, planning and so on. 'Thus the very essence of the city – the concentration of diverse relational intersections between and within such activities and elements – tends to be lost' (Amin and Graham 1997, 412).

During the 1990s many geographers were put off trying to comprehend the interconnectedness of human activity in the city by a series of epistemological objections that cut right across a 'big picture' perspective. But this is quite at odds with the position taken, for example, by environmentalists when they come to study an ecosystem. The term 'bioregion' was deliberately coined to stress the system-wide interdependence of natural and human processes operating within a territorially defined ecosystem. For geographers to vacate this kind of terrain, which would repeat what happened with environmental education in Australia, is just plain silly. As Amin and Graham (1997) point out, the city has been 'rediscovered' as the powerhouse of the globalised economy. They argue that there are three key strands to this:

- the durability of urban centrality
- the enhanced role of cities as drivers of national economic development in an age of globalisation
- the value placed upon cities as a creative milieu (i.e. 'creative cities').

But it is no easy matter to pull the elements together in an introductory textbook in a fully convincing way. First, there is a temptation to over-generalise from a handful of cases. The 'global city' hypothesis relies on the experience of just three global financial centres – London, New York, Los Angeles (Sassen 1991). As a pointer to what lies ahead for other cities, Los Angeles has definitely been oversold. According to Amin and Graham (1997, 417), 'If it "all comes together" in Los Angeles, the implication is that all cities are experiencing the trends identifiable in Los Angeles and that we do not really need to understand these processes. Therefore, unexceptional, even ordinary, cities should find a place in our analysis as well' (Amin and Graham 1997, 411).

Secondly, the textbook writer is drawing upon what are often partial and specific representations of the city and urban spaces. Accounts abound in academic articles, as well as in film and media, of the

spaces of power and centrality in the global financial centres now gentrified by 'yuppies'. Yet working-class women barely rate a mention in the agendas of feminism and cultural theory.

Typically our case studies focus on a specific aspect of place, and fail to show how the social and economic processes that make it distinctive are imbricated with other parts of the city or global economy. Geographers should be alive to:

- how 'command and control' functions in the nerve centres of global cities are linked via circuits of labour and capital to other key sectors and spaces within the information economy
- how people on the move globally, from corporate executives to guest workers, are putting pressure on local labour and housing markets, or changing the demography, and the cultural and social mosaic of cities
- how consumption patterns of an urban population dictate the inflow and outflow of materials and energy, and the extent of a city's ecological 'footprint'
- how unevenly and inequitably economic restructuring and cutbacks to welfare are impacting upon urban regions on one level, and neighbourhoods and suburbs within cities on another.

According to Amin and Graham (1997, 417–21), *partial* approaches are seldom up to making these kinds of connections. Rather, they suggest we have to begin seeing the city as a constellation of spaces interlinked by a whole mesh of relational webs and flows that coalesce, interconnect, fragment and compartmentalise. What a mouthful! The terminology gets pretty abstruse because we lack a vocabulary that can describe adequately the transformations occurring within the space economy as the capitalist system becomes more highly integrated globally. They call these interlinked nodes and places in contemporary urban regions the 'multiplex city'. Some remain place based, as is the case with the financial district, cultural zones or industrial districts in cities. But other spaces within cities rely on flows of money, information and workers that connect the locality with distant parts of the same country, or even with suppliers or markets on the other side of the globe. The essence of this perspective is that it promotes urban activity and urban existence as the combination of 'multiple spaces, multiple times and multiple webs of relations'. These all tie 'local sites, subjects and fragments into globalizing networks of economic, social and cultural change' (Amin and Graham 1997, 418).

Developing a new-generation text along these lines for students coming to urban geography for the first time would be a tall order (and may end up being totally confusing!). Rather, instead of structuring the text around the notion of a multiplex city, the connections between 'the local' and 'the global', and how they are transforming urban spaces in cities, will be made explicit throughout. With global corporations trying to harmonise their operations around the world, the 'cultural boundaries of the past are disappearing because clients want the world's best expertise … without any country or geography coming into the equation' (Roderick McGeary, worldwide chairman of KPMG Consulting, January 2000). Yet in urban communities in cities around the globe there is a growing resistance to this undermining of cultural and geographical identity. In an attempt to preserve the identity of the place where they live, citizens are becoming more and more active at a local level in place management.

A WAY OF MAKING SENSE OF CITIES

We have already noted that how one ultimately decides to represent urbanisation and the geography of cities involves a value judgement – in this case, about the educational value of the particular approach we adopt. Some of the conceptual scaffolding developed by Michael Dear and Frank Stilwell provides us with a way of making reasonable sense of cities. For Dear (1988, 270), ' "urban geography" is the study of the social, economic, and political processes as they occur in cities'. Michael Dear teaches at the University of Southern California in Los Angeles.

To unpack that further, he argues that it is these three primary processes that structure the time–space fabric of cities. Significantly, he attaches no particular order of priority to political, economic or social processes. The concept of time–space takes account of the fact that the city, or parts of the city that have a distinctive character of their own, are a complex amalgam of past, present and newly forming patterns that co-exist in the urban landscape. In as much as cities come to reflect Schumpeter's 'creative and destructive' tendencies within capitalism

over time, urban spaces/places/localities acquire their special identity gradually with the passage of time.

Systems of production, exchange and distribution are going to be at the heart of a consideration of economic processes. Chapters 3, 4 and 5 of this textbook cover the geography of wealth creation, the location of economic activities and the work performed in cities respectively. The social processes that facilitate human interaction in often densely populated and culturally diverse cities, working in combination with the housing market, hold the key to residential differentiation. Thus Chapter 6 deals with housing markets and residential location prior to the description of the social and cultural mosaic of cities in Chapter 7. In Chapter 8 the focus turns to the urban environment and the ecological outlook for cities.

Political processes are discussed in Chapter 9, which is devoted to urban politics and the management of cities. The formation of political regimes in response to the differing interest groups in society has a major bearing on the nature of cities, urban development and community affairs. While these interest groups are usually based on class, or gender, or generational, or sexual differences, there are also political coalitions and social movements that form around single issues and therefore cut across these natural affinities.

Frank Stilwell, the aforementioned political economist, who teaches at Sydney University and is interested in cities and regions, has teased out the key relationships between the processes that are steering contemporary urban change (Stilwell 2000). His aim is to show how urbanisation and city form impact – for better or worse – on the attainment of three broad social goals (efficiency, equity and environment). He is committed to developing public policy and other forms of collective action that will lead to more efficient, equitable and ecologically sustainable cities. Stilwell suggests that a reasonable synthesis can be achieved with the aid of a three-way typology (Fig. 0.3). The respective panels in Figure 0.3 represent the systems (society/economy/ecology), the institutions (market/state/community) and the spatial scales (global/national/local) affecting urbanisation and the geography of cities.

The first panel displays three interconnecting systems – economy, society and ecology. Some of society's most pressing concerns lie at the interface between these systems. In placing a value on equity

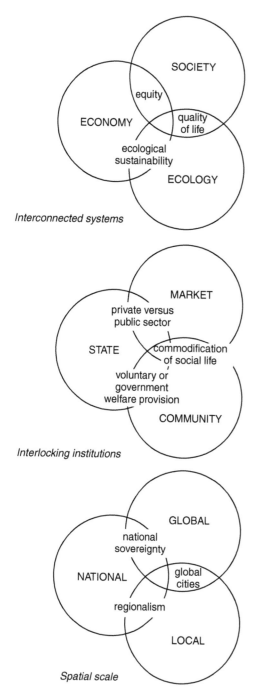

FIGURE 0.3 *Stilwell's three-way framework for understanding cities*

Source: composite based upon figures in Stilwell (2000, 16, 17, 19)

(in the distribution of income, wealth, socio-economic opportunities and life chances), ecological sustainability and quality of life, these become collision zones for the different interest groups in cities according to whether they are winners or losers from the political choices being made.

The second constellation in Figure 0.3 shows the institutions that can serve, or frustrate, the achievement of a sustainable balance between society, economy and ecology. Again, the interface between markets, the state and the community (or civil society) will always be a source of tension. Markets are regarded as efficient mechanisms for allocating resources, distributing income and rationing environmental goods. Because the consumer exercises one vote – 'one person, one vote' – markets are said to be decentralised, democratic and disinterested institutions. In practice it is capacity to pay – 'one dollar, one vote' – that determines market outcomes. Because of this, the state frequently steps in to cover for market failure. However, collective decision-making and conscious planning are prone to excesses of their own – at worst, authoritarianism and corruption – which is why it is important to have a robust and vigilant community.

The role[s] of household and of neighbourhood organisations are central here, institutions in which principles of altruism, social responsibility, and public service typically intermingle with self-interest and gain. This concept of community is notoriously difficult to define but it is an important 'catch-all' category to take account of the sources of social cohesion that are not reducible to market and state.

(Stilwell 2000, 17)

Once more, tensions arise across the interface with particular significance for the processes of urban change. Social and political conflicts can arise on any of these three frontiers. In mixed economies, the private sector/public sector split is determined by powerful pressure groups lobbying for less public ownership in favour of privatisation. The frontier between the state and community throws up different but equally complex public-choice issues about the relative roles of government and voluntary agencies in welfare provision. Lastly, the interface between community and market is often the site for struggle over the commodification of social life (i.e. the penetration of business into sport, music, the arts).

The third constellation in Figure 0.3 extends the conceptual scaffolding by introducing the issue of spatial scale. Stilwell points out that, while global, national and local processes of change are at work, and impacting upon cities and urban communities, in practice they are more likely to be inseparable. The conceptual status of spatial scale has taken on new-found importance almost proportional to the degree with which 'globalisation' in its various meanings has come to dominate in the public consciousness. It's a transformation within capitalism that is being carried along not just by economic drivers, but by an ideological groundswell too. The economic strategy of global corporations now finds willing support from politicians like the UK's PM, Tony Blair, who is convinced of the inevitability, and even desirability, of riding the tiger's tail of globalisation. In spite of this, real tensions have emerged at the global/national interface because of the challenge this poses to national sovereignty in matters of economic and social policy. More and more, governments are constrained by the ever-present fear of precipitating a flight of mobile international capital (Stilwell 2000, 18–19).

Meanwhile, at the national/local interface, growing regional disparities and regionalism are causing mounting tension between central governments and local communities. Whether it is the 'north–south' divide between south-eastern England and the rest of Britain, the 'bi-coastal' ascendancy in the United States, or the antipathy between the 'battlers in the bush' and Australia's 'urban elites', there is a big question mark over the will, let alone the capacity, of the state to work for more balanced national development (see Chapter 3). Lastly, there is the global/local interface, which is also a source of political-economic tension. Ohmae (1995) suggests that, with 'world-cities' now vigorously competing in a 'global bazaar' for the business of global corporations, the concentration of economic power in these cities could gradually undermine the integrity of nation-states. These 'bidding wars' also take place between lower-ranked cities in an urban hierarchy. When global corporations are seeking a new location for a plant or office, they play cities off against one another, and sometimes local communities within the same city (see Chapter 4). Alternatively, when a global corporation pollutes the local environment in cities, this also creates friction at the global/local interface.

Although Frank Stilwell is the first to admit that his three-way typology simplifies the real world processes that are responsible for urbanisation and the geography of cities, it does help us make sense of the broad systemic, institutional and scale effects involved. These conceptual considerations pervade the discussion throughout *Making Sense of Cities*. The other attractive feature of his approach is that it is problem oriented. For Stilwell, and for us, this is a priority:

> If one takes a problem-oriented approach to urban analysis there is no shortage of practical concerns needing illumination and resolution. The all-too-familiar parade of urban problems includes congestion, pollution, access to affordable housing, unemployment and increased social-spatial inequalities. There is an obvious need for research into the roots of these practical problems and policy prescription for their amelioration.
>
> (Stilwell 2000, 14)

FURTHER READING

All students taking a course in a specialist area of human geography like urban geography will find the omnibus *Dictionary of Human Geography* indispensable as an *aide-mémoire* (**Johnston** *et al.* 1999).

Current issues of journals like *Urban Geography*, *Urban Studies* and *Cities* are bound to contain up-to-date case studies that can be used to illustrate topics covered in this text. **Paul Knox's** *Urbanization: An Introduction to Urban Geography* (1994) makes a useful, though mainstream, companion volume written explicitly for North American students.

Australian students are equally well served by **Clive Forster's** text, *Australian Cities: Continuity and Change* (1999), which was prepared for OUP's Meridian Series on Australian Geographical Perspectives.

By contrast, *Understanding Cities*, a three-volume set of course materials prepared by a group of urban specialists teaching at the Open University (**Massey** *et al.* 1999), steps out in new directions in a determined effort to challenge students to approach cities afresh.

Susan Hanson is preparing an urban geography text for the Arnold series Human Geography in the Making, which will be worth watching out for.

1

URBANISATION PROCESSES AND PATTERNS IN AN ERA OF GLOBALISATION

INTRODUCTION

Definitions

Urbanisation is defined as an increase in the proportion of a country's population living in urban centres. When it comes to comparing country-by-country trends, measuring urbanisation presents some difficulties because there is no international agreement about the size a settlement needs to reach to count as 'urban'. Currently, the best estimate places the proportion of the world's population living in urban centres at between 40 and 55 per cent depending upon the criteria used to define an 'urban centre' (United Nations 1996, 14). Just a small adjustment in the census definitions used by populous countries like India and China could make a considerable difference to this estimate of urbanisation.

The other source of variation in statistical estimates of urbanisation arises from the different administrative boundaries that city governments adopt, and what parts of the built-up and adjacent areas of a metropolis are then counted. Table 1.1 shows how estimates of city size can vary depending upon which administrative areas are counted in or out. For example, the Shanghai Autonomous Region, with a population approaching 19 million, covers over 6000 square kilometres of some of the most fertile land in China. As well as feeding Shanghai's population, this hinterland supports a dense network of villages. The delimited boundary for 'urban' Shanghai spills over into densely populated countryside (Fig. 1.1).

The continuing sprawl of urban regions in the United States also led to an updating of boundaries in the 1980s. The so-called functional urban region takes in all those daily commuters linked by their journey to work into the greater metropolitan labour market. By comparison, statistical definitions in European countries do not yet encompass this 'functional urban region' and as a consequence understate city size. In the United Kingdom by 1990, Glasgow, Liverpool and Newcastle would all have qualified as 'million' cities if the population within their functional urban regions had counted. This means that any league table of the world's largest cities, such as Table 1.2, is always open to conjecture.

The other definitions that cause problems raise quite a different set of issues of usage. These are the bipolar categorisations 'First World/Third World' and 'North/South'. It is difficult to avoid such usage in a text like this, so we should be clear about how this is intended. Globalisation is concentrating technology, market power and wealth within a group of countries roughly corresponding to the Organisation for Economic Cooperation and Development (OECD). These are often described as the First World bloc of countries and many of them were colonial powers that left colonies in a state of economic dependency. Former communist countries – or the Comecon bloc, an economic association of the Soviet Union's east European satellites – made up a loosely described 'Second World' grouping. Before the fracturing of Euro-communism, that left a group of variously underdeveloped or less developed Third World countries in Asia, Africa and Latin America. United Nations publications now use the term 'less developed countries' (LDCs) because, as well as

TABLE 1.1 *Examples of how city size varies with boundary definitions*

City or metropolitan area	Date	Population	Area (km²)	Notes
Beijing (China)	1990	2,336,544	87	Four inner city districts including the historic old city
		c.5,400,000	158	'Core city'
		6,325,722	1369	Inner city and inner suburban districts
		10,819,407	16,808	Inner city, inner and outer suburban districts and eight counties
Dhaka (Bangladesh)	1991		6	Historic city
		c.4,000,000	363	Dhaka Metropolitan Area (Dhaka City Corporation and Dhaka Cantonment)
		6,400,000	780	Dhaka Statistical Metropolitan Area
		< 8,000,000	1530	Rajdhani Unnayan Kartipakhya (RAJUK) – the jurisdiction of Dhaka's planning authority
Katowice (Poland)	1991	367,000		The city
		2,250,000		The metropolitan area (Upper Silesian Industrial Region)
		c.4,000,000		Katowice governorate
Mexico City (Mexico)	1990	1,935,708	139	The central city
		8,261,951	1489	The Federal District
		14,991,281	4636	Mexico City Metropolitan Area
		c.18,000,000	8163	Mexico City megalopolis
Tokyo (Japan)	1990	8,164,000	598	The central city (23 wards)
		11,856,000	2162	Tokyo prefecture (Tokyo-to)
		31,559,000	13,508	Greater Tokyo Metropolitan Area (including Yokohama)
		39,158,000	36,834	National Capital Region
Toronto (Canada)	1991	620,000	97	City of Toronto
		2,200,000	630	Metropolitan Toronto
		3,893,000	5583	Census Metropolitan Area
		4,100,000	7061	Greater Toronto Area
		4,840,000	7550	Toronto CMSA equivalent
London (UK)	1991	4230	3	The original 'city' of London
		2,343,133	321	Inner London
		6,353,568	1579	Greater London (32 boroughs and the city of London)
		12,530,000		London 'metropolitan region'
Los Angeles (USA)	1990	3,485,398	1211	Los Angeles City
		9,053,645	10,480	Los Angeles County
		8,863,000	2038	Los Angeles-Long Beach Primary Metropolitan Statistical Area
		14,532,000	87,652	Los Angeles Consolidated Metropolitan Area

Source: United Nations (1996, 15). For notes on primary data, see original reference in text. Reproduced by permission of Oxford University Press

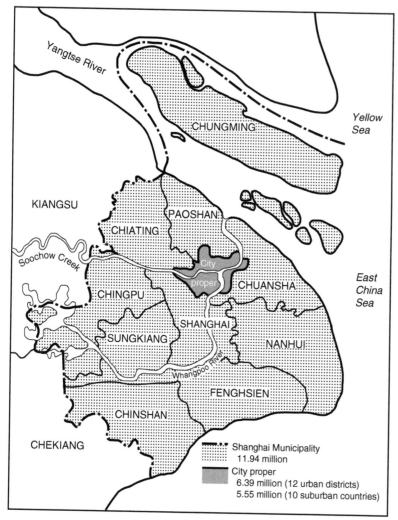

FIGURE 1.1 *The administrative areas of Shanghai for which urban statistics are calculated, c. 1980*
Source: Map of Shanghai Municipality

these geopolitical changes, uneven economic growth in the Third World is making for a more and more differentiated 'South'.

Notwithstanding the economic setback in Asia in 1987, a number of newly industrialising economies (NIEs) like Taiwan, Malaysia, Singapore, Indonesia and Thailand recorded impressive rates of growth during the 1980s. China was admitted to membership of the World Trade Organization in 2001 and is now cautiously opening up its huge market to foreign investors and goods. While South Asia and much of Africa have yet to be integrated into the new global economy in any meaningful way,

most Latin American countries remain peripheral economies with the International Monetary Fund (IMF) dictating their economic policy.

One consequence of this is that the prospects of cities and regions around the world are subject to these differences in the degree of economic development and integration of countries into the new global economy. At the same time globalisation processes are breaking down such clear-cut distinctions in many cities. In *In the Cities of the South*, Jeremy Seabrook (1996) reminds us that political and business elites in Third World countries generally enjoy a 'First World' lifestyle and are insulated from the

TABLE 1.2 *Actual (1950, 1994) and projected (2015) populations of the 15 largest cities in the world*

	1950	(million)	1994	(million)	2015	(million)
1	New York	12.3	Tokyo	26.5	Tokyo	28.7
2	London	8.7	New York	16.3	Bombay	27.4
3	Tokyo	6.9	São Paulo	16.1	Lagos	24.4
4	Paris	5.4	Mexico City	15.5	Shanghai	23.4
5	Moscow	5.4	Shanghai	14.7	Jakarta	21.2
6	Shanghai	5.3	Bombay	14.5	São Paulo	20.8
7	Essen	5.3	Los Angeles	12.2	Karachi	20.6
8	Buenos Aires	5.0	Beijing	12.0	Beijing	19.4
9	Chicago	4.9	Calcutta	11.5	Dhaka	19.0
10	Calcutta	4.4	Seoul	11.5	Mexico City	18.8
11	Osaka	4.1	Jakarta	11.0	New York	17.6
12	Los Angeles	4.0	Buenos Aires	10.9	Calcutta	17.6
13	Beijing	3.9	Osaka	10.6	Delhi	17.6
14	Milan	3.6	Tianjin	10.4	Tianjin	17.0
15	Berlin	3.3	Rio de Janeiro	10.4	Metro Manila	14.7

Source: United Nations, *World Urbanisation Prospects* (1994)

conditions that confront the residents of big cities like Bangkok, Calcutta or Dacca on a daily basis. Similarly 'Third World' conditions can now be found in global cities like New York, Los Angeles, London and even Sydney, where there is a thriving informal economic sector. To some extent the 'hidden' part of this informal economy exploits the 'sweated labour' of newly arrived immigrants.

URBANISATION AND MIGRATION

By the late 1990s, the International Labour Organization (ILO) estimated that, worldwide, there were about 120 million people on the move annually (Stalker 2000). The move into and between cities is the single most important factor behind the population shifts taking place globally. Each year, between 20 and 30 million people leave the land or small villages in the countryside to move into medium-sized towns and cities. Nevertheless, there has been some slight slowing in the overall rate of urbanisation on a worldwide basis in each of the decades since the 1950s (United Nations 1996). Importantly, how different cultures view the economic participation of women has a major bearing on the gender make-up

of the migration streams and, ultimately, on the sex ratio of urban populations. In South Asia, North Africa, the Middle East and many parts of sub-Saharan Africa, more men move to the cities and head urban households because of customary prohibitions on the right of women to work outside the home. By contrast, in the towns and cities of East and Southeast Asia, and Latin America and the Caribbean, women outnumber men and head up more urban households simply because they leave the countryside in greater numbers (United Nations 1996, 12). In an era of globalisation, many of these rural to urban migrants originate in other countries and are contracted to work in cities where there is a domestic shortage of labour (see Chapter 5).

Apart from natural increase, rural to urban migration is only one set of flows that contributes to the changing level of urbanisation. Rural to rural moves take place in the countryside and in total may exceed cityward migration. In some countries, as migrants move up the urban hierarchy in pursuit of better prospects, people moving between cities outnumber those migrating from the countryside. Yet in the United States and Australia, where about one in five people changes location each year, there is little change to the degree of urbanisation because these are inter- or intra-city moves. Urban to rural migration, which takes a number of forms, is another source of change in the size of cities. This can be temporary, when city workers return to the rural area to visit or help with the harvest. Or it can be permanent if labour conditions deteriorate in the city, or when city workers decide to live in the countryside and commute to jobs in town. There are always people returning to the countryside, even if a country is urbanising: an increase in the degree of urbanisation in any country simply means that the migrants moving into cities outnumber those leaving.

Millions of people residing in rural locations actually have jobs that they commute to in urban areas. In rural Europe and North America, these households generally have access to the same range of homes and services available in the cities. They are some of the people who have left the largest cities in the North for lifestyle-related reasons during the last inter-census period. This exodus from British, US and Australasian cities is termed 'counter-urbanisation' (Champion 1989; Boyle and Halfacree 1998).

Mega-cities

Not only will urbanisation continue into the foreseeable future; the character of settlement distribution is changing on a global scale (United Nations 1996). This is especially true of the population of those Third World countries that are growing at such unprecedented rates. It is not surprising that Third World cities mirror these trends. The average size of the world's 100 largest cities had grown to 5 million persons by 1990 compared to 2.1 million in 1950 and fewer than 200,000 in 1800. Of the 3000 million people living in urban areas at the turn of the century, proportionally more are migrating to the very biggest cities. Even though there are some commentators who object to putting these cities into the same basket, the term mega-city is now widely used to describe such gigantic cities.

Even though as many people will continue to live in rural villages and towns in Africa, Asia and Latin America, this is where most of the very biggest cities will also be in the future. By 2015, 13 of the world's 15 largest cities, which are ranked by size in Table 1.2, will be located in the South (Fig. 1.2).

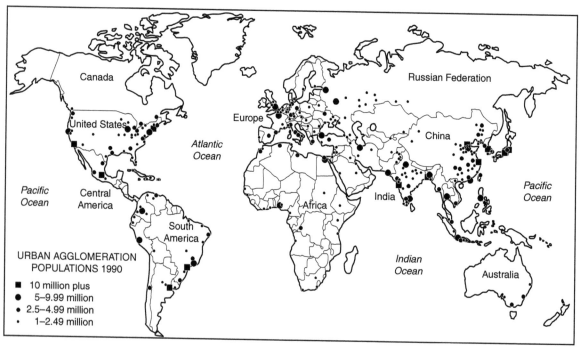

FIGURE 1.2 *The world's million-cities, including mega-cities, in 1990*
Source: United Nations (1996). Reprinted by permission of Oxford University Press

Over 60 per cent of the urban population of Asia already resides in cities like Bangkok, Shanghai and Manila, which means that most are destined to spend their lives in squalor. As Doreen Massey (1997, 100) reminds us, the First World cities that loom so large in our intellectual, and even political, debates – places like London, Los Angeles, New York and Paris – are now quantitatively unimportant in the context of worldwide twenty-first-century city life (Table 1.2).

Urbanisation is tied to economic and social development (Bairoch 1988). This chapter provides a broad description of urbanisation processes and patterns as a backdrop to the later focus on economic activity in cities. The way Chapters 1–5 are organised is admittedly arbitrary. Each of the aspects I deal with in these chapters is really an inseparable part of the same economic system. This chapter begins with a brief review of the history of urbanisation, followed by a more elaborate discussion of how globalisation is contributing to the changing nature of urbanisation. Then Chapter 2 switches to urban systems and the growth of cities. Here the focus is on the external relations and linkages formed between cities, especially the hierarchical repositioning of cities that is taking place within a world-city system. However, in Chapter 3, where we look at the geography of wealth creation in cities, it will be evident that the economic fortunes of cities are equally bound up with their responsiveness to opportunities arising in a global marketplace. In turn, that both reflects the locational requirements of business enterprises and has a powerful influence on their distribution within cities (Chapter 4). Lastly, a city is also a complex labour market, and this raises a series of important questions about the changing geography of work performed in cities (Chapter 5).

There is one further preliminary point to make. In such a brief overview of urbanisation there is always a risk of dwelling on what we observe to be common to cities in spite of their varied historical experience and cultural identity. This sense of sameness, yet difference, in urbanisation processes and patterns is mentioned by Seabrook (1996). Having grown up after the Second World War in the remnant landscape of industrial Birmingham, he decides, 'When I see the industrial and urban landscapes in Bangkok or Calcutta, or Manila now, I do not feel I am in a foreign country.' According to Seabrook, city-bound peasants in China or India experience the same dislocations and, where they can, consolations as

farm workers displaced by the enclosure movement in eighteenth-century Britain. Yet urbanisation in China or India, and other parts of the Third World, 'is not the one-way traffic it became in Britain in the early industrial period' (Seabrook 1996, 11). Most of these migrants to the city have a home place in a rural village with which they rarely sever all connections. They send remittances back to their family and return to help out at harvest time.

With the expansion of manufacturing and the beginnings of 'municipal socialism', European cities grew more prosperous and living conditions gradually improved in the last two decades of the nineteenth century. By contrast, few countries in the South enjoyed any growth in per capita income between 1975 and 2000. As a consequence, 'The number of urban dwellers living in poverty increased sharply in many countries during the 1980s and early 1990s' (United Nations 1996, 5). The main exceptions have been the newly industrialising economies in Asia, such as China, South Korea, Taiwan, Hong Kong, Singapore, Thailand, Indonesia and Malaysia. But living conditions deteriorated markedly in Southeast Asian cities following the collapse of those economies in 1997. The Chinese managed to ward off the repercussions of the Asian crisis, partly because they had not been as reliant on foreign investment to build their economy. The World Bank estimates that the number of people in Southeast Asia living below the poverty line jumped from 30 million to 60 million between 1998 and 2000. Many of those thrown out of work in the cities returned to their village to put pressure once more on local resources, not to mention the environment.

A BRIEF HISTORY OF URBANISATION, 1500–1950

For most of recorded history the world's population has been anything but urbanised. While the few urban settlements that were creatures of the great hydraulic civilisations of Mesopotamia, the Indus and China possessed the technology and standing armies to command vast empires, the populations they subdued remained overwhelmingly agrarian. In its heyday, second-century Rome had a population that probably reached 800,000 supported by a population of 50–55 million subjects (Bairoch 1988, 81–3). On the other hand, Athens, as the 'cradle of western civilisation', was a notable city-state without much territory.

Later in Europe, the colonisation of Spain by Muslim armies brought with it a transplanted urban culture. By the year 800, over a third of all the cities in Europe with 10,000 or more residents were located in Spain. By 1000, Cordoba had a population of 400,000–500,000 and Seville nearly 100,000. It took until 1700 before any other European cities – Paris and London – grew as large again. Hence it was not until the fourteenth century and the beginnings of commercial expansion that European city growth and urbanisation began to support more of the population. At this stage we will turn to some of the evidence of urbanisation and urban development between 1500 and 1950 in what were to become metropolitan powers and their colonies, leaving the question, 'How did this come about?' until Chapter 2.

Beginnings of urbanisation in western Europe

The Renaissance in Europe fostered trade links between the Italian city-states of Venice, Genoa, Milan, Florence and Vienna, and centres in the Low Countries like Bruges and Ghent, which were building a reputation for handicrafts. By 1500 there were just six European cities with populations in excess of 100,000 – though only a minority actually enjoyed the 'freedom of the city'. Otherwise the bulk of the population lived in small towns or villages and travelled out to surrounding fields to work. Then, with the rise of capitalism and nation-building during the sixteenth and seventeenth centuries, capitals like London, Paris and Madrid began to dominate the settlement system in western Europe. In times when communicable diseases like the plague could decimate an urban population, such cities depended heavily on migration from the countryside and beyond for their continuing growth. Consequently, 'Western Europe, once an area of mainly small cities with market areas limited by access from their agricultural hinterlands (classically in the German city states), became a region of rapid city growth within powerful nation states (Italy and Germany excepted)' (Lawton 1989, 2).

From the late sixteenth century onwards, these European powers explored and settled three continents. Over the next three centuries, fierce rivalries developed between Spain, England, Holland,

Portugal and France as they competed to carve up the New World and establish their own commercial spheres of influence. Russia, Germany, Italy and the United States joined in much later once they were sufficiently united on the domestic front to look outwards. The global trading system that the European powers eventually formed was tied together by a series of urban nodes made up of the imperial capitals, colonial ports and frontier outposts. Still, at the beginning of the nineteenth century it would be misguided to think that European expansionism alone was responsible for stimulating urbanisation on a global scale.

Earliest cities of Asia

As Table 1.3 reveals, Asia had more than three-fifths of the world's 100 largest cities in 1800. Historically, Asia has consistently concentrated a high proportion of the world's urban population and has always had most of the world's largest cities (Bairoch 1988). And while the size of some of these cities was undoubtedly boosted by their role as trading posts (Calcutta, Tokyo) or Treaty Ports for colonial powers (Peking, now Beijing; Canton, now Guangzhou; Amoy, now Xiamen; Tiensin, now Tianjin), others have long histories that pre-date the arrival of Europeans. Among Asian cities that owe their origins to quite different circumstances are Kyoto and Edo (now Tokyo) in Japan, Seoul and Pusan in South Korea, Karachi and Lahore in Pakistan, Delhi, Bombay and Bangalore in India, Jakarta and Surabaya in Indonesia and many of China's major cities. We will find out more about the origins of some of these cities in Chapter 2.

Industrialisation and the growth of 'great cities'

Of the 100 largest cities listed in Table 1.3 only a few exceeded 100,000 by 1800, a figure that was quite exceptional prior to the fifteenth century. Peking was the only 'million-city' in 1800. But by 1850 it had been joined by London and Paris, and by 1900 there were 16 such cities – mainly in Europe and North America – and another 27 of over half a million. Meanwhile, worldwide, the number of cities over 100,000 rose from 65 to 106 between 1800 and 1850, and nearly trebled again by 1900 (Lawton 1989, 3).

TABLE 1.3 *The regional distribution of the world's population in million-cities and the location of the world's largest 100 cities (1990, 1950 and 1800)*

| | Approximate proportion of the world's population | | | | Number of the world's 100 largest cities in | | |
| | living in urban areas | | living in million-cities | | | | |
	1950	1990	1950	1990	1800	1950	1990
Africa	4.5	8.8	1.8	7.5	4	3	7
Eastern Africa	0.5	1.7	—	0.8	—	—	—
Middle Africa	0.5	1.0	—	0.8	0	0	1
Northern Africa	1.8	2.8	1.8	3.2	3	2	5
Southern Africa	0.8	0.9	—	0.8	0	1	0
Western Africa	0.9	2.6	—	2.0	1	0	1
Americas	23.7	23.0	30.1	27.8	3	26	27
Caribbean	0.8	0.9	0.6	0.8	1	1	0
Central America	2.0	3.3	1.6	2.7	1	1	3
Northern America	14.4	9.2	21.2	13.1	0	18	13
South America	6.5	9.7	6.7	11.1	1	6	11
Asia	32.0	44.5	28.6	45.6	64	33	44
Eastern Asia	15.2	19.7	17.6	22.2	29	18	21
Southeastern Asia	3.7	5.8	3.4	5.6	5	5	8
South-central Asia	11.2	14.8	7.0	14.6	24	9	13
Western Asia	1.8	4.1	0.6	3.3	6	1	2
Europe	38.8	22.8	38.0	17.9	29	36	20
Eastern Europe	11.8	9.3	7.7	6.3	2	7	4
Northern Europe	7.7	3.4	9.0	2.1	6	6	2
Southern Europe	6.5	4.0	6.7	3.2	12	8	6
Western Europe	12.8	6.2	14.6	6.2	9	15	8
Oceania	1.1	0.8	1.6	1.3	0	2	2

Source: United Nations (1996, 20). For notes on primary data, see original reference in text. Reproduced by permission of Oxford University Press

The growth of these 'great cities' in Europe and the United States during the nineteenth century was both a necessity to support industrialisation and a consequence of it. Together with manufacturing towns they were the centres where capital and labour came together to transform raw materials like coal and iron, lumber and clay, cotton and wool, and sugar and jute. 'The rise of large manufacturing centres transformed the scale and pattern of world-cities by the mid-nineteenth century, and confirmed Europe's lead in the new urbanisation' (Lawton 1989, 3). But this was just one side of the urbanisation coin. Other

preconditions for urbanisation in Europe included the pressure of unprecedented rural population growth; the enclosure movement; mechanisation and labour-shedding in agriculture; and increases in farm productivity sufficient to feed the urban workforce.

Urbanisation in Britain led the way. In the space of 90 years, between 1801 and 1891, the proportion of the English population living in towns rose from one-quarter to three-quarters – up from 17 per cent in 1801 to 61 per cent by 1911 for towns over 20,000. Before 1800, London was the only city over 100,000, but by 1911, 44 other British cities shared that same status. Centres of British industry such as Manchester and Birmingham doubled in size between 1801 and 1831, and then again over the next 30 years. As to be expected, the redistribution of the population, along with the variable rates of urban growth, reflected the differing factor endowments of the regions relative to the rise to prominence of staple industries during the nineteenth century. So the mining, textile and clothing centres of Lancashire and Yorkshire grew fastest at first, followed by centres specialising in heavy engineering and metal manufacturing (such as Sheffield, Glasgow, Belfast, Newcastle, Birmingham and Coventry).

Colonialism and urbanisation as nineteenth-century globalisation

In recent revisions of the history of urbanisation, leading figures like Anthony King (1990a; 1990b) have developed a perspective that stresses the essentially interdependent nature of urbanisation and urban processes. From about 1500 onwards in Europe, and much later in North America, urbanisation, and the growth of the metropolitan centres that went with it, depended on the development of colonial economies. In 'world system' terms (Wallerstein 1974), the metropolitan, or colonising, powers are conceived as the core economies, and the colonies as peripheral economies. This is to say that urbanisation and city-building in Europe, on the one hand, and in far-flung colonies, on the other, had as much to do with the asymmetrical flows of labour, capital and commodities between the core metropolitan powers and the colonies as between the countryside and the growing cities within the domestic economy. Europe's core economies were dominantly a source of producer goods and investment capital, while their colonies supplied raw materials and served as markets. During the age of imperialism, colonial cities established by the Spanish, Dutch, Portuguese, British and French functioned like a hinge connecting the core and peripheral economies during the age of imperialism. Through these cities flowed capital, settlers, commodities and often an equally invasive imperialist culture.

These processes represent some of the precursors to 'globalisation' with its emphasis on the growing integration of the world's trade system, financial system, political institutions, markets, patterns of labour mobility, and cultures. In fact, two economic

historians, O'Rourke and Williamson (1999), argue that the countries constituting the nineteenth-century 'Atlantic economy' of western Europe, North and South America, and Australasia were exposed to globalisation processes every bit as dynamic as those currently transforming economies. With all the hype currently surrounding globalisation, it is worth remembering that in many respects it is building upon earlier forms of economic interdependence.

Yet there are two fundamental differences between these colonial interdependencies (King 1990a) and those that now characterise the global economy. Firstly, in their external relations prior to 1950, colonies like Australia, Indonesia or Kenya were tied to the metropolitan parent. By and large trade ties were one to one. Secondly, since 1950 there has been a progressive coming-together and fusing of international trade, investment and finance to form an integrated global economy. With the breaking-up of colonial empires and the crumbling of power blocs in the East and the West, multilateral relations are now the order of the day. These recent developments within geopolitics have helped to shape the global pattern of urbanisation since 1950.

Urbanisation in New World countries

Whereas industrialisation was the mainstay of urban growth in Britain, in the New World urbanisation had its beginning in mercantilism. 'Hinge' cities like New York, Boston, Philadelphia and Baltimore on the eastern seaboard of the United States, and like Sydney, Melbourne, Adelaide and Auckland, were facsimiles of the UK's Bristol and Liverpool. Manufacturing did not become the cornerstone of US metropolitan growth until well into the 1860s (Pred 1966, 143). Likewise, a high degree of urbanisation had been achieved in Australia long before manufacturing was significant: 'In British economic history it is possible to talk of factories giving rise to towns; in Australia towns appear to have given rise to factories' (Glynn 1970, 17). In 1861, when approximately 40 per cent of Australians lived in towns and cities, manufacturing accounted for less than 4 per cent of the gross domestic product. However, unlike the USA after 1860, when the richly endowed industrial heartland came to concentrate American manufacturing, industrial production in Australia stuck to its 'hinge' city beginnings. By 1891, allowing for some variation in census definitions, about two-thirds of Australia's population lived in towns and cities. The United States and Canada did not match this degree of urbanisation until 1920 and 1950 respectively.

Urbanisation since 1950

Estimates of urban growth can be misleading in cases where the population base is large to start with. Under these circumstances, very large increases in population have to be recorded to produce especially impressive rates of urban growth (United Nations 1996, 18). Bombay was the only city to report an annual increase in population over the 400,000 mark through the 1980s, and a large part of this was due to the extension of the city boundary between censuses. Two other urban agglomerations, Tokyo and Jakarta, grew by more than 300,000 a year between 1980 and 1990.

The boom city of Surat provides an illustration of the phenomenal growth that Asia's fastest-growing cities experienced during the 1980s and 1990s. Surat lies at the heart of the 'golden corridor' of industrialisation of Gujarat in western India. Surat's population was 471,000 in 1974, 925,000 in 1984 and over 1.5 million in 1991. This represents an increase of 64 per cent between 1984 and 1991 (Seabrook 1996, 8).

Table 1.3 summarises the broad trends characterising the global pattern of urbanisation since 1950. The first thing to note is the changing *regional* distribution of the urban population and the relative concentration or de-concentration, as the case may be, of the population in million-cities between 1950 and 1990. There is a clear distinction between trends in the South and in the North. Without exception, all subregions of Asia and Africa increased their shares of the world's urban population and of the population living in million-cities. In the Americas, Caribbean countries joined with Central and South America to mirror these trends. By contrast, the United States and Canada, along with all of Europe and Oceania, lost shares of the world's urban population and of the proportion of people living in million-cities. In fact, Europe's share of the world's urban population living in million-cities was more than halved between 1950 and 1990 (Table 1.3).

Asia leads the way in the case of million-cities. Between 1950 and 1990, Asia's share of the world's urban population concentrated in million-cities climbed from 28.6 per cent to 45.6 per cent. Put another way, this represents 42 per cent of the world's million-cities, and half of the ten largest urban agglomerations (Tokyo, Shanghai, Beijing, Bombay and Calcutta) (United Nations 1996, 75). Africa is also experiencing rapid rates of growth in shares of the world's urban population and million-cities. The rate of growth is made more impressive by the fact that, with the exception of northern Africa, the colonial powers restricted the numbers of Africans living in urban centres. At the time independence was granted in the 1950s, the colonial capitals were comparatively small but, since then, rural Africans have migrated in large numbers.

As noted, an important intra-regional shift is now under way in the Americas. By 1950, following industrialisation, the United States and Canada had most of the urban and million-city population. But by 1990 this was no longer the case. As the United Nations volume (1996, 18) points out, this shift is something of a reversion to pre-Columbian and colonial times, when most of the urban population and the major cities were in Central and South America. At the same time, this should not blind us to the fact that some of the newer cities in the southern and western states of the USA recorded growth rates between 1950 and 1990 that were among the world's fastest. For example, though much smaller in 1900, by 1990 both Miami and Phoenix had outgrown a fast-growing African city like Nairobi. The same holds for Los Angeles, which was about one-tenth the size of Calcutta in 1900, yet by 1990 they both had roughly 11 million people residing within their metropolitan regions.

The decline of Europe's shares of the world's urban population and million-cities is particularly striking. Part of the reason lies with the challenge since 1950 to the pre-eminence of great manufacturing centres like Milan in Italy, Hamburg and Düsseldorf in Germany, and Birmingham and Manchester in England. But as well as that, European countries have been among the first to face stalling, or even declining, population growth.

URBANISATION AND GLOBALISATION SINCE 1950

Although the rising level of urbanisation has been interrupted in some countries, internationally it shows no sign of abating. Continuing urbanisation since 1950 has been encouraged by political events and economic developments on the international stage. It is evident that the countries with the fastest-growing economies since 1950 are generally those with the most rapid increase in their level of urbanisation, and that the world's largest cities are heavily concentrated in the world's largest economies. On the political front, we have already noted that decolonisation in many countries was accompanied not only by a transfer of administrative powers, but by the removal of migration controls on the indigenous population and changes to the prevailing settlement pattern.

At the same time, the global economic system has been transformed from one of protected, if not entirely closed, national economies, into one where trade liberalisation has opened up most economies and where international trade, investment and finance are increasingly integrated. These processes were undoubtedly extended by the fall of the Berlin Wall in 1989 and many commentators mark the advent of full globalisation from this date. With globalisation forcing down production costs, corporations that were previously anchored within the nation-state have 'gone global' and shifted their operations from high-cost to low-cost locations (Thurow 1999, 60). The 'hyper-mobility' of capital and information, and the worldwide movement of labour, including the corporate business elite and contract workers, are also defining features of globalisation. All this has been aided by the steady fall in the cost of air travel, telecommunications and data processing between 1950 and 1990 (Table 1.4).

Some comprehension of globalisation processes is necessary if we are to make sense of urbanisation

processes and patterns during the late twentieth century. But in addition to this there are now cities that are so caught up in the global transformation that they warrant singling out for special attention as globalising cities (Marcuse and van Kempen 2000). 'Without such an understanding, it is difficult to make sense of the very rapid growth of many cities in China, the increasing concentration of the world's urban population in Asia, the slowing in growth of most major cities in Latin America (at least up to the early 1990s), the radical restructuring of the urban system in the United States and the revival of certain major cities in Europe' (United Nations 1996, 3).

We have arrived at the point where we need quickly to take stock of globalisation processes and how they can affect urbanisation and the growth of cities. (These general points will be illustrated more fully in the discussion of a functional world-city system in Chapter 2.) Nigel Thrift (1994), an urban geographer based at the UK's Bristol University, suggests that the following five processes are involved in the globalisation of the world's economic system:

- the global restructuring of production giving rise to global oligopolies
- the globalisation of finance and the consolidation of its power over production
- the global shift to knowledge industries and information processing
- the international recruiting and movement of labour (including corporate 'high-fliers' and contracted 'guest' workers)
- the rise of transnational economic diplomacy and multilateral agreements between states.

TABLE 1.4 *Falling costs of air travel, telecommunications and data processing (in US$, 1990)*

Year	Average air travel cost per passenger mile	Cost of a 3-minute phone call from New York to London	Relative cost of computing power
1950	0.30	53.20	
1960	0.24	45.86	125,000
1970	0.16	31.58	19,474
1980	0.10	4.80	3620
1990	0.11	3.32	1000

Source: IMF (1997)

THE RISE OF GLOBAL OLIGOPOLIES

The tendency towards centralisation and concentration of economic activity has been one of the most noticeable features of the forward march of western capitalism. In an era of globalisation the restructuring of productivity activity has been marked by several key developments. Firstly, it has been driven by technological advances giving birth to new industries while destroying those denied access to the breakthrough (Table 1.5). For example, the microprocessor gave the personal computer such a competitive edge over the mainframe that IBM was overtaken by the new computer industry leaders, Intel and Microsoft. Partly as a result of developments in technology, only 6 of the 25 largest corporations in 1960 remained on the list in 1997 (Thurow 1999, 59). In the 1980s, as part of the changing nature of the new global economy, Japanese, Taiwanese and South Korean transnational corporations rose to prominence. During that decade, Japan possessed the world's 10 largest banks and 315 of the top 1000 corporations, and it led in 25 of 34 technologies considered essential for a post-industrial world (Lo and Yeung 1996, 24).

Secondly, in the last 50 years or so, international trade has multiplied twelve-fold and changed from a dominance of commodities to a dominance of finance and specialised services. Transnational corporations now determine terms of trade. By the mid-1990s, just 500 corporations accounted for two-thirds of global trade, and around 40 per cent of this global

trade was between the branch plants and offices making up these corporations, regardless of where they were located around the globe (United Nations 1996, 9). But in 1983, with the emergence of Asia's newly industrialising economies, trans-Pacific trade overtook trans-Atlantic trade in absolute terms for the first time. 'Pacific Asia has become one of the fastest-growing core regions of the world, with progressive restructuring of production linkages, trade relations, and foreign investments' (Lo and Yeung 1996, 24).

Thirdly, an increasing share of global production and the services that support it is organised and planned as part of a global corporate strategy, and delivered via global corporate networks. Illustrations of how globalisation is affecting corporate operations and changing global flows of labour, markets and the face of cities abound. Here is just one from the hotel brokerage industry: the Jones Lang LaSalle Hotels group operates in 15 countries. Its CEO is based in Australia because, 'It's no longer necessary to be in an office. When your clients are everywhere you can never be in all the places you'd need to be.' Australia is very close to Asia, where 300 hotels were being built when the Asian 'meltdown' occurred: 'The majority will need to be sold, restructured or refinanced . . . Hotel markets are cyclical, the US is kicking at the moment, the UK is coming off the boil but Europe is strong, Australia has stabilised while Asia is still in a hole. Being global means that two-thirds of our operations always do well' (quoted in the *Australian Financial Review*, April 1999).

Fourthly, adoption of a global market strategy invariably involves taking over other businesses that expand market share or complement the operations of the parent company. Ironically, the relaxation of regulations in many countries restricting the foreign takeover of nationally owned assets has helped to facilitate the rise of global oligopolies, which are now powerful enough to challenge the nation-state. Cross-border mergers and acquisitions (M&As) are one of the hallmarks of full-throttle globalisation. Between 1991 and 1998, the global value of cross-border mergers increased sixfold from US$85 billion to US$558 billion (reported in the OECD *Observer*, December 1999). According to *Business Week* (December 1997), in the 1990s, 'The most powerful economic forces of our era – heightened international competition, the rise of fleet-footed

TABLE 1.5 *The changing nature of global competition*

Decade	Factor	Threatened national industries
1960s	Labour intensity	Textiles, shoes, simple assembly
1970s	Capital intensity	Automobiles, machinery, chemicals
1980s	Technology	Consumer electronics, telecommunications
1990s	Information	Financial services, media, 'systems' businesses

Source: Stopford and Strange (1991, 36)

entrepreneurs, an explosion of new information technologies and deregulation – are transforming industries from telecommunications to health care.'

These mergers, and the investment flows and jobs that these global corporations now control, can seal the fate of cities and their communities (see Chapter 3). For example, in 1998 the last European computer manufacturer, Siemens Nixdorf, was sold to Acer of Taiwan. Where does this leave European economies in the wealth creation stakes, and how will it impact upon the future income stream and jobs of Europe's cities? The 'Wimbledonisation' of the City, London's financial heart, provides another example. During the 1980s and 1990s, British firms and state-owned enterprises were sold to foreign corporations with a global orientation. Deregulation – the 'Big Bang' – created favourable conditions in which foreign players enter, and always win, the tournament. The fair play of the crowd and the skill of the grounds staff are a metaphor for the City. While the City's banks, shipping and insurance businesses earn more money abroad than any other industry in Britain, the controlling interest largely resides with foreign banks and securities houses.

Fifthly, the worldwide restructuring of production by transnational corporations has dramatically changed, and will go on changing, the international division of labour. This is being driven by the need for business to remain competitive in a global marketplace. In a global economic system with falling tariff barriers, business is forced to seek out locations where factor costs are as low as possible. These costs relate to the so-called factors of production, including the purchase of a suitable site, installation of up-to-date plant and equipment, and labour costs. Economic theory argues that, given the differing factor endowments of locations, wage rates will eventually be equalised in a 'borderless' global economic system. This opens up the possibility that the wages of low-skilled Europeans or Americans are now set in Beijing, Delhi or Djakarta rather than in Paris, Frankfurt or New York (Freeman 1995).

The 'new international division of labour' (Fröebel, Heinrichs and Kreye 1980) can adequately be explained in these terms. Between 1965 and 1990, the Third World's share of the global labour force grew from 69 to 75 per cent while, at the same time, the average number of years spent at school by workers in those countries doubled (Freeman 1995, 20). Thus the search by global corporations for

lower-cost locations is drawing these workers into the cities and special zones set up by national and city governments in their attempts to attract investment and jobs.

Technological change (Table 1.5) affects the structure of industry and the composition of the workforce in different regions and cities. Between the 1950s and the 1990s, services replaced manufacturing as the dominant employer in most of the OECD economies. During the intervening decades, access to the abundant supply of low-cost labour in the newly emerging economies of Southeast Asia, Latin America and eastern Europe forced the closing of manufacturing plants across Europe, North America and Australasia. With the loss of these predominantly blue-collar jobs from the inner cities, metropolitan growth began to slow in the older industrial regions of Europe and North America.

However, as Freeman (1995, 30) points out, one possible effect of the transition to service-based economies in the North is that, 'As more and more low-skilled western workers find employment in the non-traded goods service sector, the potential for imports from less-developed countries to reduce their employment or wages should lessen.'

GLOBAL INTEGRATION OF INTERNATIONAL FINANCE

The growth of world trade since 1945 has been accompanied by much greater financial interdependence between producers and consumers in the global marketplace and the intermediaries serving them. However, the value of international financial transactions has outstripped the growth in traded goods and services to the point where it now exceeds it by about 50 times (O'Brien 1992).

The global expansion and integration of international finance was precipitated by a series of interlocking developments in the 1970s. In 1972 the Nixon Administration broke with the international agreement (Bretton Woods) tying the US dollar as the medium of exchange between countries to the price of gold – the 'Gold Standard'. This took place against a background of offshore investment by American firms in Europe and the rise of the Eurodollar market. The Eurodollar, and now the euro, subsequently became major currencies in world money markets along with the Deutschmark and the Yen. Hence the floating of exchange rates in 1972 opened up opportunities for speculative trading on the futures market against the short- and long-term movement of currencies. By the mid-1990s the volume of business in the currency futures markets exceeded even the flow generated by daily trade in currencies (O'Brien 1992).

The oil price crises in the 1970s and their aftermath also contributed to the much wider global dispersion of capital in the international financial system. When the Organization of Petroleum Exporting Countries (OPEC) acted in unison to raise the price of oil on the world market in 1974 and 1979, this greatly increased the receipts of oil-producing states in the Middle East, Venezuela and Indonesia. As well as boosting their imports from industrial producers by about US$100 billion, between 1974 and 1976, they deposited over US$80 billion with American and European banks (O'Brien 1992). These OPEC deposits were then on-loaned to newly industrialising countries in Latin America and Asia. Following that, the growth of the Japanese economy in the 1980s generated such a surplus that the global pattern of lending was reversed for a time. Both the USA and Britain attracted major inflows of Japanese capital directed at re-investment in manufacturing and urban property development (see Chapter 3). On the other hand, yields on the Tokyo Stock Exchange siphoned investment away from other less profitable equity markets, like the City of London (Fig. 1.3).

The magnitude of foreign direct investment continued to grow throughout the course of the 1980s and 1990s. While economies in the North continue to monopolise foreign direct investment, flows to the South climbed from only a few billion dollars in 1980 to US$56 billion in 1993. By the mid-1990s, the main recipients were China with one-quarter of the foreign direct investment destined for the South, followed by Mexico, Argentina, Malaysia and Thailand. Table 1.6 illustrates the flows of foreign direct investment into and between Association of Southeast Asian Nations (ASEAN) over a three-year period in the mid-1980s. According to the Washington-based Institute of International Finance, the volume of lending to Asian countries had grown to US$100 billion in the year before the 'crash' (1996). The growth in foreign direct investment has heightened the interdependency of financial institutions and their customers around the globe to the

point where, from time to time, the debt levels of some Third World countries have threatened to bring down various European, American and Japanese banks. Foreign banks wrote off upwards of US$300 billion in bad loans in the aftermath of the Asian financial crisis.

Global monetary transactions and international financial services have also grown along with the rise to prominence of transnational corporations. Internationally traded financial services form a specialised group of activities that cater especially for transnational corporations doing business around the globe. They include: credit provision, which greatly expanded in the 1980s with the rise of trading in securities, or 'securitisation'; financial 'engineering' like arbitrage, consulting and packaging loans; insurance and hedging against future risk ('futures'); and expertise in gaining access to 'closed' markets like China and Japan.

Therefore, as the financial system became an essentially international system, cities began to fulfil specialised tasks within that system and their momentum of growth was tied to its expansion. Between 1974 and 1994, the total capitalisation of the world's stock markets expanded more than sixteen-fold in real terms from US$900 million to US$15 trillion. This rate of growth continued unabated during the second half of the 1990s. By 2000, at the height of the stock market boom in the United States, an estimated US$1.5 trillion was being traded on the world's stock exchanges each day. Cities lying in key time zones like London, New York and Tokyo began with a natural advantage once the decision was taken to deregulate national financial sectors because this enabled the world's money markets to operate continuously around the clock (Table 1.7). The changes that these developments in international finance have brought about in these cities are examined in Chapters 2 and 4.

KNOWLEDGE INDUSTRIES AND INFORMATION PROCESSING

Globalisation is also characterised by the proliferation of dense, interactive communications networks that have all but annihilated the barrier of distance for those with access to information technology (IT). In the decades ahead, existing regional and urban inequalities will be aggravated further by

Made in England. Grown in Japan.

A unique opportunity to invest in Japan's economic success.

Economic growth

Japan has the strongest economy in the world. In 1985, productivity increased by 9.8%. Exports have risen an average rate of 7.0% over the past 5 years. So now is a good time to invest in Japan.

But if you are going to invest in the 2nd largest stock market in the world, wouldn't you like to have your investment managed by Japanese fund managers who are experts on every area of the market?

Now you can. by dealing with Nomura.

Nomura's growth

Nomura is the biggest and most respected institution on the Tokyo Stock Exchange. Our dealers handle 15% of all equity trading on the Tokyo Stock Exchange. We have more assets in custody than any other Japanese financial institution – which means that Japanese investors voted with their yen when they were looking for a good investment. Here's how you can join them.

Nomura Growth Fund

The objective of the fund is capital growth by investing in Japanese equities. The Nomura Growth Fund S.A. is an open-ended investment fund listed in Luxembourg whose shares are denominated in yen; which means that your investment is based on one of the strongest currencies in the world.

If you are a professional investor or an advisor whose clients would be interested in investing in a fund, which has shown 50%* growth in sterling terms since its launch in July 1985, simply send the coupon to James Ferguson, Nomura International Ltd., Nomura House, 24 Monument Street, London EC3R 8AJ or call him on 01-283 8811 and he will send you a prospectus by return.

Please send me full details about the Nomura Growth Fund.

NAME

ADDRESS

POSTCODE

Type of Business: Professional Investor/Advisor/Other.
Please delete as appropriate.

TELEPHONE No   GW 15/5

Nomura International Ltd.,
Nomura House, 24 Monument Street, London EC3R 8AJ.

NOMURA

This advertisement is not an invitation to buy securities. *Figures correct at time of going to press.

Statistical Sources: Balance of Payments monthly · Bank of Japan   Economic Planning Agency · Japanese Government   Monthly Statistics Report · Tokyo Stock Exchange
Financial Statistics · Central Statistical Office   Guide to Japanese Investment Strategy · Nomura Securities   OECD estimates

FIGURE 1.3 *'Made in England. Grown in Japan': there was a leakage of investment in the 1980s as Japanese interest rates outstripped those offered in other money markets*
Source: unknown

TABLE 1.6 *Foreign Direct Investment (FDI) in ASEAN countries, 1986–1988 (US$m.)*

Investing country	Recipient country			
	Malaysia	**Thailand**	**Indonesia**	**Philippines**
Taiwan	517.4 (16.3)	158.6 (11.4)	964.5 (15.9)	116.2 (18.0)
Korea	34.4 (0.01)	16.3 (1.2)	119.1 (2.0)	1.2 (0.2)
Hong Kong	163.8 (5.1)	45.9 (3.3)	410.1 (6.8)	53.3 (8.3)
Singapore	349.1 (11.0)	19.5 (1.4)	267.2 (4.4)	3.2 (0.5)
Japan	813.7 (25.6)	738.8 (53.3)	1070.1 (17.6)	179.3 (27.8)
USA	336.2 (10.5)	88.9 (7.2)	859.6 (14.2)	162.5 (27.7)
Elsewhere	969.3 (30.4)	319.3 (23.0)	2371.4 (39.1)	129.6 (20.0)
World	3183.9	1387.3	6062.8	645.3

Source: Economic Planning Agency Japan (1990)

TABLE 1.7 *The world's largest stock exchanges, 1999*

	Total domestic equity	**Number of listed companies**	**Turnover (US$bn)**	**Market value (US$bn)**
1	Nasdaq	9986	5036	4829
2	New York	8655	11,160	3025
3	iX†	4693	4258	3919
4	London	3261	2868	2791
5	Tokyo	1761	4325	1937
6	Germany	1431	1389	1128
7	Euronext*	1301	2405	2038
8	Paris	709	1437	1149

†London and Frankfurt combined following their merger in May 2000.
*The merged Paris, Amsterdam and Brussels exchanges.
Source: London Stock Exchange (reported in the *Observer*, 7 May 2000)

looming gaps between the information-rich and the information-poor. By the late 1990s, OECD countries, with about 15 per cent of the world's population, had about 88 per cent of Internet users. By contrast, South Asia, with a fifth of the world's people, had less than 1 per cent of Internet users. Jeffrey Sachs, Director of the Center for International Development at Harvard, believes this new technology divide has now replaced the old ideology divide in significance (reported in *The Economist*, 24 June 2000). Investment in knowledge industries and advances in

IT sit behind the shift in the wealth-creating capacity of nations. In OECD countries, knowledge-based industries accounted for more than half of the 'rich-country' business output in the mid-1990s. According to Castells (1994) this amounted to a transition in capitalist economies from an industrial to an 'informational mode of development'. Knowledge is the new asset.

The global integration of international finance and investment would not have been possible without the fusion of computing and telecommunications.

FIGURE 1.4 *The European fibre-optic telecommunications network built by Interoute*
Source: *The Economist* (November 2000)

According to Richard O'Brien (1992), 'To a great extent the end-of-geography story is a technology story, the story of computerisation of finance.' Information technology applied to the financial services industry has obliterated delays in counting and transferring funds, altered the way traders and clients interact in markets, and vastly expanded the daily volume and velocity of money circulating in global markets. In 1990 only 100,000 computers were linked to the Internet. By 1998 there were 36 million PCs online, servicing 150 million users. By 2001 the number of people using the net had grown to 700 million; e-commerce is bound to impact upon retailing and distribution in cities in much the same way as electronic funds transfer systems have in the banking and financial services sector.

Castells (1994) expects the revolution in IT to create a more completely integrated network of cities in a united Europe and to lead to the Informational City. This is coming to fruition with the development of a fibre-optic telecommunications network in Europe (Fig. 1.4). And the nerve centres of the informational–global economy are those cities that are securing a competitive advantage in knowledge industries such as financial services, telecommunications, media, advertising and marketing. More will be said in Chapters 2, 3 and 4 about these 'global' cities, and how developments in the IT and knowledge industries are also transforming lower-order cities.

INTERNATIONAL MOVEMENT OF WORKERS, REFUGEES AND TOURISTS

The emergence of a single, worldwide market for labour, coupled with the growth in refugees and asylum seekers, and international tourism, constitute other important dimensions of globalisation. Between 1965 and 2000, the numbers of migrants on the move rose from 75 million to an estimated 120 million (Stalker 2000). Among the different categories of international migrants, the growth in the number of refugees has been the most dramatic. This reflects the increase in political instability and unleashing of nationalist movements in Africa and Europe following the cessation of Cold War tensions at the end of the 1980s. United Nations High Commission for Refugees (UNHCR) estimates suggest that, worldwide, there were 13.2 million refugees in 1996. This represents a doubling in numbers over the period 1985–92. Consequently, applications from asylum seekers wishing to resettle in Europe, North America and Australia multiplied seven-fold between 1983 and 1991. These numbers increased again with the break-up of the former Yugoslavia and the fighting in Afghanistan. Refugees display an overwhelming preference for settling in big cities where communities formed by their own kinsfolk and native speakers are already established. In US cities like Los Angeles and New York, over 40 per cent – and in the case of Miami, 75 per cent – of residents speak a language other than English at home. Flows of international refugees, along with other migrants and short-stay workers, are rapidly changing the demography of globalising cities and producing a geographical 'patchwork quilt' of ethnically and culturally distinct neighbourhoods (see Chapter 7).

One of the hallmarks of globalisation has been the growth of a single international labour market for the professional and managerial staff that transnational corporations compete for. Global corporations recruit internationally and periodically transfer staff with high-level management and technical skills from city to city around the world. Leyshon and Thrift (1997) see this 'footloose' corporate elite forming a transnational business class. Typically, these highly qualified and exceedingly well-paid workers have no place of permanent domicile, and are likely to identify more with the global corporate culture than with local cultures. Japanese corporations represent a notable exception: in 1988, there were 83,000 Japanese working in overseas branches.

The expansion of global banking and financial services over the last two decades has created a worldwide network of offices that are usually staffed at senior levels by overseas personnel with international experience (Beaverstock 1996; Beaverstock and Smith 1996). The emergence of a transnational corporate workforce is helping to transform globalising cities in at least four ways. Firstly, their salaries and conditions are the result of bargaining in a global market, but the presence of these highly paid corporate staff in global cities like London, New York, Tokyo, Los Angeles, Paris, Singapore and Sydney puts upward pressure upon local wage rates and house prices. Secondly, as short-stay expatriates in a foreign country, their loyalties are to the global enterprise rather than to local causes. Thirdly, Richard Sennett, who is Professor of Sociology at the London

School of Economics, contends that these global 'free-riders' in the media and finance sectors tend to pay most of their taxes elsewhere if they can get away with it (reported in *Guardian Weekly*, 4–10 January 2001, 11). Fourthly, they are the bearers of an international business culture that exists apart from the parochial, 'indigenous' culture of cities.

The worldwide movement of contract and illegal migrant workers is also advancing globalisation in regional and big city labour markets (Sassen 1988). An estimated 4 million workers from countries like India, Pakistan, Bangladesh, Sri Lanka and the Philippines were working on contract in the cities of the oil-rich Gulf States at the beginning of the 1990s (Hugo 1997). The Asian 'miracle' in the 1980s and early 1990s was greatly assisted by foreign labour. Migrant workers in Hong Kong built much of the US$20-billion airport at Chek Lap Kok, Malaysia's huge road network and Petronas Towers in Kuala Lumpur, as well as Japan's Winter Olympics Village in Nagano. Similarly, hundreds of thousands of female domestic servants from countries like the Philippines, China and Indonesia were employed by the burgeoning middle class in Japan, Hong Kong and Singapore. In 1980, a total of 1 million migrants worked in Japan, South Korea, Malaysia, Singapore, Hong Kong, Thailand and Taiwan. By 1997, the year of the 'crash', the number had grown to more than 6.5 million throughout the region. In East Asia, most of these overseas contract workers were hired in the Philippines, China, South Korea, Malaysia, Indonesia and Thailand. But with the collapse of Asia's fastest-growing economies at the end of the 1990s, these contract workers and illegal job seekers faced deportation.

International tourism is yet another means by which the new global economy is being increasingly integrated. In some countries in the South, spending by overseas tourists now rivals export income and foreign direct investment as contributors to economic growth. Between 1977 and 1989, the volume of annual international tourist arrivals in Thailand increased from 1.2 million to 4.8 million (Krongkaew 1996, 305). During the same period, income from tourism grew from 4.6 million baht in 1977 to 96.4 million baht in 1989. However, international tourism is a mixed blessing for Third World cities: it creates valuable jobs in the service sector, but often at the expense of distorting investment in urban infrastructure and overtaxing the environment.

CREATION OF MULTILATERAL MECHANISMS FOR LIBERALISING WORLD TRADE

The integration of global trade and money markets was facilitated by a sea change in government policies after the Second World War. When globalisation took hold in the 1970s, the way was smoothed by an economic regime imposed by the World Bank and the IMF (some would say serving as an extension of US foreign policy). Exchange rates were pegged to the US dollar and this gave a measure of stability to world trade. The IMF was created at the 1944 Bretton Woods Conference to act as the world's central banker, the lender of last resort to member states. The system worked reasonably well until 1972, when the then President of the United States (Nixon) uncoupled the dollar from gold so as to fund the mounting deficit caused by the Vietnam War.

Throughout the 1980s and 1990s, the World Trade Organization (WTO) and its predecessors orchestrated the lowering of protectionist trade barriers among member states. However, the process of freeing up international trade and financial transactions worked to the advantage of global corporations as much as to that of nation-states. There is a general view that global corporations would have sheeted home this advantage if nation-states had signed the Multilateral Agreement on Investment (MAI) in 1998, but France led a group of countries that were not prepared to forfeit their sovereignty in these matters and the MAI has been placed on hold indefinitely.

Countries borrowing from the IMF are made to comply with a package of measures, which have been caricatured as 'one size fits all'. These include: the imposition of high interest rates to tighten the money supply; cutting public expenditure and selling off state-owned enterprises; freeing up the financial markets; and the elimination of import restrictions. These measures take effect with varying degrees of severity, and ultimately have repercussions for urbanisation processes and cities. In 1982 the Mexican government reneged on its debts with private-sector banks, precipitating a crisis throughout Latin America that threatened the western banking system. The IMF stepped in and introduced a series of austerity measures to restructure bad debt, control inflation and eliminate the budget deficit. The effect upon economic growth and living conditions in cities like Mexico City can be compared to 'going cold turkey'. Public transport, housing and sanitation programmes came to a halt and out-of-work public servants were often forced back to the countryside. On the other hand, with the signing of the North American Free Trade Agreement (NAFTA) in the mid-1990s, small Mexican coffee, corn and sugar producers faced competition from large-scale producers in Canada and the United States, some of whom benefit from agricultural support schemes and subsidies. The countless thousands of farm families displaced by 'free trade' end up heading for the cities, or, if they can successfully get across the border, to Los Angeles.

The interventions of the IMF and the World Bank – most recently in Asia – and trade liberalisation have triggered similar 'aftershocks' in the countryside and cities around the world. Furthermore, this restructuring has left very few regions or cities unscathed over the last three decades:

While multinational companies are freed even more to roam the globe . . . more people leave the land and make for the cities. Yet they cannot go to just any city. 'Free trade' does not go so far as that. The US border and the EU border still remain closed to such people from outside. (This is, of course, one reason why New York and Los Angeles, London and Paris are not higher up the projected size-rankings of the cities of the new millennium.) And so it is that yet more people arrive in unprepared, polluted, Mexico City.

(Massey 1997, 103–4)

In an era of globalisation, urbanisation has to be understood increasingly in these terms. All the processes that have been attributed to globalisation are impacting, directly or indirectly, upon global urbanisation processes and patterns. The shifts occurring in the sectoral composition of modern post-industrial economies (Table 1.5) are translating directly into new regional and urban landscapes. There are at least three main transformations taking place in the global settlement system: firstly, Castells (1989) envisages a transformation of the space of flows within urban networks; secondly, Sassen (1991) believes that a distinctively 'global city' is emerging at the apex of the world urban system; and, thirdly, Friedmann and Wolff (1982) were the first to speculate about the formation of a necklace of globally linked cities stretching around the world. Subsequently, Peter Taylor, a geographer at the UK's Loughborough University, set up the Globalisation and World Cities (GaWC) Research Group and Network to monitor the formation of the world-city network. In Chapter 2, closer attention is paid to these transformations of urban and regional space.

FURTHER READING

The volume reporting on human settlements globally and prepared on behalf of the **United Nations** (1996) by a cast of respected urban scholars is the single best compilation of materials on urbanisation.

A special issue of *Urban Studies*, edited by **Dieleman and Hamnett** (1994), on processes of globalisation and social regulation canvasses their impact on the Dutch urban system.

Similarly, a volume edited by **Lo and Yeung** (1996) examines emerging forms of urbanisation in Pacific Asia.

Globalizing Cities (**Marcuse and van Kempen** 2000) presents a series of case studies that were commissioned to compare the impact of globalisation on the geography of cities like New York, Calcutta, Frankfurt, Rio, Tokyo, Singapore and Sydney.

2

URBAN SYSTEMS AND THE GROWTH OF CITIES

INTRODUCTION

Definitions

This chapter deals with the external relations and linkages formed between cities and how their size and rank within the urban settlement system ultimately governs urban growth and decline. The history of urbanisation shows that cities do rise and fall, and not necessarily just along with the growth and decline of empires.

Towns and cities connected in this way constitute an urban system of nodes and networks. The transport and communications networks connecting towns and cities provide carriageways for flows of people, money, goods and ideas moving between nodes in the urban settlement system. Historically these goods being exchanged between cities were products not available locally in the 'hometown' area. For example, with the spread of Islam across black Africa in the eighth century, only salt and gold could bear the costs of transport on the 3000–4000 km journeys that Arab traders undertook. By about AD 1000 West Africa had developed a reputation in Baghdad as the 'land of gold' and cities like Kano and Koumbi were quite sizeable. The gradual expansion of trade in Europe and the rise of cities were based on the exchange of wine and salt from the South in return for the textiles of Flanders. And now the worldwide movement of goods is dominated by container ship and jumbo aircraft. This gave rise in the 1960s and 1970s to highly specialised goods-handling facilities in port cities for the global container trade like Rotterdam's Europoort,

Singapore, Shanghai and airports like Newark, Changi and Frankfurt.

Travellers might be traders, judges or ministers on circuit, pilgrims, artisans and journeymen, or students. Geoffrey Chaucer's *Canterbury Tales* is a famous Middle-English epic poem relating the experiences of one such group of assorted travellers on their way from a Kentish village to Canterbury Cathedral. In the past, the speed of travel defined the extensiveness of urban systems. Figure 2.1 illustrates how travel times from New York City shrank between 1830 and 1857. Now, on a daily 'round-trip' basis, 'bullet' trains carry people who work in business and government, and even students, between destinations, e.g. Tokyo–Nagoya–Osaka in Japan, and Hamburg–Amsterdam–Cologne–Bonn–Munich in Europe. In much the same way, airlines provide 'walk-on' shuttles for commuters in high-density corridors connecting cities like Los Angeles and San Francisco on the US west coast, or New York City and Washington DC, or the London Docklands and cities in the European Union.

Meanwhile in the realm of ideas and money, advances in telecommunications have reached the point where the flow of information between even the most distant cities is now instantaneous (Cairncross 1997). By the early 1870s, the telegraph was being used for the first time to communicate foreign exchange rates and to transfer funds between New York and London. The first credit card was developed by Diners Club for use in the United States in the 1950s. In his book on global financial integration, O'Brien (1992) tells of a major upgrade to a City of London trading floor that was undertaken by one

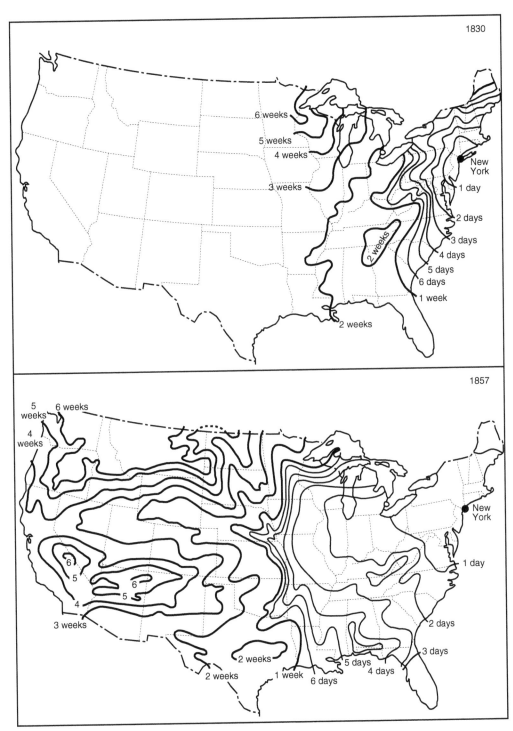

FIGURE 2.1 *Travel times from New York in 1830 and 1857*
Source: Cronon (1991). Used by permission of W.W. Norton & Company, Inc.

of the big banks simply to gain a 30-second lead over its rivals.

CITIES AS NODAL REGIONS

Each town node, or city, within the urban system also has its own sphere of influence. It is a 'central place' for the surrounding population. As service centres in their own right, towns and cities provide business and professional services, and schooling and health care to the surrounding population in the countryside. Towns and cities are also a hub for political and administrative functions, cultural and religious life, and community social interaction.

The tributary area around a city is variously defined as an 'urban sphere of influence' or 'hinterland'. It could take the form of a trade area, an administrative jurisdiction such as a service delivery

region, or a water catchment. The delimitation of the tributary area around cities occupied the attention of urban geographers in the 1950s, following on from A.E. Smailes's pioneering work (Smailes 1947). In a classic paper published in 1955, H.L. Green selected seven criteria for demarcating the divide between the catchments of New York City and Boston: the fall-off of commuter rail patronage; the volume of trucking attracted by the two cities; newspaper circulation; long-distance telephone traffic; the holiday destinations of New Yorkers and Bostonians; the residential addresses of company directors based in NYC and Boston; and the 'big city' affiliations of local banks. Figure 2.2 shows where the divide lies between New York and Boston when these criteria are combined into a composite boundary.

The city of Paris is the classic nodal region. It has so dominated the economic, political and cultural life of France over the centuries that the provinces

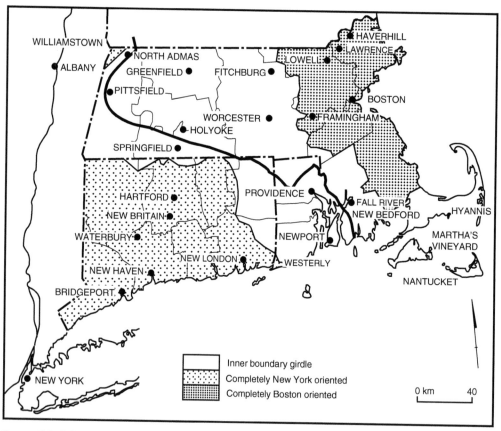

FIGURE 2.2 *Spheres of influence and overlapping trade areas in the north-eastern United States*
Source: Green (1955). Reproduced by permission of Economic Geography, Worcester, MA

pose none of the same threats to national sovereignty currently faced by other European countries. Not all cities develop the kind of nodal region that has been so important to the historical development of Paris: religious centres that attract pilgrims (Mecca, Damascus) or cities that are the seat of government (Islamabad, Canberra, Brasilia) tend not to, but nor have other cities that might otherwise have been expected to. As Jane Jacobs (1984) points out, both Glasgow and Edinburgh with their combined economies have not managed to create a dense, rich mixture of urban and rural activities within the space that lies between them. Even though Marseilles is a large seaport and has no real competitor in southern France, it has no city region to speak of. The same goes for Dublin, Belfast, Cardiff, Liverpool, Lisbon, Madrid, Zagreb, Rio de Janeiro or Montevideo. In the United States, the size of Atlantic City and Seattle belies the extent of their hinterlands. And among the developing cities of the Pacific Rim, Manila's sphere of influence does not reach as far into the Philippines' countryside as might be expected.

According to Jacobs (1984, 46–7), what these cities lack is the capacity for import replacement: where the city's economy is actively involved in import-replacing production, it is bound to develop strong relations with its surrounding region. On the other hand, 'When a city at the nucleus of a city region stagnates and declines, it does so because it no longer experiences from time to time significant episodes of import-replacing' (Jacobs 1984, 57–8). This distinction between the basic and non-basic sectors of an urban economy is examined further in Chapter 3.

Models of town/country and inter-city interaction

A German gentleman farmer, Johann Heinrich von Thünen, is associated with one of the first attempts to formalise the interaction between a town, its surrounding countryside and the pattern of rural land use that might be expected to evolve under a given set of conditions. The first edition of his book *The Isolated State* was published in 1826 (see Chisholm 1979). In his own district in northern Germany, he noticed that the intensity of farming activity fell away with increasing distance from the township of Mecklenburg. On this basis he proposed a simple model of the underlying economic constraints to represent the way land use and agricultural productivity are arranged around an urban centre. Assume a completely isolated region – the equivalent in those days was a small Germanic state – in which sat a single city on a featureless, but uniformly fertile plain. What farmers choose to plant or graze on the surrounding land at any given interval from the town centre depends on two variables: how much townsfolk are prepared to pay for different produce and how much it costs to transport the produce into the town market.

Figure 2.3 portrays the zonal arrangement of agriculture that von Thünen expected to find around any given town centre. So in a belt just outside the town, farmers compete for land on which to produce heavy, bulky or perishable consumables like fruit and vegetables, or milk and poultry. Because of the

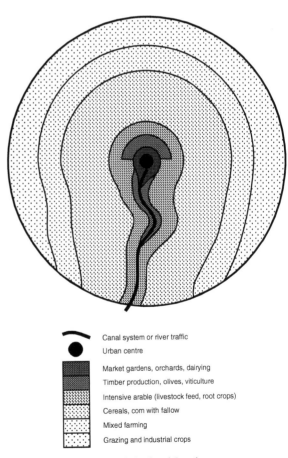

Canal system or river traffic

Urban centre

Market gardens, orchards, dairying

Timber production, olives, viticulture

Intensive arable (livestock feed, root crops)

Cereals, corn with fallow

Mixed farming

Grazing and industrial crops

FIGURE 2.3 *Von Thünen's 'Isolated State'*
Source: adapted from Chisholm (1979)

value of these products, producers can pay higher rents that those engaged in cereal cultivation or other forms of pastoral activity (sheep or beef). But, significantly, because of the value of firewood and the cost of cartage in pre-rail Europe, in von Thünen's model this second zone also contains woodlots scattered among the fields of grain and sheep. Beyond that lay uncleared forests, which were a source of deer and wild boar, or pelts and fur. This far from town, there was no point in trying to clear and farm the land because the cost of getting the produce to market exceeded the costs of production.

Notwithstanding the special place that von Thünen's model occupies in urban geography, stand-alone cities in splendid isolation on a featureless plain represent the unlikely case! On fertile plains in the New World towns jostled for market share; while along rivers and canals, transport costs were much more elastic. While this inevitably leads to distortions of the model in the real world, it need not negate the underlying principles specifying market relations between town and countryside. In Cronon's estimation, although the flat glacial plains of Illinois may not have been entirely featureless, and the city by the lake may not have been entirely isolated, the nineteenth-century geography of Chicago's market 'mimicked uncannily the pattern von Thünen had first predicted in 1826':

Beyond the central city lay the zone of intensive agriculture, filled with orchards, market gardens, dairy farms and feed lots; beyond it the zone of extensive agriculture, with its farms raising mainly wheat and corn; beyond it the zone of livestock and lumber production; and beyond it the zone of hunters, where fast-disappearing game species were opening up new niches for cattle, to say nothing of farmers, miners, and lumbermen. Each element of this new market geography had its roots in the original ecosystems that had assigned pine trees to the north woods and bison to the Great Plains. But each was no less affected by its distance from the city and its ability to pay the transport costs of getting there.

(Cronon 1991, 266)

Gaining the upper hand in hinterland trade meant making inroads into, and even capturing, a rival's suppliers and customers: 'All western cities served as markets for their hinterlands, but Chicago did so with greater reach and intensity than any other' (Cronon 1991, 148). By the 1860s, through the efforts of its industrialists and boosters, Chicago started to overtake St Louis at the confluence of the

Missouri and Mississippi rivers. As the up-river trading partner of New Orleans, St Louis had been a major trade centre for nearly three-quarters of a century. But as the railway network fanned out into the West, Chicago became both the new gateway between East and West, and the wholesale market for the entire mid-continent. With the advent of the grain elevator, wheat and corn from all over Iowa and Illinois were processed and packaged for shipment to Europe. Trees from the forests of upper Michigan and Wisconsin arrived by raft on Lake Michigan and then left Chicago as milled lumber on railroad flatcars for cities on the Eastern Seaboard. Refrigeration enabled Chicago's meat-packing industry to overcome the hurdle of shipping dressed meat – initially bison, then beef cattle and pigs – over long distances. A similar story of commercial rivalry can be told for Sydney and Melbourne despite the presence of a natural boundary like the Murray River. The Victoria and New South Wales rail and news distribution networks encroached over the border with the intent of capturing trade from the neighbouring state.

After the Second World War, the rise of mass vehicle ownership and trucking extended the inter-penetration of trade areas and even the commuting fields of adjacent towns and cities. By 1960, if national parks and reserves are excluded, most parts of the continental United States lay within daily commuting range of metropolitan areas and smaller urban centres. Aided by the expansion of the Interstate Highway System in the 1950s and 1960s, these overlapping commuting fields give workers access to jobs in any one of a number of cities in densely settled parts of the USA (Fig. 2.4).

The vast body of descriptive work undertaken in the 1950s on urban spheres of influence was followed by an attempt actually to predict the nature of the interaction between two urban centres and where to locate suburban shopping centres (Berry 1967, 40–58). Drawing on the Newtonian notions of gravity and the intervening field between two masses, William J. Reilly (1931) suggested that the strength of the interaction between two attracting magnets – such as cities or shopping malls within large cities – and the forcefield or trade area that separates them depends on the population of each city, or drawing power of the respective malls. According to Reilly, competing cities attract shoppers from a given rural region in direct relation to their respective size and

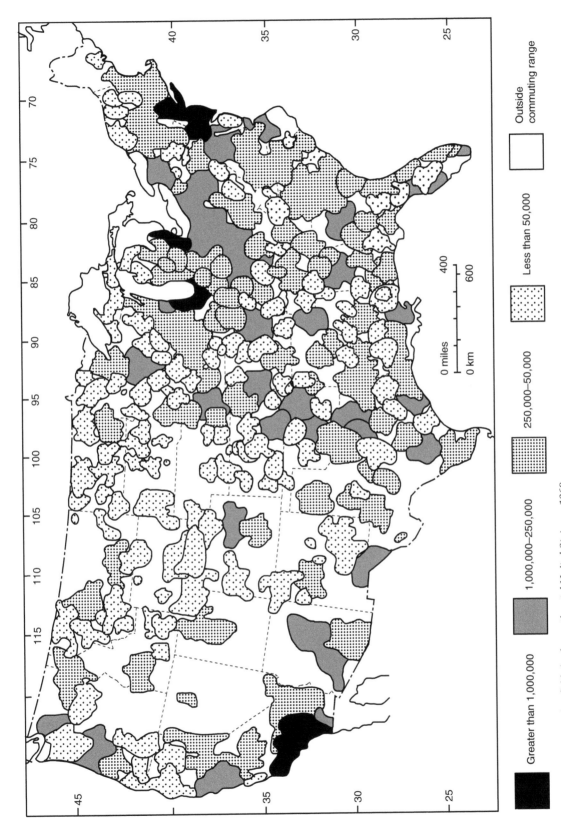

FIGURE 2.4 *Commuting fields in the continental United States, c. 1960*
Source: Berry and Horton (1970, 43). Reproduced by permission of Prentice-Hall

Greater than 1,000,000

1,000,000–250,000

250,000–50,000

Less than 50,000

Outside commuting range

0 miles   400
0 km   600

inversely in relation to the square of the distance to be travelled in getting to them. Or in the case of retailing in large metropolitan areas, where shoppers may be equidistant between similar-sized malls, Reilly's 'law of retail gravitation' has to be relaxed somewhat to allow for the probability that they will shop around more.

The most sophisticated version of this essential relationship between rural and urban service centres and their tributaries is the location-allocation model. Various kinds of location-allocation models have been developed to help draw the boundaries of electoral districts, fire and ambulance service areas, hospital catchments and school districts in the USA in the 1970s as part of an attempt to integrate racially segregated neighbourhoods and suburbs.

Central place systems

An urban system is also a nested hierarchy of villages and hamlets, small towns and county seats, provincial or regional centres, and cities of national significance. Figures 2.5 illustrates the way rural service centres at the bottom of an urban hierarchy are spaced so as to deliver services to farming communities. What struck Brush and Bracey (1955) at the time is the similarity in the spatial geometry of the urban hierarchy at this level despite the differences in population densities, accessibility and settlement history of south-west Wisconsin and southern England. In both places, lower-order, or the smallest service centres, are spaced between 5 and 10 km apart while higher-order centres are at 12 to 16 km intervals on average.

In the case of distributive trades and services, this can give rise to a central place system with a highly regular spatial geometry. In fact, the size and spacing of individual towns and cities within a pure central place system is so consistent that it has generated a body of urban theory known as 'central place theory'. Working from first principles, the economist August Lösch (1941) derived this expected hierarchical structure of market towns, or central places, deductively. The geographer Walter Christaller (1933) took another tack. He built up his formulation after observing the size and spacing of market towns in part of southern Germany. In this classic example of a central place system, Christaller concluded that the regular layout of the settlement system was due to the evenness of agricultural productivity and population densities within the region. The province of Szechwan in China provides another example of a classic central place system; but in this case it has its origins in the periodic marketing system that traditionally operated in rural China (Skinner 1964).

Among the numerous studies of central place systems undertaken by urban geographers during the 1960s and 1970s, the research of Brian Berry was perhaps the most systematic. He compared the geography of settlement systems in the US mid-west and in suburban Chicago (Berry 1967). A south-western corner of Iowa in the cornbelt was chosen because this state is regarded as satisfying the assumptions of central place theory more nearly than any other region in North America. Two counties in South Dakota provided 'tests' of central place theory for population densities supported by farming in the rangelands and wheatlands. In the suburbs of Chicago, Berry examined the geography of major regional centres, smaller regional centres, community centres and neighbourhood centres (Fig. 2.6). Figure 2.7 summarises the systemic relationships that tend to apply across urban systems. These relationships seem to hold for different-sized centres at their corresponding level in the hierarchy. For each level of the hierarchy, the size of those centres tends to be proportional to the population served, the density of that population and the extensiveness of trade areas.

City-size distributions

As a consequence, the urban systems of some countries have evolved in such a way that the relative size and position of cities appear to remain fairly constant over a long period of time. Figure 2.8 shows that, broadly speaking, the numbers of different-sized cities in the US urban system has borne the same log-linear relationship since 1790. This relationship between the size of each city and its rank within a country's urban system was first noticed by Zipf in 1949. When the population of each city in the urban system is plotted on log-log graph paper, the relationship conforms to a rank-size distribution. For example, if the largest city in an urban system has a population of 10 million, the tenth-ranking city ought to have a population of 1 million and so on.

The 'rank-size rule' does not require cities to hold their exact position in the distribution. Cities are changing rank in the city-size distribution along with the redistribution of population within countries. Sunbelt cities in the USA, like Houston and Dallas, Miami or Orlando, Phoenix or San Diego, moved up in the distribution between 1940 and 1990, while other cities, like Savannah, dropped several places (Fig. 2.8). On present trends, by 2025 southeastern Queensland, including the Greater Brisbane region, is expected to leapfrog over Melbourne, the second-ranked city in Australia's urban system.

Cities at the top of the urban hierarchy are referred to as 'primate cities' when they completely overshadow other cities in an urban system and distort the rank-size distribution. In Argentina, Buenos Aires is more than ten times the size of Rosario, the next largest city. In France, the population of Paris is

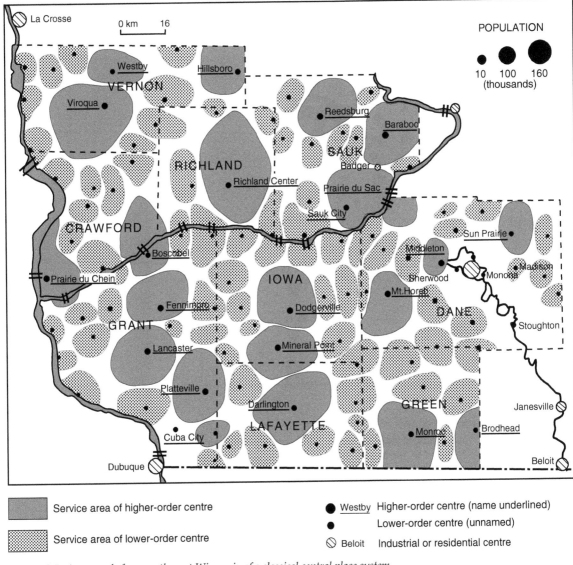

FIGURE 2.5 *An example from south-west Wisconsin of a classical central place system*
Source: Brush and Bracey (1955). Reprinted by permission

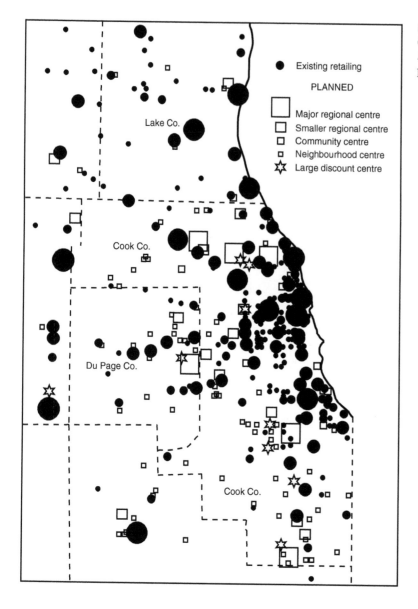

Existing retailing

PLANNED

Major regional centre
Smaller regional centre
Community centre
Neighbourhood centre
Large discount centre

Lake Co.

Cook Co.

Du Page Co.

Cook Co.

FIGURE 2.6 *Central places in suburban Chicago*
Source: Berry (1967). Reprinted by permission of Prentice-Hall

eight times that of Marseilles. And notwithstanding Thailand's national and regional development policies, Bangkok has continued to grow rapidly at the expense of cities in other parts of the country.

THE EVOLUTION OF URBAN SYSTEMS

The notion of a world-city system is a comparatively recent one that has taken hold along with the coming of globalisation. Prior to that cities evolved either as reasonably independent city-states, each with its 'respective hinterland entirely separate from each other' (Bairoch 1988, 24), or as one of a loose-knit constellation of urban centres seeking mutual advantage from trade or cultural ties. Historians like Toynbee (1970) believe city-states to be distinctive on the grounds that they exhibit the trappings of states, while also supporting a comparatively sophisticated urban culture. City-states first appeared in Mesopotamia around 3200–3000 BC, and faded with

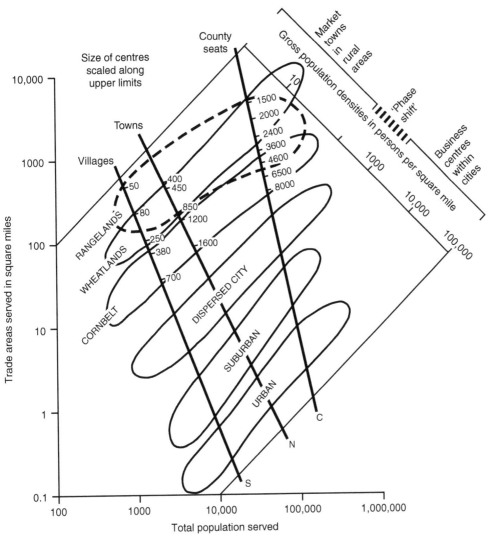

FIGURE 2.7 *Some of the regularities observed for market centres in different urban systems in the United States: trade area sizes, populations served and population densities show remarkable regularity at differing levels of the urban hierarchy*
Source: Berry (1967). Reprinted by permission of Prentice-Hall

the unification of the German and Italian city-states in the middle of the eighteenth century. Since Hong Kong rejoined the People's Republic of China, the only surviving city-states are the Vatican, Singapore and Monaco.

While they cannot compare with the degree of global integration that now binds together 'global' cities into a world-city system, there have at times in the past been loose-knit, and geographically widespread constellations of cities linked by long-distance, cross-border trade, or cultural and religious forms of affiliation. As Bairoch (1988, 20) points out, it is not correct to talk in terms of 'international trade' because these links between early cities pre-date the existence of real nations. None the less, the very first archaeological 'digs' on early city sites like Jericho unearthed evidence of long-distance trade from the earliest times. Pottery shards from the

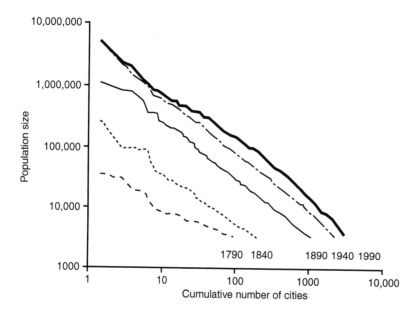

FIGURE 2.8 *The rank-size distribution of US cities according to their changing size over time*
Source: Knox (1994). Reprinted by permission of Prentice-Hall

more advanced cultures around the Mediterranean have been found in Europe and pre-date the emergence of the continent's first cities. When the Arab Empire was expanding between the seventh and tenth centuries, Islamic cities were linked by the one religion. Alternatively, the Hanseatic League, a loose-knit confederation of Baltic and North Sea towns, was formed during the thirteenth century for mutual protection and to promote trade.

River basins and hydraulic city systems

The origin of the first real cities remains a hot topic of debate among urban scholars that we cannot really hope to do justice to here. Trying to generalise about the experience of cities that evolved over a time span roughly equivalent to four millennia (3500 BC–AD 500) is hazardous, not least because what we know about the cities of antiquity is based on a very incomplete archaeological picture.

Which came first? Agriculture or settled life? It has always been assumed that an agricultural surplus is necessary to support an urban population. 'Even the fostering of non-agricultural occupations heightened the demand for food and probably caused villages to multiply, and still more land to be brought under cultivation' (Mumford 1961, 42). But in the late 1960s, Jane Jacobs challenged the prevailing wisdom in a book called *The Economy of Cities*

(Jacobs 1969). She proposes that the sedentary, non-agricultural activities that mark urban life come before rural development.

Where did the first urban settlements take shape? In the Middle Eastern Fertile Crescent where they were close to the sites of animal and plant domestication in south-west Asia? Logically, the evidence of early trade disposes archaeologists to 'the role of diffusion, rather than simultaneous growth, in accounts of the origins of urbanization' (Bairoch 1988, 20). Or did pre-contact urban settlement evolve independently elsewhere in Asia and the Americas? A plausible case can be made for the spread of urbanisation from a single Middle Eastern source throughout other parts of Asia, and then on to Europe and Africa. Cities first appeared in China about 1500 years after the Middle East, and 1000 years after the Indus. Chinese scholars have unearthed evidence for the existence of at least 20 large cities in northern China before 221 BC. The historical geographer Paul Wheatley (1971) remains convinced that the Chinese were very effectively insulated from other centres of urban evolution. It is possible that the pre-Columbian societies of Meso-America also developed their own forms of urban settlement.

In his book *The Preindustrial City* Gideon Sjoberg (1960, 27–31) lists three pre-conditions for the emergence of cities. The first is a favourable 'ecological' base, by which he means climate and soils capable

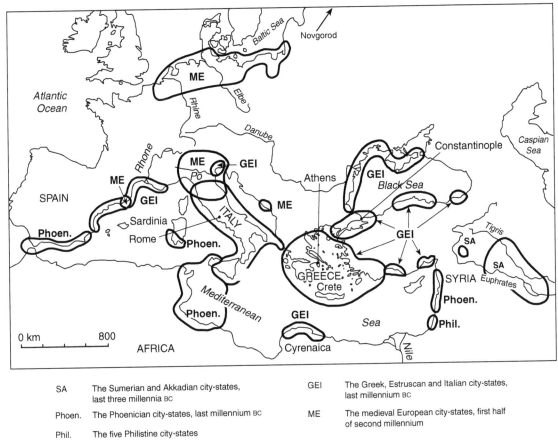

SA	The Sumerian and Akkadian city-states, last three millennia BC	GEI	The Greek, Estruscan and Italian city-states, last millennium BC
Phoen.	The Phoenician city-states, last millennium BC	ME	The medieval European city-states, first half of second millennium
Phil.	The five Philistine city-states		

FIGURE 2.9 *Early city-states in the Mediterranean and Asia Minor*
Source: composite based on figures in Toynbee (1970). Original maps by Reginald Piggot, reproduced by permission of Oxford University Press

of supporting the development of plant and animal life. An adequate, year-round water supply harnessed by irrigation techniques was crucial to the rise of the urban civilisations – though the earliest cities in Meso-America never developed the large irrigation and flood control schemes that fed and protected cities in the great river valleys of Asia (Fig. 2.9). The second requirement for settled life is the possession of reasonably sophisticated technologies for increasing agricultural yield. This includes the ability to domesticate grains like wheat in south-west Asia, barley in northern China and maize in Meso-America. This freed up city workers to concentrate on the administrative, planning and educational tasks that form the basis of the third requirement for urban development. Without some form of political structure and ruling class, the organisational means

for levying taxes, mobilising slave labour, regulating the grain trade, or coordinating the smelting of iron and bronze simply did not exist.

For these reasons, it is most likely that the first cities sprang up in a few great river valleys in exceptionally favoured regions: the Tigris-Euphrates (Eridu, Ur, Babylon); the Nile (Thebes, Memphis, Alexandria); the Indus (Harappa, Mohenjo-daro); and the Hwang Ho (Hsien Yang, Lo-yang, Kaifeng). 'Even without animal manure, the rich silt deposited at flood time guaranteed crops almost a hundred times greater than the original seed: sometimes two or three crops a year' (Mumford 1961, 71–2). These extensive river systems also greatly increased the capacity to move grain from afar to feed the urban population; although, in the case of China's northern cities it took the construction of the Grand Canal to link them with the rice bowl to the south (Fig. 2.10).

Pre-Columbian societies in the New World, including the Aztecs, Mayas and Incas, built as many as seven cities with populations approaching 50,000, and 25–30 cities with populations of 20,000–50,000. At the beginning of the sixteenth century, Tenochtitlan, the capital of the Aztec Empire, had an estimated population of between 150,000 and 200,000, and ruled a hinterland population of 400,000 (Bairoch 1988, 62–6). Yet these societies never developed irrigation beyond a little terracing, and lacked the plough, the wheel and proper metallurgy. As well as these handicaps, the Incans never developed writing beyond a simple numbering system. They also failed to provide for formal education or cultivate an intellectual elite. This ultimately set the limits to their expansion. Therefore, even if new evidence emerges to challenge the south-west Asian sites as the first cities, their importance as a crucible of urbanisation will remain.

Islamic cities

'Islam was implanted in pre-existing urban worlds' (Bairoch 1988, 371). By the eleventh century the Maghreb, Egypt, Persia and Spain were all societies ruled in accordance with the Koran; yet, in their own way, they were geographically, historically and culturally distinctive. The spread of Islam was spearheaded by Arabs who either took over the cities they conquered or established new ones as military encampments (Basra, Kufa, Fostat). Most of the cities they founded were in the Maghreb and along the coastline of black Africa. Baghdad was established by decree to be the new capital of the Muslim world and the Caliph was installed in AD 765. Very soon it had a population of 400,000, and by AD 930 it had reached a million. But invading armies can also use trade routes and Baghdad was sacked on numerous occasions.

There is no doubt that Muslim dominion greatly enhanced the economic development of those cities it adopted. By AD 1000 this Islamic urban system comprised between 40 and 50 cities connected by long-distance trade routes, both maritime and overland. The populations of six to eight of these cities probably exceeded 100,000. Spain possessed about 30–40 per cent of Europe's urban population (i.e. in cities >20,000). Cordoba, with half a million

residents, was Europe's largest city, whereas Venice, the next largest, had a population of only 40,000–60,000 (Bairoch 1988, 372–4).

City-states (Athens, Rome, Constantinople, Ch'in cities, Edo)

Toynbee (1970, 44) defines a city-state as a state in which one city is so superior in terms of population and power relative to any other cities in the urban hierarchy that its paramountcy is indisputable. He distinguishes four main types of city-state (Toynbee 1970, 40–66) and Bairoch (1988, 25) adds a fifth: the African Hausa culture that existed between the fifteenth and nineteenth centuries was organised along city-state lines. In the Middle East and Europe, the Sumerian and Akkadian cities rose to prominence between 3000 and 2000 BC; the Phoenician and Philistine cities began around 1000 BC; the Greek, Etruscan and some Italian cities also date from that time; the city-states of medieval Europe emerged elsewhere in Italy from the seventh century on (Milan, Venice, Florence, Siena, Treviso, Padua, Vicenza, Verona), and in northern Europe during the thirteenth century. Constantinople was the capital of the Byzantine Empire and for a time the world's largest city. Toynbee (1970, 145) also suggests that the four dynasties that ruled over northern China from 200 BC to about AD 1420, and Japan's Tokugawa regime (1603–1867), were not unlike city-states in their political organisation.

Greece achieved a high level of urbanisation – perhaps 25 per cent – due to the commercial role of city-states like Athens, Sparta, Heraclea and Syracuse (Fig. 2.9). Of course, some of those living in the city were farm workers who made the daily trip out to the wheatfields or olive groves. Despite this, Bairoch (1988, 77) estimates that a city-state like Athens needed to import enough wheat from the colonies to support between one-quarter and one-third of its population. Merchants established trading posts, grain stores and ports in Greek colonies, which eventually stretched from France (Marseilles) to Asia Minor. Manufactured goods and olive oil were traded for agricultural imports, especially grain. By about 500 BC, the Greek city-states were 'the trader for the whole of the Mediterranean basin' (Bairoch 1988, 76).

Rome was the uncontested centre at the heart of the Roman Empire (Fig. 2.9). The empire was essentially

a body politic composed of city-state cells administered by fiat from Rome. They were all deprived of
their sovereignty but allowed to manage their own
domestic affairs (Toynbee 1970, 58). At its height in
the second century AD, Rome possibly had a population of more than one million, which was not
approached again by other (mainly Chinese and
Muslim) cities until the period between the seventh
and ninth centuries AD. By this time the empire
embraced 50–55 million subjects whose main role
was to feed the capital. According to Bairoch (1988,
83), 'Rome received much and furnished little.' Late
in the fourth century, with the Roman Empire in
decline, Constantine transferred the capital to
Constantinople, which was more central to the eastern half of the empire and straddled a major sea lane
(Fig. 2.9). The fall of Rome dates from 476.

For three centuries (roughly 360–650), Constantinople, capital of the Byzantine Empire and at the
crossroads between Christian Europe and Muslim
Asia Minor, was the largest city in the world
(400,000–600,000) and still figured prominently until
about 1250. It was 'simultaneously the empire's chief
port, royal seat, administrative capital (headquarters of a highly bureaucratised and thus heavily
manned administration), religious capital, and

industrial and commercial center' (Bairoch 1988,
367). Such a large city with very limited water supplies depended on food imports from as far away
as Egypt. The Byzantium urban system included
four to seven other cities between 30,000 and 50,000
(Jerusalem, Antioch, Ephesus, Edessa), and perhaps
ten or more of 20,000–30,000. Unlike Constantinople,
these secondary cities largely drew their food supplies from their hinterland.

For all but 600 of the 2200 years since China was
first unified, the country has been governed by a
series of dynastic empires with northern capitals –
Hsien Yang, Ching-chao, Ch'ang-an and Peking,
now Beijing (Fig. 2.10). These centres of imperial
authority upheld the Chinese state. Apart from the
official and state-run monopolies in salt, iron, copper, weapons, tax or tribute grain, and foreign trade,
these cities also quartered local merchants, artisans
and coolies. Imperial supervision reached down the
urban hierarchy to the level of the county seat. These
small cities, ranging between 10,000 and 50,000 in
size, were the centres of an administrative district,
or *hsien*. Thus Chinese Emperors extracted tribute
grain from each *hsien*, and installed an imperial
magistrate, civil mandarins and militia in their
midst to keep them subdued.

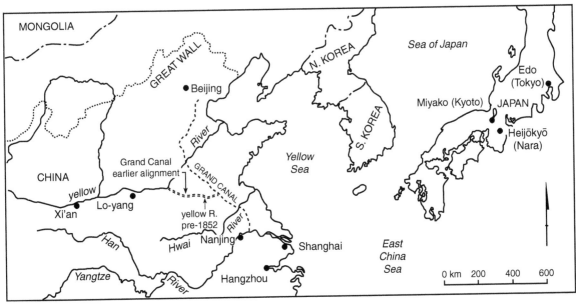

FIGURE 2.10 *Early city-states in northern China and Japan*
Source: composite based on figures in Toynbee (1970). Original maps by Reginald Piggot, reproduced by permission of
Oxford University Press

Between 221 BC, when the Ch'in State annexed six other local states, and the end of the thirteenth century, when a warlord moved his capital to Peking from Mongolia, feeding these imperial capitals posed an ongoing problem (Toynbee 1970, 72–6). Because China's two main navigable rivers, the Yangtze and the Hwai, flow east to west, these natural waterways were no use for transporting food from China's 'rice bowl' in the south to cities in the north. It took about 1500 years before Peking was finally connected by a vast inland system of canals to the grain terminals of Hangchow (Fig. 2.10). Bairoch (1988, 44) estimates that, by the fifth century BC, China already had four to six cities with populations between 100,000 and 200,000, while, in their time, Sian, Kaifeng, Peking, Hangchow and Nanking have each been the largest city in the world, with populations reaching a million in the second millennium AD.

Japan's two earliest cities, Nara and Miyako (present-day Kyoto), were successively the south-western capitals of Honshu, the main island. Then, when the economic and social centre of gravity shifted to the north-east during the twelfth century, the seat of political power moved with it: at first to Kamakura (1185–1333); and then, much later, to Edo on the Kwanto Plain (Fig. 2.10). At its peak in the second half of the eighth century Nara's population approached 150,000, while Miyako had 100,000 residents by the ninth century and eventually reached 200,000–250,000.

Later on, the Tokugawa shoguns (1603–1867) established the seat of government at Edo, on the site of present-day Tokyo. Steadily the regime exerted control over most of the principalities in Japan. Then they moved the Emperor with his imperial household from Kyoto to Edo (Fig. 2.10). He was installed in the massive fortress-palace that still dominates the centre of Tokyo to reign as the regime's figurehead, but played no part in governing Japan (Toynbee 1970, 113–14). Thereafter they compelled *daimyos* and their nobles from the provinces to reside for part of the year, and their families the year round, in Edo. This secured their loyalty and made for a bustling and prosperous capital. For these reasons, and despite the emergence of other market and port cities, the level of urbanisation in seventeenth-century Japan probably did not exceed 5–8 per cent and was dominated by just one or two cities. Some estimates place the population of Edo (Tokyo) as high as one million during the

Tokugawa era (Sjoberg 1960, 81). Tokugawa Japan was a model city-state.

Merchant city systems

Urban life re-appeared in Europe after the Dark Ages along with the revival of trade and the growth of population (Pirenne 1948). The places where traders met to exchange their merchandise had to be secure. In an unruly countryside overrun with vagabonds looking to plunder a medieval caravan, it made sense to seek out the protection of a feudal baron or an ecclesiastical overlord. For example, when medieval German warlords pushed east of the Elbe, they built walled towns to protect them from hostile Slavs. Likewise, as Swedish traders developed commercial links with Byzantine suppliers, their backers established palisaded depots at regular intervals along the great rivers of eastern Europe.

But, as trade increased, and travelling merchants grew more numerous, those who could not find quarters inside the walls of the city, or burg, had to camp outside. These encampments were called *faubourgs* – outer burgs – and gradually took on a quite different mercantilist complexion from the ecclesiastical or fortress way of life of the walled city population (Pirenne 1948). The *faubourg* was the forerunner of the much more specialised merchant city that evolved along with the expansion of trade in Europe. Bairoch (1988, 161) describes this as a muddled series of steps from the fortified castle to the burg, from the burg to the *faubourg*, from the *faubourg* to the merchant's city.

The 'merchant city' system grew up around the overland trade between northern and southern Europe. From the beginning of the eleventh century, economic expansion and the rise of cities in Europe rested on two major foundations (Bairoch 1988, 126–7). The first was the continuation of the ancient trade between Europe and Asia, where merchants went in search of spices and other luxuries not produced in Europe. This greatly enhanced the fame and wealth of a number of well-positioned Italian cities, which held a virtual monopoly over trade with the East. Of the five European cities with populations over 100,000 around 1300, three were Italian (Venice, Florence and Milan), whilst, in northern Europe, the signing of an accord in 1241 between the cities of Lübeck and Hamburg guaranteed protection to shipping moving between the Baltic Sea and

URBAN SYSTEMS AND THE GROWTH OF CITIES

the North Sea. This pact became the basis of the Hanseatic League, a federation of about 25 towns scattered across the North German Plain. As well as trade, Hanseatic League cities like Bremen and Rostock developed strength in ship-building, which capitalised on the abundant supply of timber in northern Europe.

The other foundation of European economic and city growth, which saw even greater development during the thirteenth and fourteenth centuries, was the trade between the different regions of Europe, especially cities in the north and south. Spices, dyes, silks and other goods from the Middle East and beyond, and carried by caravan across central Europe, funnelled northwards through Italian cities like Venice, Genoa and Naples. On the return journey, merchants carried woollen goods and elaborately worked metal and wood products from the rest of Europe for distribution throughout the Middle East and Asia.

In the Low Countries to the north and down the length of the Rhine, cities like Bruges, Ypres, Ghent and Cologne attracted and trained the craftsmen who transformed raw materials like linen, clay, timber and metals into objects for the court and wealthy burgers. By the middle of the fourteenth century, there was a population of about 220,000 engaged in textile production within a 30 km radius of these first three Flemish towns. Also, by this stage in the Netherlands, Delft was exporting earthenware and Leyden made black broadcloth for export (Bairoch 1988, 168). Trade between European cities also led to the creation of fairs in the central part of the continent where the merchants met. Cities like Frankfurt, Leipzig and Champagne on one side of the Alps, and Milan, Geneva and Lyons on the other, became sites of trade fairs several times each year throughout the thirteenth century.

Europe's colonial urban systems

European maritime expansion, exploration and conquest, and the founding and settlement of colonies, relied upon developments in ship-building, navigation and geography. Also, printing was re-invented in Europe in 1440, followed a couple of decades later by the development of the blast furnace. These technological advances in Europe all paved the way for the cultural revival, or Renaissance, in the latter half of the fourteenth century. But it was not until the sea

routes to Asia and the Americas were discovered (or rediscovered in the case of some parts of Asia) in the last decade of the fifteenth century that the future of Europe's merchant cities was assured.

With the growth in international trade – silver, gold, sugar and cotton from the Americas; spices, tea, precious stones and rubber from India, China and the East Indies – that followed, commerce came to dominate the European economy in the sixteenth and seventeenth centuries. And, as the fortunes of Europe's rival colonial powers waxed and waned, the centre of gravity of commerce shifted from cities arrayed around the Mediterranean to those with access to the Atlantic sea-lanes:

From Venice and Genoa the heart of commercial Europe, and consequently the centre of financial services and transport, moved first to Lisbon and Seville, and finally to Amsterdam and London (not forgetting the important interlude of Antwerp). And this shift carried in its wake the formation of new networks of commercial cities.
(Bairoch 1988, 132)

But as well as this, in each European country, particular cities favoured by international trade also gained from the centralisation of political control and government functions that accompanied the rise of the nation-state and expansionist colonial empires. Such was the dominance of these capital cities within the European urban system during the sixteenth and seventeenth centuries that only nine of them accounted for a third of the total increase in the continent's urban population. Between 1500 and 1700, the populations of the following capitals increased from 500,000 to 2 million: Amsterdam, Copenhagen, Dublin, Lisbon, London, Madrid, Paris, Rome and Vienna (Bairoch 1988, 183). At the same time, the growth of these impressive cities should not be allowed to cast a shadow over the 130 lower-order cities (>20,000) that also partly owed their existence in 1700 to non-local trade (Bairoch 1988, 198).

The rapid increase in urbanisation that took place in Britain over such a short time-span during the nineteenth century was fuelled by an expanding population, but even more vigorously driven by the rise of manufacturing. Between 1830–40 and 1900, the urban population grew at a rate of about 2 per cent annually. About 1840 the level of urbanisation in England passed 40 per cent, and by 1880 had climbed to 68 per cent.

The concentration of industrial production in regions favoured by their comparative advantage in manufacturing brought with it marked changes in Britain's urban system (Table 2.1). The spinning, dying and weaving of cotton and wool were centred in the mill towns of Lancashire and Yorkshire respectively. By the mid-nineteenth century, Lancashire was dominated by Manchester, which was linked by canals to the regional port of Liverpool. On the other side of the Pennines were the mill towns of Leeds, Bradford and Sheffield. In the Midlands, Birmingham (light engineering and machine tools), and, in the Black Country, Stoke (pottery) rose to prominence. This came at a cost to a number of other English provincial centres, as Table 2.1 makes clear. The use of water power in textile manufacturing almost to the middle of the nineteenth century, and then the movement of iron smelting to the coalfields, meant the death-knell for pre-industrial market towns like Norwich, Exeter, York, Colchester and Coventry. Among the leading urban centres in Britain before the Industrial Revolution, only port cities like London, Bristol and Newcastle held their own (Table 2.1).

King (1990a, 139) points out that by the beginning of the nineteenth century the ten largest industrial and port cities in Britain were, in fact, those that were tied together economically not only as centres of production and transhipment, but as the origin and destination for the flow of migrants and goods to the colonies. These ten British cities in descending order of size were: London, Edinburgh, Liverpool, Glasgow, Manchester, Birmingham, Bristol, Leeds, Sheffield and Plymouth. Hence it is now generally accepted that 'British industrial urbanisation – its degree, location, and distinctive built environment – was, within the larger world economy, strongly influenced by and, in places, dependent upon a colonial system of production: it was largely produced by it

TABLE 2.1 *Major urban centres in England before and during the course of the Industrial Revolution (population in thousands)*

	Population (000)		
	1700	**1800**	**1850**
London	550	860	2320
Major urban centres before the Industrial Revolution			
Norwich	29	36	67
Bristol	25	61	150
Newcastle	25	33	110
Exeter	14	16	16
York	11	16	35
Colchester	8	10	12
Coventry	7	16	36
Total for these seven cities	119	188	426
Major urban centres in the first phase of the Industrial Revolution			
Birmingham	10	71	230
Liverpool	6	76	422
Manchester	9	81	404
Leeds	7	52	185
Sheffield	8	45	141
Bradford	4	13	100
Stoke	3	22	65
Total for these seven cities	47	360	1547
Total urban population	880	2100	8000

Source: Bairoch (1988, 254). Reproduced by permission of University of Chicago Press

and cannot be understood except as part of it' (King 1990a, 148).

For their part, colonial cities were 'planted' with the express purpose of exploiting the territories annexed by the European metropolitan powers. Table 2.2 lists some of the major cities established as part of the British colonial urban system. These cities were usually situated on the coast to expedite the transhipment of the minerals and farm produce flowing out of the colonies. In the case of the African colonies, these port cities often supplanted inland centres and broke up indigenous urban systems. In turn, they served as the entry points for foreign settlers, investment and the manufactured goods that were often dumped on the local market. Hence there is a striking contrast in the rail networks developed to serve the urban systems of European countries on the one hand, and their colonies on the other: in Europe, they radiate outwards from centrally located capital cities like London, Paris, Madrid, Vienna and Moscow; whereas in the colonies, they funnel progressively towards a port terminus.

Thomas (1972) shows how the pattern of urban growth in former colonies like Australia, the United States, Canada and Argentina coincides with the extra outflow of farmworkers and capital when the industrial economies of western Europe temporarily stalled. O'Rourke and Williamson (1999) estimate that about 60 million Europeans migrated to the New World after 1820, or about one-sixth of the total population. Between 1870 and 1914, British investors placed 5.27 per cent of total GNP in overseas lending. This level has not been approached since (King 1990b, 19). English and Scottish banks invested heavily in land development, railroad-building and public utilities in the American mid-west and Argentina. Four British pastoral companies exported 75 per cent of the wheat produced

TABLE 2.2 *Key cities forming Britain's colonial system*

Colony	Date of incorporation		Major city		Population in 1900 (000)
	pre-1800	pre-1900	pre-1800	pre-1900	
EUROPE					
Gibraltar	1704		Gibraltar		25
Malta	1800		Valetta		60
Cyprus		1878	Nikosia		13
ASIA					
Aden		1838	Aden		42
India	1612-		Calcutta		1027
			Bombay		776
			Madras		509
				Delhi	209
Ceylon	1796		Colombo		127
Straits			Penang		
Settlements	1785			Singapore	512
		1819		Malacca	†
		1843		Hong Kong	254
Sarawak		1888		Kuching	†
AFRICA					
Cape Colony	1815			Cape Town	79
				Port Elizabeth	23
				Durban	27
Natal		1856		Pietermaritzburg	18
Transvaal		1900		Pretoria	12
				East London	7

TABLE 2.2 *(continued)*

Colony	Date of incorporation		Major city		Population in 1900 (000)
	pre-1800	**pre-1900**	**pre-1800**	**pre-1900**	
Rhodesia				Salisbury	†
Gambia	1664			Bathurst	14
Sierra Leone	1787			Freetown	†
Gold Coast/Ghana		1868		Accra	†
Lagos		1861		Lagos	33
Nigeria		1886			
East Africa/with Uganda		1888		Mombasa	†
Zanzibar		1888		Zanzibar	30
Mauritius		1810		Port Louis	65
NORTH AMERICA					
Canada	1623–			Montreal	268
				Toronto	208
				Quebec	69
				Ottawa	60
				Halifax	41
Newfoundland	1583–			St Johns	31
SOUTH AMERICA					
British Guiana		1803		Georgetown	49
Falkland Islands		1833		Port Stanley	†
WEST INDIES					
Bermuda	1609			Hamilton	2
Bahamas	1670			Nassau	†
Jamaica	1629			Kingston	†
Leeward Islands	1626–			St John (Antigua)	†
Windward Islands	1605–			St George's/Grenada	†
Barbados	1605			Bridgetown	†
Trinidad & Tobago	1797			Port of Spain	34
Honduras	1783			Belize	7
AUSTRALASIA					
New South Wales	1787			Sydney	99
Victoria	1787			Melbourne	478
Western Australia		1829		Perth	38
South Australia		1836		Adelaide	162
Queensland		1859		Brisbane	121
Tasmania		1803		Hobart	†
New Zealand		1841		Wellington	47
				Auckland	67
Fiji		1874		Suva	†

† Data not available.

Source: based on King (1990a, 141–2). Reproduced by permission of Routledge

in Argentina. In Australia, rail, road and bridge-building accounted for three-quarters of new capital formation in the 1880s, much of which was funded by British investors. In fact, the flow of British capital into Australia in the second five years of the 1880s was such that capital imports made up almost half of the gross domestic capital formation (Butlin 1964, 135).

Hence O'Rourke and Williamson (1999) argue that the development of New World colonies and cities during the nineteenth century was carried along by early globalisation processes. The technological breakthrough in railways, steamships, refrigeration and the telegraph reduced the costs of moving migrants and commodities. For the first time: productive activity was able to move closer to low-cost locations; capital was able to shift more easily to centres offering higher rates of return; migrant workers were able to seek out places promising better wages (see Chapter 3).

METROPOLITAN EXPANSION AND COUNTER-URBANISATION

During the second half of the twentieth century, urban systems in Europe and North America that had been shaped by earlier modes of transportation began to take on a rather different and much more complicated geography. Jean Gottmann (1961), an urban geographer, was one of the first scholars comprehensively to document this new geography of urbanisation in a famous monograph called *Megalopolis: The Urbanized Northeastern Seaboard of the United States*. Gottmann's classic study reveals the extent to which the economies and commuting zones of great cities like Boston, New York, Philadelphia, Baltimore and Washington DC began to overlap and coalesce to form a formidable 'urban agglomeration' on the east coast of the United States (Fig. 2.11). Another urban geographer (and planner), Peter Hall (1966), suggests that the same 1950s and 1960s phenomenon was beginning to transform the urban systems of Europe (south-east England, Randstat-Ruhr) and Japan (Tokyo-Osaka) as well.

After the Second World War, the suburbanisation of population and jobs, together with increasing car ownership and freeway construction in North America, Australia and New Zealand, pushed the commuting zone of already large cities further and

further outwards to created expanded metropolitan regions (Fig. 2.4). The outward extension of metropolitan regions in North America and Europe continued during the 1980s and 1990s, as manufacturing declined and living conditions deteriorated in the core areas of older industrial cities. The 1990 US census revealed for the first time that a majority of Americans lived in suburbs, up from one-quarter in 1950 and one-third in 1960. In fact, 19 of the country's 25 fastest-growing 'cities' between 1980 and 1990 were actually suburban offshoots in the Greater Los Angeles region (Moreno Valley, Rancho Cucamonga and Irvine), Phoenix (Mesa, Scottsdale and Glendale), and Dallas (Arlington, Mesquite and Plano). Three of the ten fastest-growing counties in the USA were on the edge of Atlanta.

In Europe, suburbanisation has been more restrained – though some city workers have always commuted to work from outlying villages throughout the European countryside. In the Netherlands, for example, the aspirations of middle-class families are leading to higher levels of car ownership and suburban development on the edge of cities like Amsterdam and Utrecht. Outer Paris, on the other hand, is more working class. During the 1960s and 1970s inner-city housing was cleared to make room for projects and monuments of national cultural significance, like the Georges Pompidou Centre and the Opera-Bastille, and the workers were decanted to new towns built on the periphery of Paris.

In addition to moving to suburbs on the expanding edge of US, British and Australasian cities, increasing numbers of city dwellers are now choosing to migrate further afield to the smaller towns and rural areas that are attracting post-industrial jobs and can offer high-amenity living environments to newcomers. The redistribution of people and jobs now taking place in conjunction with this 'urban–rural shift' is called 'counter-urbanisation' (Champion 1989). Sometimes this reversal of the long-standing trend that has seen the populations of western countries concentrating in the largest urban centres is also known as the 'population turnaround'.

The economic and social shifts underlying counter-urbanisation are also playing their part in reshaping the geography of urban systems in post-industrial societies like the United States, Britain and Australia. Many of the highly educated white-collar workers in the 'knowledge economy' live and work nearby in emerging urban corridors like California's Silicon

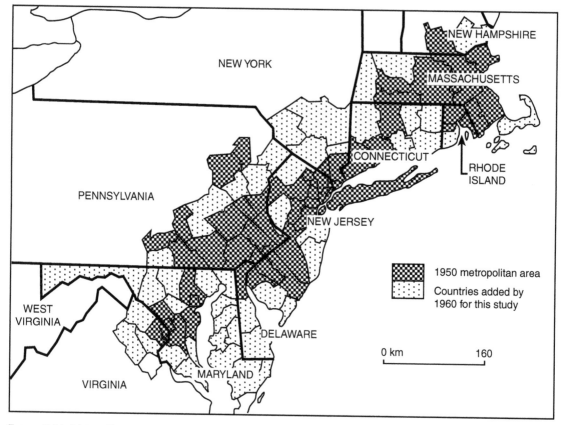

FIGURE 2.11 *Metropolitan expansion on the Eastern Seaboard of the United States, called 'Megalopolis' by Jean Gottmann*
Source: Gottmann (1961). Reproduced with permission from the Century Foundation

Valley or the beltway around Washington DC, the M4 corridor between Bristol and London in the UK, or the corridor linking the Gold Coast, Brisbane and the Sunshine Coast in south-east Queensland, Australia. This is powered in part by new communications technologies, which have freed back offices and call centres to locate away from big cities and permit 'teleworking' from a home office.

Significantly, counter-urbanisation gained momentum as increasing numbers of middle-class families bypassed the suburbs to seek refuge in rural villages or on 'lifestyle blocks' and 'hobby farms' in semi-rural locations. Between 1990 and 1994, more than 1.1 million net migrants moved into rural areas and small towns in the US countryside, most of them from the inner and middle suburbs. Some of these households are making a shift in conjunction with retirement (Boyle and Halfacree, 1998), but others contain members who commute into jobs in the

suburbs or the 'edge cities' that have sprung up in the last decade or so at the periphery of the larger metropolitan areas (see Chapter 4). According to John Kasarda, one of America's leading demographers, 'It's not just the old move to the suburbs, it's the exurbs and beyond' (quoted in Kotkin 1996, 15).

Metropolitan expansion and coalescence, together with counter-urbanisation, are also occurring against a backdrop of regional change that is helping to reconfigure the urban system in countries like the United States, Britain, Australia and New Zealand. Marked regional shifts are also affecting inter-country urban systems across large regions such as western Europe. The 2000 census in the United States revealed a 13.2 per cent gain in the nation's population in the preceding decade. With the US population standing at 281.4 million, this represented the biggest jump in 30 years. The regional shifts of investment activity and labour, which favoured cities in the south and

TABLE 2.3 *British conurbations 1961–2001*

Region	Population in thousands				
	1961	1971	1981	1991	2001 (projected)
Greater Manchester	2710	2750	2619	2571	2560
Tyne and Wear	1241	1218	1155	1130	1114
Merseyside	1711	1662	1522	1450	1386
Greater London	7977	7529	6806	6890	7215

Source: Power and Mumford (1999)

the west throughout the 1970s and 1980s, continued into the 1990s. Together, southern and western states like Georgia, North Carolina, Florida, Texas, Arizona, Nevada, Colorado and California accounted for over three-quarters of the increase (reported in *Guardian Weekly*, 4–10 January 2001, 25). These states are drawing people and jobs from the North and mid-west. Cities in the north-east lost 1.5 million people to other parts of the USA between 1990 and 1994. New York City suffered a net domestic out-migration of more than 861,000 in the same period. According to Kotkin (1996, 15), the mainly 'urban–rural shift' is 'largely to heavily white enclaves such as central Florida, the southern Appalachian hill country as well as the edge cities around the Research Triangle in North Carolina and Atlanta'.

Britain's urban system is also undergoing change. Net out-migration continues to lower the populations of Greater Manchester, Tyne and Wear, and Merseyside (Table 2.3) as young people migrate from the north to job opportunities in south-east England. After dipping in the early 1990s, the population of the south-east grew by 70,000 a year during the rest of the decade. An Office of National Statistics (ONS) report (Wilson 1999, 7) on population growth and redistribution in Britain predicts that the south-east will grow by a further 12.8 per cent over the next two decades (from 7.9 million in 1996 to 8.9 million by 2021), while the population in the north-east is expected to fall from 2.6 million in 1996 to 2.5 million in 2021. A metropolitan region like the Merseyside may lose nearly 10 per cent of its population (down from 1.4 million in 1996 to 1.3 million in 2021). As a result, a growing number of homes are being abandoned in the north. About 120,000 tenants vacate council housing in northern

cities every year. By contrast, over the same period, population growth is expected to approach 30 per cent in a band from Bristol to Cambridge. London will grow by 9.4 per cent if population increase matches the forecast rise from 7.1 million to 7.7 million between 1996 and 2021. Although not enough to meet the demand, in March 2000 the British government approved plans for 215,000 new dwellings in the south-east, and a further 115,000 in London over the following five years. The population and employment trends behind these projections involve a continuation of the exodus from Britain's inner cities coupled with the drift from large metropolitan areas to prosperous rural subregions.

In Europe, expanding cities exist in a corridor across southern Germany, northern Italy and southern France (Cheshire 1995). But there is also evidence to suggest that counter-urbanisation might have temporarily slowed in parts of western Europe and the United States between the 1980 and 1990 censuses (Champion 1992; Frey 1993). A number of cities in the northern states of the USA at least managed to halt their loss of population during the decade (Sommer and Hicks 1993). This is mirrored to some extent by urban trends in Australia (Hugo 1996). During the 1970s the rate of metropolitan growth stalled and was overtaken by non-metropolitan growth for the first time since the 1920s. Then in the 1980s it is possible to detect a slowing-down of the turnaround in Australia (Fig. 2.12). In the 1990s the pace of rural restructuring quickened in Australia further depleting less favoured regions. At the same time regions with high amenity values on the east coast of Australia between northern New South Wales and central Queensland continued to attract a greater share of international tourists, retirees,

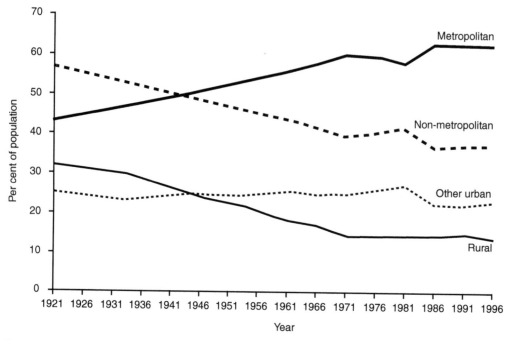

FIGURE 2.12 *Counter-urbanisation in Australia*
Source: Newton and Bell (1996). Reproduced by permission of P.W. Newton and M. Bell

workers in service and knowledge industries, and welfare recipients (for whom it makes good sense to move from high-cost cities).

Meanwhile, Sydney, Melbourne, Brisbane, Adelaide and Perth continued to expand. Notwithstanding the coastal 'sunbelt' urbanisation, these cities slightly increased their share of the country's population between 1986 and 1996. Throughout the 1980s and 1990s, Australia's metropolitan centres continued to attract disproportionate numbers of overseas migrants, especially Sydney, and to expand – though at slower rates than in the 1970s – as more affluent city commuters opted for rural living (Hugo 1996). This newish low-density, lifestyle-driven development beyond the suburban fringe is referred to as ex-urban or peri-urban development (see Chapter 4).

Based on trends in urban and regional change between 1960 and 2000, variations in regional restructuring are likely to grow more marked across the space economy of post-industrial societies. As business and government, and workers, retirees and welfare recipients, respond to the growth and decline of different regions, some cities will prosper and improve their position within the urban hierarchy, while others will lag behind and slip back in the rankings. Both Cheshire (1995) and Meijer (1993) have documented the changing position of cities within the European Community's urban system during the 1970s and 1980s.

Emerging urban corridors in Asia-Pacific

The urban corridor development that has hitherto been a feature of metropolitan expansion in the industrialised regions of North America, Europe and Japan is about to be eclipsed by the conurbations that are now emerging in the Asia-Pacific region. All the following are destined to be mega-cities, and some will join the ranks of 'global city' status within the next two decades. Indonesia's Jaotabek is the conurbation formed by Jakarta, Bogor, Tangerang and Bekasi, and had a population of 20 million by the mid-1990s (Dick and Rimmer 1998). With the functional integration of Hong Kong, Shenzhen, Guangzhou, Macau and Zhuhai in southern China, the Zhujiang Delta region could eventually support a population of 40 million (Lo and Yeung 1996, 12). On the east bank of the Huangpu River, Shanghai, already with a population of 15 million, is constructing the Pudong New Area. By 2000, Pudong, with a population of two million, was linked by subway to Shanghai and was close to opening a giant airport that dwarfs anything in Asia. The dynamic 'city-state' of Singapore is fast expanding its economy into the adjacent regions of Johore State (Malaysia) and the Riau archipelago (Indonesia). The population of the Seoul–Pusan corridor in South Korea exceeded 26 million by 1990.

While it does not apply to all Asian cities, 'region-based urbanisation' is qualitatively different from the older forms of urban agglomeration found in Europe and North America in three main respects. Firstly, rather than being burdened with outdated transport and communications infrastructure, these urban corridors are equipped with new-generation technologies like bullet trains, fibre-optics and super-airports that give their cities a comparative advantage in global markets. Japan's Kansai region is dominated by a corridor that links Osaka–Kobe–Kyoto. In the mid-1990s, Malaysia unveiled ambitious plans for a multimedia 'superhighway' of 750 km, connecting Kuala Lumpur to Sepang airport (Forbes 1997, 460). Two new cities within the advanced communications corridor will build up the country's IT industries. Secondly, a considerable number of these dynamic growth points in the Asia-Pacific region comprise clusters of cities that are capitalising on the cross-border flows of finance, information, labour and materials that have become the hallmark of borderless economies. Examples include the 'growth triangles' of Singapore–Johore–Riau (Macleod and McGee 1996) and the Hong Kong–Zhujiang Delta (Chu 1996). Thirdly, in these border regions, regional and city governments have assiduously pursued economic integration and complementary regional development strategies to boost the international competitiveness of the growth triangle (Macleod and McGee 1996, 424).

'GLOBAL CITIES' AND THE WORLD-CITY SYSTEM

According to Friedmann and Wolff (1982), the globalisation processes discussed in the previous chapter are creating a new kind of extra-national urban system based on the global strategies of transnational corporations. The cross-border flows of

goods, finance and people between 'global' cities are responsible for the formation of a world-city system. In particular, cities that are well placed to dominate the business of circulating capital and finance around the globe have gained the ascendancy in this international urban system. In King's view (1990b, 24–32) it is the global control function and the geopolitical leverage this gives to truly global cities that set them apart from the other contenders. London, New York and Tokyo are regarded as the pace-setters in this respect (Sassen 1991). More and more of the business transacted by the corporations based in global cities is directed towards international rather than domestic markets. Cities that qualify for membership of the world-city system are likely to be: the headquarters for a sizeable number of global corporations; major centres of international finance; centres for the worldwide recruitment of labour; gateways for immigrants; the base for international institutions and agencies; the hubs for several international airlines; homes to universities and institutes of international standing.

A variety of attempts have been made to classify and group cities within the framework of an international urban hierarchy. But, in spite of the reasonably high degree of agreement about the defining criteria, indicators for all the candidate cities are often difficult to come by. Friedmann's (1986) classification represents the original attempt to construct a world-city hierarchy (Table 2.4). The criteria he uses to indicate world-city status are: major financial centre; base for world and regional headquarters of multinational corporations; international institutions; rapid growth of business/services sector; important manufacturing centre; major transport node; population size. Alternatively, a classification put together by a group led by the British geographer Peter Taylor measures the respective strength of 263 cities in accounting, advertising, banking and law (Taylor 1999). This is based on the presence, or absence, of 69 global corporations in each of these cities. What they come up with is the world according to GaWC, or the Globalisation and World Cities Research Group (Fig. 2.13).

However, the tendency to fall back on pin-pointing the headquarters locations of multinational corporations underestimates 'the complexity, diversity, and interdependent nature of global networks, thereby marginalising important nodes of globalisation in

TABLE 2.4 *Friedmann's world-city hierarchy*

Core countries				Semi-periphery countries			
Primary		**Secondary**		**Primary**		**Secondary**	
London[†]	***	Brussels[†]	*				
Paris[†]	**	Milan	*				
Rotterdam	*	Vienna[†]	*				
Frankfurt	*	Madrid[†]	*				
Zurich	*		*			Johannesburg	*
New York	***	Toronto	*	São Paulo	*	Buenos Aires[†]	***
Chicago	**	Miami	*			Rio de Janeiro	***
Los Angeles	***	Houston	*			Caracas[†]	*
		San Francisco	*			Mexico City[†]	***
Tokyo[†]	***	Sydney	*	Singapore[†]	*	Hong Kong	**
						Taipei[†]	*
						Manila[†]	**
						Bangkok[†]	**
						Seoul[†]	**

[†] National capital.
Note: Population size categories (recent estimates referring to metro-region)
* 1–5 million   ** 5–10 million   *** 10–20 million
Source: Friedmann (1986, 72)

developing regions' (Godfrey and Zhou 1999, 269). For example, league tables of the headquarters locations of the 500 largest multinationals produced by both *Fortune* and *Business Week* in 1996 gave greater prominence to US cities than is really warranted. In this way the inherent bias in world-city studies 'perpetuates well-established Eurocentric views of the global economy under the guise of objective data' (Godfrey and Zhou 1999, 269). But, by including first-level subsidiaries, or the regional offices, of the global corporates as well, Godfrey and Zhou (1999) show that other influential Asian cities like Hong Kong, Singapore and Seoul join Tokyo in the top ten.

The bases upon which the status accorded some global cities rests have also been questioned by Ann Markusen, a prominent American regional economist. As she points out, by emphasising the similarities that have been the hallmark of the rise of London, New York and Tokyo to world-city status, promoters of the 'global city hypothesis' 'tend to downplay the roles that national systems of cities and the conditions of host national economies

play in their success' (Markusen 1999, 875–6). In the case of New York, unlike either London or Tokyo, the city is only one of half a dozen major specialised centres in a nation with a 'multi-polar and progressively flattening urban hierarchy'. By this she means that New York has to compete within its national space with Boston, Washington, Chicago, Los Angeles and San Francisco, each of which variously serves as a 'capital' – educational, political, industrial, high-tech, military, industrial – in its own right (Markusen 1999, 876). Not only is New York surprisingly under-represented in many producer services categories, but this position is being steadily eroded *vis-à-vis* other US cities.

CONCLUSION

One of the strongest impressions to emerge from this historical survey of urban systems is of the continuing importance for city growth of the political, economic, and sometimes cultural, linkages connecting places. Another is of the extent to which the

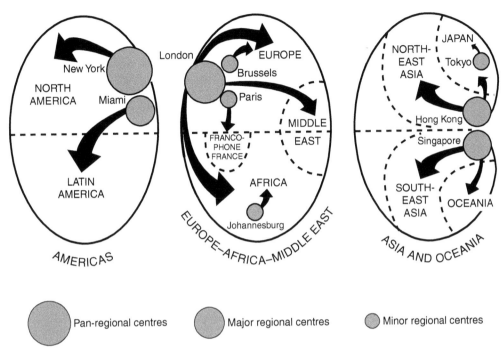

FIGURE 2.13 *The world according to GaWC (the Globalisation and World Cities Research Group)*
Source: Taylor (1999); available on the Globalisation and World Cities Research Group website: www.lboro.ac.uk/
departments/gy/research

long-distance flows and interaction between the cities forming these earlier urban systems are the actual forerunners of the world-city system, with its 'global cities' and emerging urban corridors. While the degree of integration and global coverage achieved by the more imperialist city-states and by Europe's colonial powers fell well short of what global corporations are capable of today in their network structure and connectedness, these earlier urban systems were not unlike today's world-city system. What has changed historically is the nature of the relationship between powerful cities and the state.

Principal centres of the city-state were instruments of imperial power and quite parasitic with respect to the demands they placed on their hinterlands. This is the forerunner of the core-dependency model perpetuated with the rise of mercantilism. Again, metropolitan cities achieved supremacy by serving Europe's imperial powers (core) and by tying their commercial livelihood to subservient colonies (periphery). Still, they evolved in the service of the state: 'The reinforcement of the power and executive

activity of the state and the growth of international trade' were the two factors chiefly responsible for the rise of Europe's most powerful cities (Bairoch 1988, 183).

Now core–periphery, state–city relationships are in the process of being transformed by globalisation. A select 'club' of what are none the less intensely competitive global cities has emerged, and these cities are less and less answerable to the state. This is not to say that they will ever assume the full range of powers and functions that states perform. But, with decolonisation, urban core-dependency relations have been re-aligned along a predominantly global–local axis. It is in urban affairs that global corporations operating within a global marketplace are replacing the state as the chief arbiters of the future of cities.

FURTHER READING

Lewis Mumford's (1961) account of the progress of urban civilisation remains unsurpassed. Though written in the late 1950s, it raises many of the 'big' questions about urban growth and the impact of cities upon the environment.

Morris (1974) is a well-illustrated guide to the history of urban form before the industrial revolution. One of the self-appointed high priests of the Los Angeles 'school of urbanism' prefaces his latest attempt to portray southern California as the prototype for future urban systems with an extended digression on the First and Second Urban Revolutions (**Soja** 2000).

Sjoberg (1960) remains one of the most authoritative works on The *Preindustrial City*.

Bairoch (1988) traces the rise of cities from the perspectives of economic historian and demographer.

Two monographs by **King** (1990a and 1990b) link urban development and the growth of urban systems with colonialism and the gradual integration of the global economy.

Of their time, **Jean Gottmann's** (1961) study of the Northeastern Seaboard of the United States and **Sakia Sassen's** (1991) characterisation of the 'global city' were influential in redefining urban systems.

Lo and Yeung (1996) contains case studies foreshadowing the emergence of urban corridors within Asia.

3

GEOGRAPHY OF WEALTH CREATION IN CITIES

INTRODUCTION

Employment opportunities and living conditions are bound up with the prosperity of cities. And, within a city, the neighbourhood or suburb in which a household can afford to live may be equally critical to access to work and quality of life (see Chapter 6). The composition of a city's economic base has a major bearing on the range of occupations on offer and access to jobs. Urban areas where inner-city jobs were once plentiful for female machinists in textiles, footwear and clothing, or for male stevedores or assembly line workers, may now be engaged in bidding wars to attract the suburban call centres and back offices that hire predominantly female 'pink-collar' workers. Displaced workers who are unable to retrain and get work locally are often urged to migrate to other regions where labour is in demand. Indeed, there is much greater labour migration between cities and regions in North America than in Europe or Australasia (see Chapter 5).

Is the city in which you are based in the economic doldrums? Or is the city in which you live as economically dynamic as Singapore, Shanghai, Atlanta, Vancouver or Sydney? Why are some cities depressed and the population falling, while other city regions with buoyant economies are attracting population? Moreover, the economic welfare of cities no longer requires population growth under certain conditions:

In a mature information society it will be possible for a city's economic growth to occur without necessarily being accompanied by population growth. This results from the fact that the information-intensive industries, unlike their more labour-intensive counterparts of earlier eras, are not limited in their growth and development by the confines of a local labour market or by a local market catchment area associated with sales.

(Newton *et al.* 1997, 7)

Yet this is also a narrow First World perspective that is restricted to technologically advanced economies. In many Third World cities, population growth is outstripping economic growth and the capacity of the urban economy to provide for newcomers.

These issues raise two broad sets of questions that bear upon wealth creation and quality of life in cities. The first relates to the role that cities play in national economic development. How significant are the economies of cities in the larger context of a country's economic system? What part do cities play in the creation of wealth? How does the creation of wealth in cities, or value adding, relate to the growth path of economies?

The second broad set of questions relates to how a city goes about earning its living? Why are some cities in decline in an otherwise prosperous economy when others are booming? Is the growth or decline of an individual city part of a secular long-term trend, or is the pattern of economic downturn and rebound more cyclical in nature? In a global era of open economies and heightened competition for market share, how can urban managers better position cities both to capitalise upon and to ride out the vicissitudes of trade and currency movements, inflation, fiscal crisis and falling wages rates? All these externally driven forces within the global economy can deeply affect the economic welfare of cities and levels of urban well-being (Mera 1998). What strategies for improving and expanding the productivity of a city's economic base, or what defensive strategies, are available to those with the task of managing cities? How do cities attract value-adding activities in sectors of the economy with the fastest rates of growth? What strategies have enabled cities like Singapore, Birmingham, Barcelona and Toronto to adapt more successfully to the challenges of globalisation and restructuring? Or is it just that those cities and others like them are naturally endowed with the mix of assets and human capital that makes for greater resilience in the long term?

There is an inescapable sense in which each of these vital questions is both open and contingent. Open because the economic strategies that are available in the pursuit of urban development and well-being have been periodically revised with a shift in public policy. And contingent because, as Cohen (1999) reminds us, the national economies in which cities are embedded are at differing stages of development and technological sophistication. First World cities are energetically trying to develop strategies that apply the 'four Cs' of Kenichi Ohmae, the Japanese economist and information theorist: cities, communications, capital and competition (Ohmae 1995). Alternatively, in Cohen's (1999) view, the 'four Ps' – people, pollution, poverty and politics – deserve a much higher priority in the development planning of Third World cities. He also points to even more obvious differences between cities: in the 1990s the budget of Osaka (US$50 billion) was twice the annual lending of the World Bank to 130 developing countries.

While theories about wealth creation in cities abound, it is important to keep Ann Markusen's (1999, 871) advice firmly before us: 'good theory must encompass both process and institutions, both structure and agency.' This is to emphasise that we must take care at all times not to reify 'the city'. By reify, I mean the tendency to attribute an innate capacity to the city as an object, or as a set of abstract processes. It permeates the 'competitive cities' debate between two economic gurus, Paul Krugman and Michael Porter, and their supporters (Lever and Turok 1999). Krugman maintains that 'competitiveness' is an attribute of firms and cannot be applied to cities or regions, let alone countries or continents. Porter, on the other hand, argues that cities and

regions do indeed compete – though not necessarily in the way that nation-states or businesses do. Rather,

they do compete for mobile investment, population, tourism, public funds and hallmark events such as the Olympic Games. They compete by, for example, assembling a skilled and educated labour force, efficient modern infrastructure, a responsive system of local governance, a flexible land and property market, high environmental standards and a high quality of live.

(Lever and Turok 1999, 791)

Entrepreneurs and bureaucrats, acting on behalf of enterprises or institutions, mobilise these potentialities. In conceiving of the city as a 'growth machine', Molotch and Logan (1987, 51–98) stress that the economic fortunes of cities are tied up with the deals done between city managers and business investors, and the complex politics and inter-bureau competition of state agencies (see Chapter 9). Furthermore, since the late 1980s a lot of attention in economics, as well as in urban and regional studies, has been devoted to the role that urban-industrial clusters can play in nurturing globally competitive local industries (Norton 1992).

THE ROLE OF CITIES IN NATIONAL ECONOMIC DEVELOPMENT

The role that cities play in national economic development has been hotly debated since the problems of urban over-concentration and congestion in cities surfaced in the 1960s. At the time, research findings emphasised the diseconomies that began to mount with an increase in city size, as well as the economic and social costs of unbalanced regional growth (Neutze 1965). In part, this harks back to a concern that 'over-sized' cities might become economically parasitic by draining human talent and resources from a dependent national hinterland. In those Asian and African countries where the level of urbanisation has surged ahead of the growth in per capita income it has been suggested that perhaps 'over-urbanisation' is responsible for holding back economic growth.

'Over-urbanisation'

Rising levels of urbanisation and the rapid growth of large cities are often viewed as problematic either because governments are unable to keep up with the incessant demand for new infrastructure and services, or because they fail properly to enforce pollution control and other regulations that ensure minimum standards in densely populated urban areas. During the 1980s the urban growth rate in African countries typically approached 5 per cent a year. The comparable rates for Asia and Latin America and the Caribbean were less than 4 per cent and close to 3 per cent (United Nations 1996, 25). One basis for the 'over-urbanisation' thesis is that the levels of urbanisation in the South are associated with much lower levels of per capita income. At the stage when levels of urbanisation were comparable in Latin America and Europe, per capita GNP was only a third in many Latino economies. Supporters of the 'over-urbanisation' thesis maintain that economic growth, agricultural productivity and industrialisation in many developing economies have not been sufficient to sustain the growth of their cities.

There are other commentators who are not so perturbed by the rates of urbanisation reported for countries that seem unable to cater adequately for their urban populations. They are inclined to point to the even higher rates of population growth experienced by cities in the United States between 1820 and 1870 (an average of about 5.5 per cent annually) and Japan during the 1930s (around 6 per cent a year). They also object to the implication that the growth path of the advanced economies, including the pattern of urban development in the North, is somehow exemplary. It is certainly the case, where public investment and pricing policies have favoured the urban at the expense of the rural sector in the South, that more migrants have left the land and been drawn into the cities.

Also, economic progress, relative to levels of urbanisation in parts of Africa, Asia and Latin America, was set back by a series of debt crises in the 1980s and 1990s. While per capita incomes fell by about 10 per cent in many Latin American countries during the 1980s, poverty in the cities rose by 17 per cent. In the late 1990s, urban poverty in Asian countries like Indonesia rose from about 11 per cent to 17 per cent. Yet this was somewhat less than expected owing to the shift of urban Indonesians from formal- to informal-sector employment, and back to the villages from which they had migrated. Unlike the post-industrial economies, 'generating jobs and income in the long climb out of recession ... is likely

to be more successful for the poor if those jobs are in low skill, low capital-intensive sectors such as construction' (Cohen 1999, 73). This tends to contradict prevailing nostrums, which state that the path to prosperity in a global economy involves competing in higher-end jobs and productivity gains. Even in those Latin American and Asian countries where economic growth has recovered, substantial urban poverty still persists.

It can be shown statistically for the period 1950–2000 that 'there is a close relationship between the level of urbanisation and per capita income (measured by GDP per capita)' (United Nations 1996, 28). But this should not be viewed as a general prescription for all developing countries because they differ so much with respect to the dominance of agriculture, physical factors such as size and topography, the level of security in the countryside, cultural norms and government policies towards human settlement.

Cities as national economic 'generators'

A number of countries were relatively successful in lifting their economic output during the 1970s and 1980s. These are countries that experienced a considerable spurt in urbanisation as well. They include newly industrialising economies (NIEs) like Hong Kong, South Korea, Taiwan and Singapore in the 1970s, followed by Malaysia, Thailand and Indonesia in the 1980s. Moreover, the fact that major cities generally concentrate more of a nation's total economic output than of its population lent growing support in the 1990s to the view that dynamic and competitive cities play a key role in economic growth.

This realisation that cities function as economic 'generators' within the domestic economy (Jacobs 1984) helped to overcome the particularly negative attitudes towards cities that prevailed in policy circles when 'de-industrialisation' was causing plant closures and high unemployment in the North. Structural analyses of inner-city decline in the United States and the UK elicited a policy response that accepted the inevitability of restructuring and cautioned against spending to prop up ailing urban areas. But, when the amount of physical capital sunk into urban areas is taken into account, there is an economic price to be paid for abandoning cities in decline. The regional economy suffers from falling investment and consumption, while a drop in revenue from export earnings undermines national economic performance. Current thinking recognises that, in standing by while the infrastructure and productive capacity of cities are run down, governments are placing their economies at a competitive disadvantage in an era of globalisation (OECD 1996).

Economic transition and technological innovation

Cities have always been a locus for productive activity. Table 3.1 sets out the series of technological innovations and social developments that have paved the way for the transition from 'pre-industrial', to 'industrial', and now to 'post-industrial' modes of production and urban structure; although there is an ongoing debate over the extent to which the shift from an industrial to a post-industrial, service-oriented economy is really based upon the overtaking of manufacturing output by services. Much high-wage and high-value-added service employment is actually devoted to providing services to manufacturing – hence the term 'producer services'. All the same, the distinction between 'pre-industrial', 'industrial' and 'post-industrial' is a very handy one

because it identifies the key technological shifts impacting on wealth-creating activity in cities.

In pre-industrial times, the development of tools and water power played a vital part in lifting agricultural productivity sufficiently to provide surpluses for off-farm workers living in towns (Jacobs 1984). The labour force of these early economies was dominated by direct employment in agriculture, with city workers making a relatively modest contribution to GDP. The Industrial Revolution changed all that. The substitution of steam power and mechanised equipment for hand-operated tools increased the capital-intensity of manufacturing and gave birth to the factories and mill towns of northern England and the industrial heartland of the United States. Later, with the widespread adoption in the 1940s and 1950s of Henry Ford's moving assembly line and Taylor's principles for organising workers (i.e. Fordism), cities were able to mass produce and distribute capital goods and consumer durables to increasingly urbanised and middle-class consumers.

The shift from a dominance of blue-collar manufacturing jobs to white-collar service jobs in economies like the United States has given rise to speculation about an approaching era of 'post-industrialism', and the crossing of a notional 'industrial divide'.

TABLE 3.1 *Transition to an information society*

Transition factor	Societal transitions		
	Agricultural	**Industrial**	**Informational**
Industry location	Dispersed	Centralised	Centralised with decentralisation
Industrial process	Handicraft	Mass production	Flexible specialisation
Economic engine	Human muscle	Machines	Human knowledge
Product	Customised	Uniform	Personalised
Work conditions	Informal	Formal	Team
Dominant mode of interaction	Face to face	Hierarchical line management	Information networks
Type of information transfer at work	Verbal	Paper	Electronic
Market orientation	Local	National	Global
Commuting pattern	Dispersed	Focused	Dispersed
Transport network	Minimal grid/ribbon	Radial	Extensive grid
Transport mode	Private, walk	Public rail	Private, car

Source: Newton *et al.* (1997, 8). Reproduced by permission

Piore and Sabel (1984) believe that this amounts to nothing less than a transition in manufacturing from 'Fordism' to 'post-Fordism'. As well as the increasing numbers of jobs in data processing, designing expert systems, manipulating symbolic code, e-commerce and advanced management functions (Reich 1993), since the 1970s a significant transformation has been taking place within the manufacturing sector of the advanced industrial economies. This has seen the introduction of more flexible and adaptive manufacturing systems to augment, rather than completely replace, mass-production lines. Even though their operations are still dominated by assembling components, the latest manufacturing plants turning out consumer durables like cars and white goods make use of flexible systems to link robotics with computerised causeways. Economies of scale will be replaced by 'economies of scope' as the software that controls machines assumes more importance than hardware in manufacturing.

At this stage, the adoption of more flexible ways of organising production and labour has progressed much further in small to medium-sized firms that are operationally suited to capturing the benefits of 'flexible specialisation'. These operating systems feature batch production of small quantities of highly differentiated, and even customised, items. The use of computers for stock control and 'just-in-time' scheduling allows companies to shorten the product cycle and respond more quickly to shifts in consumer tastes. By comparison with firms that are vertically integrated and control each phase of production from the inputs all the way through to the intermediate goods and components to the finished product, flexible specialisation promotes 'vertical disintegration'.

Moreover, 'as productive processes become increasingly complex in advanced industrial societies, the largest reserve of economic opportunities will be in organising and coordinating productive activity through the process of information handling' (US Office of Technology Assessment 1995, 12). For example, all 10 cutting-edge technologies with the potential to boost productivity and create jobs in the future rely upon the exploitation of microprocessors to manage and manipulate information (e.g. lasers, simulation of virtual reality, genomics, integration technology, biotechnology, smart products, nanotechnology, bionics, global positioning systems, micro-machines).

Table 3.2 illustrates the sectoral shift in employment that took place between 1954 and 1991 for a number of major Australian cities. The employment data have been subdivided into manufacturing, services and information groups in order to highlight the growing importance of information-processing activities within the urban economy. With few exceptions, the cities with a higher than average concentration on information-processing activities appear to enjoy superior economic well-being if the employment-to-population ratio is accepted as a reliable guide (Table 3.2).

As cities make the transition from primary production to secondary production, and then on to the tertiary and quaternary activities that are the hallmark of the 'new information-intensive, services economy' (Reich 1993), they are generating more and more of the national wealth. In Australia's case, farming accounted for 20 per cent of GDP in the 1950s; 20 years later it had slumped to less than 10 per cent of GDP; and finally, another 30 years on, it had fallen to less that 5 per cent of GDP.

Four main factors contributed to the rise in the productive capacity of cities in the late twentieth century.

- Resource-based primary production accounts for a diminishing share of export income and job creation relative to expanding, city-based sectors of the economy like elaborately transformed manufactures (ETMs), producer services, educational exports and tourism.
- Within the industrial sector of the 'knowledge economy', production is becoming less energy- and labour-intensive, and more dependent on expert systems and the processing of information. Large cities provide a stimulating and synergistic intellectual milieu, where institutions of higher learning and R&D labs cluster, and innovative ideas in media and fashion are exchanged.
- Because of the high degree of interaction between firms using flexible systems in production, including the sharing of orders, facilities, technical information and machine tools, and marketing and trade services, it is not uncommon to find them clustering in cities – though outside the traditional industrial districts with their organised labour and congestion costs.
- Profits generated from international trade in commodities, finance and business services, and

TABLE 3.2 *Employment trends by major industrial sector, 1954–1991: major urban centres*

Urban centre	1991 population (000)	Industry sector (% persons employed)						Employment-to-population ratio
		Manufacturing		Services		Information		
		1954	1991	1954	1991	1954	1991	
Sydney	3698	37	14	41	39	20	39	0.44
Melbourne	3153	40	18	39	37	19	37	0.44
Brisbane	1327	28	13	46	40	23	39	0.43
Adelaide	1062	36	15	43	36	18	39	0.43
Perth	1197	25	11	50	40	22	39	0.42
Newcastle	432	41	15	39	40	12	33	0.39
Canberra	315	6	3	34	29	55	60	0.50
Wollongong	239	45	20	34	36	9	33	0.39
Gold Coast	274	13	9	59	50	17	31	0.38
Hobart	185	26	11	46	37	25	43	0.41
Geelong	152	46	22	39	38	12	32	0.39
Townsville	115	23	8	54	40	20	44	0.44
Darwin	77	6	5	48	39	40	46	0.46
Toowoomba	83	25	12	32	41	38	38	0.39
Launceston	94	30	14	49	40	18	34	0.40
Ballarat	82	37	14	41	38	18	40	0.36
Cairns	84	21	8	57	49	16	34	0.43
Australia		28	13	39	38	17	36	0.42

Note: ASIC classification: Information sector (communications, finance, property, business, etc., public administration, R&D community services).
Source: Newton *et al.* (1997). Reproduced by permission

technology are replacing productive activity *per se* as the driving force in the global economy. As a result, a few select cities have acquired an unrivalled economic and strategic influence over capital flows and the location of new investment around the world.

HOW DO CITIES GENERATE THEIR INCOME?

The 'economic base' of a city determines the prosperity of its citizens. By concentrating productive activity, a city provides a living for its residents in a very direct sense – and also indirectly, inasmuch as, these days, governments provide various forms of income support to the economically inactive in society. The industrial composition, or mix of economic

functions located in a city, holds the key to how well the urban economy is performing. 'Functional classifications of cities' can provide a useful analytical tool in helping to understand why cities with a particular industrial profile, or economic base, are performing well or poorly. In the 1960s and 1970s, urban geographers like Robert Smith (1965) put a lot of effort into classifying cities according to the types of economic function they supported. But this has largely been superseded now by more sophisticated techniques like discriminant analysis, which is used, for example, by Paul Cheshire (1990) to explain the performance of the European Community's major urban regions between 1971 and 1988. Alternatively, Markusen and Gwiasda (1994) use shift-share analysis to compare the economic performance of a number of major US cities, based on their functional specialisation.

Conceptually, this interest in the economic performance of cities with differing degrees of functional specialisation harks back to more traditional forms of economic base analysis, which in turn applies the rudiments of trade theory. This states that employment growth occurs fastest in those countries, or cities, that capitalise on their comparative advantage in producing the goods and services in greatest demand on the world market. This view of how cities grow and prosper is called 'export base theory'.

Export base theory

For a long time economists and geographers have debated why it is that some urban economies perform better than their counterparts, or prove to be more resilient during a cyclical downturn. In the 1920s, the New York Regional Planning Committee drew a distinction between the 'primary' economic functions, like processing materials sourced within the region for export, and the 'ancillary' functions that concentrate on serving the needs of the local population. Homer Hoyt (1939), an urban economist, refined this further when he suggested that the terms 'basic' and 'non-basic' be used to describe the activities that generate urban growth as distinct from the provision of services. In the 1950s 'economic base analysis' was a veritable growth industry in its own right. In all of this, the work of geographer Gunnar Alexandersson (1956) on 'city-forming and city-serving production' and economist Charles Tiebout (1956) on the 'urban economic base reconsidered' stand out as classics.

In essence, cities grow or stagnate according to the ability of their businesses to earn export income. In a practical sense, these earnings are 'basic' to the growth and prosperity of an urban economy. Naturally some of the goods and services produced by city firms are purchased locally. While this activity might be considered to be 'non-basic', the revenue also returns to circulate within the urban economy. Much of this activity helps to replace the purchase of imports and makes a net contribution to local prosperity. But import replacement does not have the same powerful stimulatory effect locally as the earnings on goods and services exported outside the city region. Tiebout (1956) was the first to recognise that this gives rise to a local-multiplier effect when workers in 'export' industries spend their wages to purchase local goods and services.

As a city-state, Singapore is crucially dependent upon exporting goods and services for its livelihood. Singapore has been run by exceptionally smart managers, initially under the tutelage of Lee Kwan Lew, its first Prime Minister. Its economic development strategy has been continuously adjusted to keep ahead of its main economic rivals in what is a very dynamic region. For example, in the 1970s, with other low-wage production zones in Southeast Asia catching up, it decided that the economic base had to be transformed to increase both labour output and the value added in manufacturing. In a series of measured moves the government phased in higher wage rates. (This represents a form of heresy to orthodox economists!) Industrialists struggling to maintain their returns to investment were left with little choice but to retrain their workers and upgrade their plant and equipment. By 1990, Singapore had 41 programmable robots for every 10,000 manufacturing jobs. At the time, this index of 'robot density' in a selection of other advanced economies ranged between 172 for Japan and 63 for Sweden, to 29 for Germany and 19 for the United States. This period has since become known as Singapore's 'second industrial revolution' (Macleod and McGee 1996, 431).

Then, once more in the mid-1980s, with wage rates edging ahead of productivity gains in manufacturing, Singapore's economic managers realised that financial and business services would need to become the vanguard sector if Singapore were to retain its competitiveness. This conclusion was reached by an Economic Commission established in 1985 to analyse the economy's structural weaknesses. The present strategy involves converting Singapore into an 'intelligent island' founded upon the twin pillars of leading-edge telecommunications and a highly educated, computer-literate society. By 1986, Singapore's expenditure on IT per S$1000 of GDP was double that of Japan. With very impressive government-run agencies and public utilities like Telecommunications Singapore and Singapore Airlines, the city-state has been well placed to capitalise upon the export potential of these information-based services throughout the region and beyond.

Import replacement

A city can also strengthen its economy by substituting 'home-made' goods and services for imports. Because the availability of materials and the size of the local market are bound to impose some limits to import substitution, the amount of added productivity that can be achieved rests with the ingenuity of local entrepreneurs. With the lowering of protective trade barriers in countries like Australia and New Zealand since the mid-1970s or so, import substitution has become dreadfully unfashionable. Yet Jane Jacobs (1984, 38–58) makes a compelling case for the stimulatory role of import-replacing investment. The multipliers include the:

- investment capital that import-replacing brings into the city
- extra jobs and local spending created by the process
- boost to local suppliers and the building industry
- generation of profits for ploughing back into local businesses.

Significantly, during the 1980s the Australian and New Zealand governments deliberately opened up the domestic markets to competition from cheaper imports in an effort to accelerate structural adjustment. The effect has been to force cities to restructure manufacturing, and generate export earnings along lines that apply mainly to North American and European cities. This has drastically eroded the economic base of a number of smaller Australasian cities that previously earned their income from import-substitution and the export of semiprocessed

primary products. With the lowering of the average effective rate of assistance for the manufacturing sector from 25 per cent in the mid-1970s to a projected 5 per cent by 2004, a number of industrial suburbs in Sydney, Melbourne and Adelaide are expected to lose more jobs. For example, in 1997, a forecast indicated that Adelaide would lose 9393 jobs given a 'worst-case' scenario. This would be the outcome if the car-makers Mitsubishi and General Motors-Holden closed their plants in South Australia (Table 3.3). With the takeover of Mitsubishi by Daimler-Chrysler in March 2000, the eventual loss of the Mitsubishi jobs moved a step closer. On 30 September 1999, the 'Big Australian' BHP, or Broken Hill Proprietary Ltd, closed its Newcastle steelworks, effectively leaving the one-company town without its economic mainstay. This marked the end of a long process of adjustment to intermittent recession (in 1981–2 and 1992–3) and a mounting glut in steel on the world market. As a result, the Newcastle workforce was progressively 'downsized' from a peak of about 12,000 in the late 1970s, to just 3000 in 1997 when BHP announced the planned closure of the steelworks.

Shift-share analysis

A methodology called 'shift-share analysis' was developed in the 1960s to throw more light on the processes responsible for the growth or decline of

TABLE 3.3 *Where jobs would be lost if Mitsubishi or GM-Holden closed their Adelaide car plants (estimates for late 1990s)*

Suburb	Postcode	Mitsubishi	GM-Holden	Total
Morphett Vale	5162	766	5	771
Paralowie/Salisbury	5108	30	586	616
Elizabeth/Hillbank	5112	10	443	453
Andrews Farm/Blakeview/Craigmore/One Tree Hill/Smithfield	5114	8	400	408
Aberfoyle Pk/Flagstaff Hill/Happy Vlly	5159	344	12	356
Davoren Pk/Elizabeth	5113	5	348	353
Hackham/Hackham W/Onka Hills	5163	311	—	311
Brahma Lodge/Salisbury	5109	16	289	305
Hallett Cove	5158	238	4	242
Reynella	5161	203	1	204

Source: Centre for Labour Studies, University of Adelaide

urban regions. One of its great attractions is the detail it can reveal about industry structure and regional dynamics as separate sources of employment change in cities. Several economists with a special interest in regional science and working for the Resources for the Future Inc. (Dunn 1980) subsequently applied this very useful technique to 171 urban regions in order to chart economic development between 1940 and 1970 in the United States.

Shift-share analysis makes use of matrix algebra to break down the individual contributions of each industrial sector – 31 for the United States in the Dunn report (1980) – to employment change in an urban region. The model accepts that there are three main components of employment change within an urban labour market. Firstly, there is the 'national growth effect', which would exactly reproduce the rates of national growth in each industrial sector if all the component industries in a city formed a mirror image of the national economy, and only if these 'home town' industries grew at the same rate as their national counterparts between 1940 and 1970. Secondly, there is an 'industrial mix effect', which identifies how much of the employment growth in an urban region can be attributed to its industrial structure over and above national trends, sector by sector. It helps to throw light on whether or not a city region is handicapped by poorly performing industries. Thirdly, there is a 'regional share effect', which allows for the fact that a local industry may be relatively more or less competitive than the sector is nationally. For this reason, it's sometimes referred to as the 'regional shift' or 'competitive effect'.

Markusen and Gwiasda (1994) use a modified shift-share analysis to assess the impact of domestic, as opposed to international, demand upon the performance of the manufacturing sector in six major US cities between 1977 and 1986 (Table 3.4). San José, in the heart of Silicon Valley, and Boston both benefited from the stimulus of domestic spending on defence (missiles, armoured vehicles, computers, semiconductors) and international demand for computers. These extra jobs due to industry mix helped to offset those jobs lost to import competition (Table 3.4). The strong domestic demand for manufactured goods in the case of Los Angeles, and for producer services in the case of Washington DC, meant that they were able to withstand import penetration even though their export performance in manufacturing was unimpressive during the 1980s. Likewise, both New York and Chicago lost jobs in export-dependent manufacturing, and, despite the reasonable domestic demand for output from these two cities, failed to capture their 'expected' shares of new and relocating activity as measured by 'competitive shift' (Table 3.4).

Localisation and agglomeration economies

The extra part that 'localisation' and 'agglomeration' economies can play in the location decision-making of the firm was first formalised by the great

TABLE 3.4 *International manufacturing shift-share components, selected cities, 1977–1986*

MSA	Net job change	Industry mix (domestic demand)	Industry mix (exports)	Industry mix (imports)	Competitive shift
San José, CA	64,874	175,276	37,710	−29,070	28,110
Los Angeles, CA	25,022	129,764	691	−7439	5356
Washington DC	20,770	13,189	846	266	14,994
Boston, MA	16,333	96,013	16,338	−17,248	18,158
New York, NY	−125,985	37,241	−6822	−18,922	−87,342
Chicago, IL	−160,099	22,036	−4037	2869	−123,566

Source: Markusen and Gwiasda (1994); compiled from unpublished Department of Commerce Data on international trade. Reproduced by permission of Blackwell Publishers

English economist Alfred Marshall (1920). Simply put, localisation economies exist where there is a cluster of same-sector businesses and employees deriving benefits from their geographical proximity. The city's economic base is much more specialised owing to the presence of only one or two industrial sectors. Agglomeration economies are found in highly urbanised regions where there are many more industry groups together with a complex array of economic and social institutions. For this reason they are sometimes called 'urbanisation economies'. In a classic paper, Chinitz (1961) compared the respective difficulties and opportunities faced by a specialised company-town like steel-based Pittsburgh, with a diversified economy like New York.

Entrepreneurs that can capitalise on these localisation and agglomeration economies get a price advantage, or make savings over and above any efficiency gains achieved from internal improvements to the operation of the firm. Economists also call these external economies, or positive externalities, because they are not part of the inner workings of a business. The benefits of industrial clustering extend to the strong magnetism for other firms once a cluster gets going: 'A concentration of rivals, customers, and suppliers will promote efficiencies and specialisation' (Porter 1990, 157). Also, the possibility of attracting inward investment and talented people is greatly enhanced by the presence of other firms in the same industry. The presence of a skilled workforce is also crucial. A study of European urban regions undertaken by the Centre for Economic Policy Research (Braunerhjelm *et al.* 2000) estimates that a 10 per cent increase in average years of schooling in the local workforce is capable of raising labour force participation rates by between 22 and 45 per cent. Typically, industrial clusters have a strong localised labour market where workers with skills in high demand move from firm to firm, but remain committed to the region. Low rates of out-migration foster the evolution of a strong local cultural identity and shared industrial expertise.

Figure 3.1, which was drawn up by Paul Knox (1994), another urban geographer, summarises the chain reaction whereby productive activity in a city feeds off itself. It sets out the self-reinforcing multipliers, and 'backward' and 'forward' linkages between new and existing firms within a city. A backward linkage is created when a new firm exploits a niche to provide local firms with components or services in short supply. On the other hand, a forward linkage is created when an out-of-town firm establishes a local presence in order to take advantage of locally produced goods or services. They become an intermediate input into the newcomer's operation as part of the assembly, or packaging, or distribution functions. If a city is successful in attracting newly linked industries, this in turn creates a threshold of activity large enough to support ancillary functions like maintenance and repair, recycling, security, staff training, and even technical and laboratory support for product development. Beyond a certain scale, these multiplier effects translate into localisation economies that further enhance the competitiveness of a city as a prospective location for an expanding firm. With continuing urban-industrial growth, a city's economic base generally becomes more highly diversified, even though it may be especially strong in a number of core industrial sectors. Cities like Osaka, Birmingham, Melbourne and Minneapolis-St Paul come to mind.

The Tokyo–Yokohama region provides a prime example of a conurbation that has gained economies with each stage of its development. In the 1960s it was a magnet for the whole Japanese economy, drawing investment and labour from the other regions. The next stage, in the 1970s, involved a deliberate strengthening of Tokyo as the Northeast Asian hub for transnational corporations, finance and telecommunications. This necessitated a huge land reclamation scheme in Tokyo Bay to accommodate the twenty-first-century infrastructure and corporate facilities that a global city must possess. Finally, during the 1980s, Tokyo's corporate leaders began to decentralise its manufacturing capacity by seeking out lower-wage locations around the Asia-Pacific Rim. Nevertheless, Tokyo has managed to hold on to its dominant share in micro-electronics and small-scale machining industries (Markusen and Gwiasda 1994, 184). Tokyo's comparative advantage grew commensurate with the emergence of the Japanese economy as a global economic power. By 2000 Tokyo was a giant agglomeration containing more than 30 million people who manage to function daily without too much inconvenience owing to the extensiveness and efficiency of the regional mass transit system.

On the other hand, poorly managed urban growth in cities like Athens, Bangkok, Sydney and Auckland can begin to backfire and produce diseconomies of

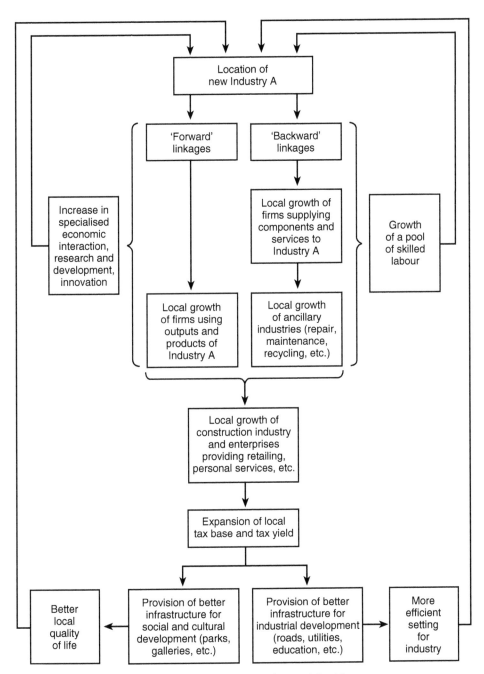

FIGURE 3.1 *The linkages between firms that spur economic growth in cities*
Source: Knox (1994). Reprinted by permission of Prentice-Hall

scale, or agglomeration diseconomies. These include the costs to firms of exorbitant land prices, expensive hotel accommodation, traffic delays and congestion, inefficient ports, solid waste treatment and the higher taxes levied to combat these diseconomies. Table 3.5 illustrates the respective costs of locating a plant in a number of international cities in the late 1990s.

TABLE 3.5 *Locational cost benchmarking study: Melbourne and 10 other Pacific Rim cities*

Comparative index criteria	Melbourne	Auckland	Bangkok	Jakarta	Kuala Lumpur	Los Angeles	Osaka	Singapore	Shanghai	Hong Kong	Seoul
Industrial land	100	304	311	286	723	358	7944	3644	306	1988	658
Factory construction	100	132	100	62	117	186	348	216	228	391	86
Factory construction time	100	75	163	138	263	150	138	263	263	88	200
Office rental	100	101	146	131	180	208	369	538	392	653	180
Marketing director salary	100	93	75	102	100	119	163	152	48	123	120
Accountant salary	100	85	70	72	71	154	208	161	23	159	118
Electronic engineer salary	100	80	65	71	63	135	153	120	19	85	110
Process worker salary	100	119	14	22	26	154	166	83	18	81	122
Frequency of international flights	100	41	74	34	41	117	50	98	34	110	57
Air freight	100	102	117	175	129	187	338	225	130	146	161
Sea freight cargo	100	137	86	88	68	149	87	76	78	75	71
Ship turnaround time	100	96	144	109	64	88	64	72	168	100	72
International calls	100	131	157	184	121	137	152	108	204	89	119
Local calls	100	161	197	101	171	139	183	429	94	193	59
Electricity	100	134	138	251	183	160	327	137	154	155	125
Gas	100	305	188	64	397	229	321	703	136	808	204
Water	100	134	118	275	92	111	890	161	24	115	61
Waste water disposal	100	116	59	186	8	62	430	237	10	—	4569
Industrial waste disposal	100	127	443	52	100	475	709	165	2	16	336
Hotel accommodation	100	140	126	126	102	191	189	108	188	271	161

Source: **Business Victoria** (1997)

CITY GROWTH AND 'NEW INDUSTRIAL DISTRICTS'

Alfred Marshall's pioneering ideas have been taken a step further as a way of accounting for the emergence of 'new industrial districts' in a number of urban regions where specific conditions are met (Marshall 1920). As such, they represent a real break from traditional manufacturing organised along Fordist lines. This new respect for industry– city ties was rekindled in the 1980s by Piore and Sabel (1984) in their book *The Second Industrial Divide*. They were the first to suggest that Marshall's 'industrial district' was something of a forerunner to emerging post-Fordist manufacturing zones with their reliance on small-batch or flexible production systems, vertical disintegration and spatial linkages between localised clusters of small and medium-sized firms. These ideas were taken up and expanded upon by a group of Californian urban and regional geographers, led principally by Allen Scott (1988), with help from Michael Storper and Richard Walker (1989):

Thus Porter and the regionalists came independently to the same conclusion. To the regionalists, post-Fordist competition, highlighted by flexible systems and quick responses, bolsters the economic base in resurgent, generative cities. To Porter, anything that speeds the information flow becomes a source of competitive advantage. Either way urban industrial clusters hold centre stage.

(Norton 1992, 162)

What is drawing attention to these new industrial districts is their apparent resilience and adaptability in the face of global restructuring. Small, flexible and innovative firms embedded within a regionally cooperative system of industrial governance and management are a recurring theme in these newly forming industrial zones.

Attention initially focused on regional clusters of mature craft centres in central Italy – the so-called Third Italy – and southern Germany, where researchers noted that regional competitiveness benefited from a cooperative approach by firms to training, product development and marketing (Sforzi 1987). In Italy's Emilo-Romagna region, cities like Bologna (food processing and packaging machinery), Prato (woollen textiles and garments), Parma (foodstuffs), Carpi and Modena (knitwear, woodworking machinery), Sassuolo (ceramic tiles) and Arezzo (jewellery) spring to mind. In southern

Germany: chemicals are concentrated in Frankfurt and Ludwigshafen; automobiles and machine tools in Stuttgart, Munich, Ingolstadt, Neckarsulm and Regensburg; surgical instruments and cutlery in Tuttlingen; jewellery in Pforsheim.

The concepts of 'flexible specialisation' and 're-agglomeration' have since been drafted into use by Scott (1988) and Storper and Walker (1989) to help account for the formation of new clusters of predominantly post-Fordist industries both in, and around the edges of, some North American and European cities. Silicon Valley and Orange County LA, in California, along with the M4 corridor between London and Reading, and Cambridge in England, as well as the Scientific City in the southernmost portion of the Greater Paris region, are invariably cited as the most fully developed examples of post-Fordist industrial spaces.

However, the promotion of 'flexible specialisation' and the 'neo-Marshallian' industrial district, or its Italianate predecessor, as somehow paradigmatic of post-Fordist industrial complexes has drawn criticism from several researchers (Gertler 1988; Lovering 1990). From one direction, the 'high-tech' North American districts are said to be only superficially similar to their Italianate progenitors in that they cannot compare in their approach to managing the workforce or labour welfare. Italian cities have a long tradition of active political engagement based on strong unions and well-organised labour, whereas the enterprise culture of Silicon Valley and Orange County is resistant to unionism. And, coming from another direction, in a recently completed study of fast-growing 'second tier cities' in the USA, Brazil, Korea and Japan, Ann Markusen and her colleagues find that flexible specialisation is not pervasive, even among this set of relatively new, 'untainted' and highly competitive locales (Markusen *et al.* 1999).

Markusen (1996) identifies at least three additional variants of the 'new industrial district' and pronounces two of them – 'hub and spoke districts' and 'satellite industrial platforms' – more dominant in the United States than either of the other two; that is, the 'Marshallian industrial district' and 'state-anchored industrial districts'.

Marshallian industrial districts

Diagram A in Figure 3.2 portrays the main characteristics of a Marshallian industrial district to set it

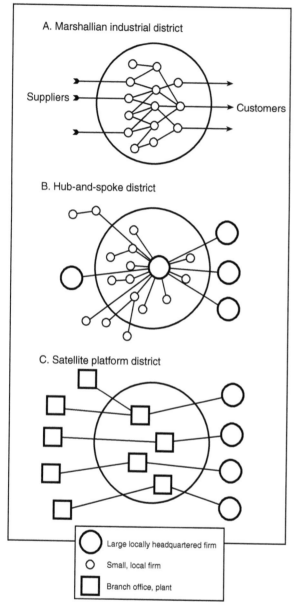

A. Marshallian industrial district

Suppliers

Customers

B. Hub-and-spoke district

C. Satellite platform district

○ Large locally headquartered firm

○ Small, local firm

□ Branch office, plant

FIGURE 3.2 *The Marshallian industrial district and its variants Source*: Markusen (1996). Reproduced by permission of *Economic Geography*

industrial district' typically comprises small, locally owned firms that make investment and production decisions locally, do a lot of business among themselves, often share industrial expertise and work together to build up a local talent pool of workers with specialised skills. And frequently, often with the help of regional agencies or trade associations, they consciously 'network' to solve problems of cycles and overcapacity, or to respond to the need for even greater flexibility or shifts in fashion.

The Marshallian district also encompasses a relatively specialized set of services tailored to the unique products/industries of the district. These services include technical expertise in certain product lines, machinery and marketing, and maintenance and repair services. They include local financial institutions offering so-called 'patient capital', willing to take longer-term risks because they have both inside information and trust in the entrepreneurs of local firms.

(Markusen 1996, 300)

Apart from the European and North American candidates put forward by the promoters of 'flexible specialisation', Markusen suggests that the southern sector of Tokyo, which takes in Yamahama and Kangwan (a south side district in Seoul), also possibly qualify as 'neo-Marshallian industrial districts'.

Hub-and-spoke industrial districts

There are fast-growing regional economies where a number of key firms and/or facilities act as anchors or hubs to suppliers and other subsidiaries that spread out around them like the spokes of a wheel. In the USA, Seattle, central New Jersey and Anaheim/ Santa Ana are good examples, as are Toyota City in Japan, Ulsan and Pohang in South Korea, and San José dos Campos in Brazil. In Seattle, the economy is organised around Weyerhauser as the dominant resource-sector company, Boeing as the dominant industrial employer (commercial aircraft and military/spacecraft), Microsoft as the leading services firm, the Hutchinson Cancer Center as the magnet for a swarm of biotechnology firms, and the Port of Seattle as the transport hub (Markusen 1996, 302).

Satellite platform industrial districts

The satellite platform is characterised by a congregation of branch facilities of externally based multiplant

apart from three other possibilities identified by Markusen (1996). She drew these diagrams to highlight differences in firm size, inter-firm connections inside and outside the region, and dominant market orientation. The business structure of a 'Marshallian

firms. These have often intentionally been funded by central or regional governments to stimulate growth in laggard regions. An outstanding high-end case in the USA is the internationally admired Research Triangle centred upon Raleigh and Durham in North Carolina, which groups the research institutes of a number of large multinational corporations. In South Korea, Kumi provides a low-end textile and electronics platform, while Ansan groups together otherwise unwanted polluting industries. In Japan, Oita and Kumamoto are reasonably successful technopoles. A state-sponsored import/export zone in Brazil provides the platform for development in Manaus.

State-anchored industrial districts

Many of the fastest-growing industrial districts in the United States and elsewhere owe their performance to the location of a major government facility (Markusen 1996, 306–7). Such cities often doubled, or occasionally better than trebled, their employment growth between 1970 and 1990. Military bases and academies, or weapons labs, account for the phenomenal post-war growth of US cities like Santa Fe, Albuquerque, San Diego and Colorado Springs, while defence plants contributed dramatically to the growth of Los Angeles, Silicon Valley and Seattle. State universities and/or state capitals explain the prominence of cities like Madison, Ann Arbor, Sacramento, Austin and Boulder among the fastest-growing cities in the United States. Denver owes much of its post-war growth to the second-largest concentration of federal government offices in the nation. In Japan and South Korea, the government research complexes at Tsukuba and Taejon respectively have fuelled regional growth. In Brazil, Campinas has a top-flight university, while the government-sponsored military facility at San José dos Campos accounts for the region's strength in aerospace.

Lastly, it should be said that this typology of new industrial districts does not preclude a mixture of all of the above. Part of Los Angeles and Silicon Valley, for instance, have 'Marshallian' districts largely given over to electronics or software engineering, but also anchor 'hub' firms like Lockheed Space and Missiles, Apple and Hewlett Packard, as well as hosting large 'platform-like' branch plants of US, Japanese, Korean and European corporations. Further, over

the years both regions have been the recipients of massive defence contracts that have continued to fuel development in missile systems, electronics and communications technology.

DIVERSIFICATION AND URBAN GROWTH

In the literature on urban and regional development, connections have been drawn between growth and the diversification of a city's economic base. But cause and effect are not at all clear. On the one hand, there are reasonable grounds for assuming that the larger a city gets, the greater the industrial mix. Thus urban growth begets diversification (O'Donoghue 1999). On the other hand, Thompson (1965) turns this on its head somewhat. He proposes that once a city attains a certain size, the structure and diversity of its economic base will ensure further growth and even insulate it against absolute decline. That is to say, the diversity of a city's economic base – the range and mix of industrial sectors – is a key factor in determining how solidly any urban region grows, or copes with structural adjustment, or rides out cyclical fluctuations in the broader economy. Diversification could well act to stimulate growth by drawing in workers who cannot find the same variety of employment opportunities in cities with a narrower range of industries.

O'Donoghue (1999) is interested in finding out whether the converse applies. Does a specialised economic base increase the vulnerability of cities to instability and even decline? He takes a period in recent UK economic history (1978–91) that corresponds with extraordinarily severe job losses in the manufacturing sector. In a 13-year period, 2.6 million workers lost their jobs in manufacturing, representing a fall of 36 per cent since 1978. The period was marked by a deepening economic trend that hollowed out ageing core-manufacturing sectors over a 30-year period, 1951–81 (Table 3.6). The plant closures and redundancies can be traced to the changing organisation of industry at the firm level and the antiquated nature of many UK factories (Massey and Meegan 1982).

Large plants with worn-out assembly lines in congested industrial regions could no longer compete with newly installed technology in the NIEs. Hence the largest conurbations in northern England

TABLE 3.6 *Employment trends in major UK industries, 1951–1981*

Industry	Employees (000)				% change 1951–81
	1951	1961	1971	1981	
Agriculture, forestry and fishing	1126	855	432	360	−68
Mining and quarrying	841	722	396	332	−61
Food, drink and tobacco	727	704	770	632	−13
Chemicals and allied trades	435	499	438	395	−9
Metal manufacture	616	626	557	326	−47
Engineering and electrical goods	1601	2031	2028	1730	+8
Ship-building and marine engineering	277	237	193	144	−48
Vehicles	735	838	816	636	−13
Other metal goods	458	525	576	428	−7
Textiles	986	790	622	363	−63
Leather, leather goods and fur	78	60	47	31	−60
Clothing and footwear	676	546	455	313	−54
Bricks, pottery and glass	314	321	307	216	−31
Timber, furniture	326	304	269	227	−30
Paper, printing and publishing	515	605	596	493	−4
Other manufacturing industries	264	295	339	265	negligible
Total: manufacturing	9975	9958	8841	6891	−31
Construction	1388	1600	1262	1132	−18
Gas, electricity and water	357	377	377	340	−5
Transport and communications	1704	1673	1568	1440	−16
Distributive trades	2689	3189	2610	2635	−2
Insurance, banking, finance	435	572	963	1233	+183
Professional and scientific services	1524	2120	2916	3695	+142
Others, miscellaneous	3485	3519	3379	3993	+15
Total: service sector	11,582	13,050	13,075	14,468	+25

Source: Mounfield (1984, 142); extracted from Central Statistical Office Annual Abstract of Statistics

and Scotland suffered disproportionately from 'de-industrialisation', while the south-east picked up the service-sector growth. The only employment growth in the UK manufacturing sector between 1951 and 1981 was in 'engineering and electrical goods' (Table 3.6). Otherwise, the majority of new jobs created throughout this period were in the services sector – 'insurance, banking, finance', 'professional and scientific services' – where small, flexible firms were in the ascendancy. The divided 'north/south' pattern of industrial disinvestment in the UK continued to the end of the 1990s. While northern cities still concentrate on mainly traditional industries like textiles and engineering, in the twentyfirst century the south-east's expanding share of the UK's GDP is based on diversification into communications and electronics.

Significantly, the larger cities in the north of England and the Midlands were hardest hit by the decline of the UK manufacturing. Yet these were urban economies that were initially more diverse than small industrial centres (O'Donoghue 1999, 553). Table 3.7 documents plant closures on Merseyside during the 1970s and 1980s. Clearly size gives no real protection against the economic restructuring that took place in the UK in the 1980s. O'Donoghue (1999, 561) also found that, irrespective of the centres involved, the growth of service-sector jobs

TABLE 3.7 *Major job losses on Merseyside, 1975–1979 and 1980–1985*

Name of firm	Number of redundancies	Year
a. 1975–79		
British Leyland	3750	1976–8
Dunlop	2600	1979
Plessey Telecommunications	2400	1975–9
Lucas Victor	2400	1978–9
Mersey Docks and Habour Board	2280	1975–9
Cammell Lairds (British Shipbuilders)	1900	1978–9
Courtaulds	1600	1976–8
Thorn EMI	1600	1976
Spillers	1227	1978
BICC	1180	1977
GEC	1000	1978
KME	740	1979
Western Ship Repairers	675	1978
Tate & Lyle	600	1976
Bird's Eye Foods	450	1978
Meccano	425	1978
Other notified losses of 100–399 jobs	2795	—
Sub-total (1975–79)	27,422	
b. 1980–85		
United Biscuits	4000	1982–3
Ford Motor Company	3500	1980–5
Liverpool City Council	3000	1981
Plessey Telecommunications	2830	1980–4
Tate & Lyle	1700	1981
Pilkingtons	1700	1981
Mersey Docks and Harbour Board	1520	1980–1
Lucas Aerospace	1177	—
Associated Biscuits	1164	—
BL	1100	1981
Pressed Steel Fisher (BL)	1100	1981
Courtaulds	900	1981
GEC	700	
Barker and Dobson	647	—
Northgate Group	600	—
Cadbury Schweppes	500	—
CBS Engineering Co.	487	—
Metal Box	470	—
Lyons Maid	450	—
Cousins' Bakeries	450	—
S. Reece and Sons	420	—
Tillotsons	411	—
Other losses of fewer than 400 jobs*	25,348	—
Sub-total (1980–85)	54,174	
Total (1975–85)	81,596	

* Includes losses of over 400 jobs not publicised (especially losses by contraction over a period of time).
Sources: compiled by Merseyside Socialist Research Group and Townsend; original source unknown

helped to diversify the economic base of those cities. Lastly, he found that specialisation definitely accompanied the growth of certain British cities between 1978 and 1991. But, as developments in 2000 demonstrated, even the competitiveness of some of these cities can be undermined by an overvalued currency or overproduction. Ford closed its car assembly plant at Dagenham, with a loss of 2000 jobs, while the Rover plant at Longbridge, Birmingham – a closure that threatened 8000 jobs – was saved by a last-minute rescue package. Furthermore, none of the 'transplant' factories opened in the 1980s by Japanese car-makers at Sunderland (Nissan), Derby (Toyota) and Swindon (Honda) were trading profitably in 2000 (reported in *The Economist*, 6 May 2000, 34).

The economic fortunes of 12 US cities

We can gain further insight into the impact of industrial mix upon the economy of cities by examining the performance of 12 US cities during the decade 1977–87. This is based on an analysis undertaken by two urban geographers, Jane Pollard and Michael Storper (1996), of three groups of fast-growing industries – intellectual capital industries, innovation-

based industries and variety-based industries. The employment growth reported by 12 metropolitan economies, representing different phases of urban-industrial development in the United States, is the measure of performance used (Table 3.8).

Pollard and Storper assembled the three groups from the official US classification of industries by combining three-digit SIC codes. Intellectual capital industries are characterised as 'high-wage, non-production occupations' such as banking, securities and broking, laboratories, data processing, consulting, legal services and educational services. US employment in this group grew by about 152 per cent between 1977 and 1987. Innovation-based industries are those employing 'high proportions of highly skilled, technical labour'. They typically include communications equipment, electronic components, scientific equipment, aviation, missiles and space vehicles. Employment rose nationally by 140 per cent for this group between 1977 and 1987. The third group of variety-based industries are defined on the basis of their 'diversity of products'. This group comes closest to providing a reasonable indication of diversification at the city region level. These range from food, textile, paper and chemical processing to

TABLE 3.8 *Employment change in the United States and 12 selected cities, 1977–1987*

	Employment			Employment change (%)		
	1977	1982	1987	1977–82	1982–87	1977–87
Atlanta MSA[†]	707,698	878,247	1,233,837	+24.1	+40.4	+74.3
Boston MSA	1,624,239	1,977,900	2,364,226	+21.7	+19.5	+45.5
Chicago–Gary CMSA*	2,985,227	3,003,671	3,250,941	+0.6	+8.2	+8.9
Dallas–Fort Worth–Arlington CMSA	1,056,227	1,373,159	1,666,154	+30.0	+21.3	+57.7
Detroit–Ann Arbor CMSA	1,525,403	1,436,511	1,760,620	−5.8	+22.5	+15.4
Los Angeles–Long Beach CMSA	2,647,263	3,130,772	3,546,393	+18.2	+13.2	+33.9
Minneapolis–St Paul MSA	755,348	910,582	1,070,778	+20.5	+17.5	+41.7
New York–New Jersey CMSA	3,721,328	4,012,812	4,407,798	+7.8	+9.8	+18.4
Phoenix MSA	392,619	547,499	778,148	+39.4	+42.1	+98.1
Pittsburgh–Beaver Valley CMSA	787,526	818,590	813,049	+3.9	−0.6	+3.2
San Diego MSA	398,969	552,699	734,813	+38.5	+32.9	+84.1
San Francisco–Oakland–San Jose CMSA	1,684,019	2,103,528	2,472,910	+24.9	+17.5	+46.8
United States	64,975,580	74,297,252	85,483,804	+14.3	+15.0	+31.5

[†]Metropolitan Statistical Area; *Consolidated Metropolitan Statistical Area
Source: Pollard and Storper (1996); extracted from US Bureau of the Census. Reproduced by permission of *Economic Geography*

metal products and machines. Although these job numbers increased by only 88 per cent during the 1980s, they consistently contributed more than their share of value-adding nationwide (Pollard and Storper 1996, 7). Variety-based industries were also intended to serve as a pointer to the tendencies to re-agglomeration that the supporters of post-Fordism seek to detect in urban transformations. Taken together, the three groups of industries account for about one in five jobs in the US economy.

On the basis of their analysis, Pollard and Storper (1996) sort the 12 US cities into six categories depending on the key drivers of metropolitan growth.

I. High growth, single specialisation: the new high-technology regions

Dallas, San Diego and Phoenix combine high overall employment growth (Table 3.8) with strong specialisation in innovation-based production, backed up by the intellectual capital sector. Computer programming and software design, data processing, management consulting, aircraft and parts, electronics, and research and development laboratories all feature prominently in these cities.

2. High growth, multiple specialisations: the older high-skill, high-technology centres

San Francisco, Boston and Los Angeles are older cities with somewhat different industrial histories, which none the less achieved better-than-average overall employment growth (Table 3.8). Each contains a favourable mix of dynamic industries, though Los Angeles, despite its strength in movie-making, has fewer variety-based industries. For Los Angeles, aeronautics and space equipment, and electronics, were all major employers. San Francisco and Boston stand out in software design and development, office and computing equipment, engineering and scientific laboratories, non-commercial educational, scientific and research establishments, and electronics.

3. The non-specialist growth regions

Atlanta and Minneapolis–St Paul emerge as cities with high overall growth rates but without the degree of specialisation in any of Pollard and Storper's 'dynamic industry' clusters that export base theory expects. Employment growth in the variety-based industries also went against the downward national trend in both cities.

4. The low-growth, single-specialisation regions: headquarters alone are not enough

New York and Pittsburgh are similar to the extent that throughout the 1980s they both suffered disproportionately from declines in routine manufacturing, including variety-based production. And while New York, with the greatest concentration of financial and business services in the USA, experienced some narrowly based growth in intellectual capital in the 1980s – security and commodity brokers, investment activity and management consulting – it was 'not sufficient to carry the burden of the regional economy's employment needs. The problem is that New York has not developed compensating sources of diversity' (Pollard and Storper 1996, 17).

5. Low growth, multiple specialisations: unwinding a multitextured economic fabric in Chicago

Like most northern metropolitan areas in the United States, the Chicago–Gary Consolidated Metropolitan Statistical Area (CMSA) suffered from below-average overall employment growth in the 1980s (Table 3.8). The city appears to be holding on to its headquarters and advanced services functions without managing to arrest the loss of variety-based or routine manufacturing, or strengthen its capacity to innovate.

6. A non-specialised, low-growth economy: Detroit

The Detroit–Ann Arbor CMSA does not possess real specialisation in any of the dynamic industry clusters holding the key to employment generation. During the decade under consideration, jobs in routine manufacturing – much of it in automobiles – fell by 20 per cent. Evidence of limited overall job growth (Table 3.8), modest investment in intellectual capital and manufacturing in decline is suggestive of resignation to economic restructuring without much capacity building in the dynamic sectors of the economy.

Four main findings emerge from the Pollard and Storper study of the industrial composition of urban economies and metropolitan growth in the United States. The first is that, rather than becoming a motor of the American economy, as some of the literature on post-Fordism and industrial districts might suggest, 'variety-based' manufacturing in the USA is actually in relative decline. The second affirms the credence given to the post-Fordist and informational economy arguments about the propulsive effects of innovation-based production:

The most consistent pattern we detect is the link between specialization in innovation-based employment and overall regional employment growth. Almost all the regions with the greatest overall growth rates were specialized in innovation-based industries in the 1980s. In other rapidly growing regions, specialization was, at the very least, stable.

(Pollard and Storper 1996, 19)

The third finding relates to the ubiquity of the growth recorded by the intellectual capital industries – they advanced virtually everywhere. Fourthly, the causal relationships between the industrial base – whether it is more or less specialised or diversified – and employment growth in cities are much more complex than previously thought. Atlanta and Minneapolis–St Paul posted respectable rates of growth (Table 3.8) despite lack of specialisation in the dynamic industry groups scrutinised by Pollard and Storper (1996). By the same token, having prospered because of their relative strengths in these specialised clusters of dynamic industries, will San Diego, Phoenix and Dallas encounter limits to growth in the future because their economies are not sufficiently diversified?

CITY-BUILDING AND CAPITAL FORMATION

The property market of cities represents a huge storehouse of capital funds. Enormous wealth and also many jobs are created in the process of building and running cities. City-building plays a dominant role in capital formation in all economies, rich or poor. As we shall soon see, economists have only recently come to appreciate how the flow of investment funds into urban property and housing markets can accentuate the cyclical fluctuations of regional economies and feedback into the national economy. Urbanists refer to the physical stock of cities created by human endeavour as the 'built environment'. Yet we might as well say that cities have been 'man-made' because, as Susan Fainstein (1994) shows

in her book on property, politics and planning in London and New York, in the past city builders have overwhelmingly been men.

Public expenditure on physical assets and infrastructure goes to build and operate urban transport systems, water and sewage, electricity and gas, hospitals, schools and colleges, facilities, prisons, and some council or state housing. Private investment funds industrial estates and factories, and commercial buildings (and the land they are built on) like downtown office blocks, regional shopping centres, storage and warehousing, privately owned and operated hospitals and schools, and owner-occupied housing. Of course, there is always some 'cross-over' between public investment and private-sector spending owing to the usual need for governments to lever funds from private developers. Understandably, leverage, or the ratio of private investment to public underwriting, has become a key concern of city governments when deciding which development projects to support in their town.

Investment opportunities and financial flows

In a global money market, bankers and fund managers, and now day-traders buying and selling on the Internet, scour the globe for investment opportunities that exceed the promise of returns in their own domestic markets. This has been a temptation since money and currencies were first created (Leyshon and Thrift 1997). But what has changed are the volume, acceleration and destination of the financial flows that are currently moving around the globe on a daily basis. A ranking of the world's largest stock exchanges based on total turnover in 1999 immediately following the merger of London and Frankfurt is displayed in Figure 3.3. But, to be realistic, any estimate of the magnitude of these flows is going to date immediately owing to the constant expansion of money.

Pension fund managers, the managers of property trusts, bankers lending for project development, want to know which markets are 'hot' or where the most secure long-term yields can be obtained. This is the profit motive that drives investment and underpins the 'accumulation of capital' in all market economies. Even in communist China, where deviationists used to be accused of being 'capitalist roaders', markets and the profit motive, including a stock

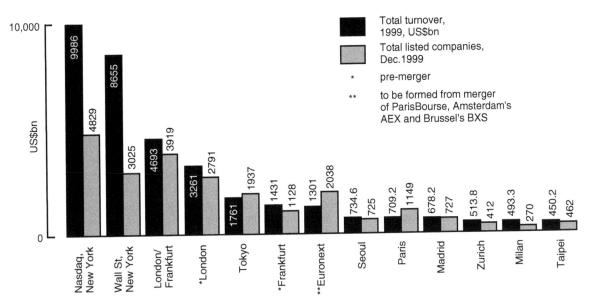

FIGURE 3.3 *The world's largest stock exchanges*
Source: London Stock Exchange

exchange, are now acceptable means of deciding investment priorities. Yet, as the recent experience of Shanghai shows (see below), the logic of the market does not always prevail in administered property systems.

The geographical destination of these financial flows – or, more formally, the 'circulation of capital' – holds the key to the economic growth of cities and regions. And, because investment opportunities and returns to investment (profits and capital growth) vary from place to place geographically, regional and urban development is intrinsically uneven. As a result of the assessment by the capital markets of the differing investment opportunities from place to place, capital is preferentially flowing to some cities, and locations within them, and withdrawing from others. For example, there has been a secular change in the relative profitability of 'smokestack' industries. Much of this industry was located in the core areas of old cities devoted to industrial production. When profit levels fall beyond a certain point, the incentive to re-invest and modernise plant and equipment is no longer there, so owners begin to disinvest or divest, which starts the process of de-industrialisation.

De-industrialisation has been part of a characteristic pattern of restructuring throughout the so-called rustbelt manufacturing regions of western

Europe, the north-east of the United States and south-east Australia. When corporations shut down unprofitable plants, rather than undertaking a programme of modernisation, to set up in another location – possibly offshore – this is described as 'capital flight'. About 1.2 million American workers lost their jobs in 1982, a recession year, as a result of plant closures and capital flight.

De-industrialisation provides a good illustration of the process of 'creative destruction' at work. This expression was coined by a famous Austrian economist called Joseph Schumpeter to convey the idea that it is a necessary, and therefore unavoidable, part of the dynamics of capitalism. But the process need not stop there: when industries pull out of old inner-city areas and neighbourhoods are abandoned, site values can fall to the point where the land is so undercapitalised that banks are prepared to lend to developers wanting to redevelop the area (although this usually has to be accompanied by the granting of some pretty favourable planning concessions by City Hall too). Investment can 'see-saw' backwards and forwards over time as the prospects of urban regions change, and it can flow in a reverse direction once the risks and profitability of redeveloping CBD sites, or old city neighbourhoods, become acceptable to developers.

Harvey's conceptualisation of urban development

The urban geographer David Harvey (1978) has set out a framework that helps to make sense of the way in which financial flows in the economy feed the urbanisation process, shape the restructuring of cities and determine the form of the built environment. His conceptualisation of the circulation of capital (Fig. 3.4) and how it gets allocated to fund property development in cities is based on a reading of Marx's *Das Capital*. He calls this the 'production of the built environment'. While there are aspects of the application of neo-Marxist theory in urban geography that have been well and truly criticised, even rejected, Harvey's framework contains some very useful insights that help us to understand building cycles and the pattern of investment and disinvestment in cities.

To begin with, he develops Marx's notion of circuits of capital. The 'primary circuit' of capital is placed across the central axis in Figure 3.4. It contains the linkages between wealth-generating activities, notably industrial production, on the left-hand side of the figure, and consumption and the supply of labour to industry on the right-hand side of the figure. The 'secondary circuit of capital' in the upper half of the figure combines those sectors that help to expand economic output. They include capital markets, infrastructure and the built environment, and household assets and savings. For example, cities with efficient roads, railways, airports, container terminals, telecommunications, waste disposal systems or cheaper house prices enable local industrialists to increase their competitiveness. The 'tertiary circuit of capital' in the lower half of the figure includes the spending normally coordinated by the state in support of economic output. It covers corporate and government spending on technology and science, including R&D, and 'social investment' dedicated to improving the quality of labour (schooling and training, welfare support, public health education, child-care supplements, rehabilitation and so on).

This raises two key questions. Under what conditions is capital switched from one circuit to another? That is to say, what determines the timing and location of new investment in cities? Secondly, why are

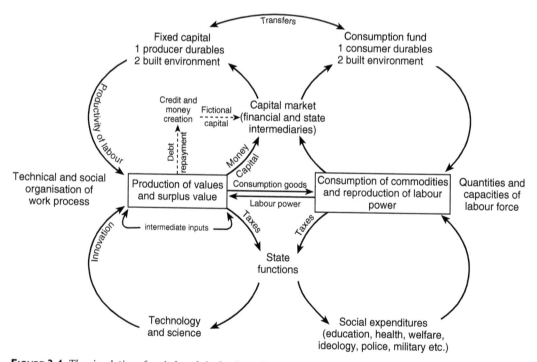

FIGURE 3.4 *The circulation of capital, and the funding of infrastructure and property development in cities*
Source: Harvey (1978). Reproduced by permission of Blackwell Publishers

some sectors of the urban property market favoured by investors at the risk of a glut in, say, office or condominium submarkets, when at the same time they are unresponsive to demand for suburban housing blocks or low-rent housing?

In answer to the first question, Harvey suggests that excess funds are invariably switched from one sector or geographical location to another in conjunction with 'crisis conditions'. At times of overproduction and falling profits in industry, which Harvey describes as a 'crisis of overaccumulation', idle reserves are switched through the financial system into the property markets of cities. This invariably causes a 'property boom' when the supply of land, or commercial and residential stock coming on to the market in the secondary circuit, exceeds demand. Property buyers might be responding to a tightening of credit such as happened in Australia in 1974 (this is called a crisis of 'underconsumption'), or, in an era of floating exchange rates, to shifts in the relativity of currencies. For example, Southeast Asian investors who bought condominiums in Sydney and on the Gold Coast with Singapore dollars in the early 1990s subsequently quit the market so as to cash in on the strengthening Australian dollar, which reached a five-year high of US80 cents in mid-1996.

For Harvey, the property boom in 1972–4 came as confirmation of the connection between the 'crisis' in British industry and the switching of capital into more profitable outlets like south-east England's property market. A fairly similar story can be told for Australia's cities in 1972–4 when there was a massive flow of speculative capital into the fringe land market and over-building of office accommodation (Badcock 1984, 146–50).

But the switching of capital between sectors and locations does not need to be driven by crisis. Corporations flush with funds buy real estate to diversify their investment portfolio, or perhaps because property is outperforming other sectors of the economy. The splurge on so-called telcos and high-tech shares in the United States in the first half of 2000 was consistent with the transition to the knowledge economy. 'Information rich' shares outperformed 'blue chip', and 'bricks and mortar' stock. US companies in the top-performing 50 per cent disposed of 24 per cent of their fixed assets between 1986 and 1996. Corporate investment in IT and people pushed expenditure on property from second to fourth place on company balance sheets in the United States. If the fascination with technology continues to bite into the demand for real estate, rental income and property values in prime commercial markets could begin to suffer.

An article that appeared in the *Guardian* on the eve of the new millennium asked, rather pointedly, 'Can Wall St survive?' The explosion of online information, trading and investment sites as part of the Internet start-up culture on the west coast is beginning to make inroads into the turnover and fees of trading floors and investment banks around the world. Estimates predict that 37 per cent of US bonds will be traded electronically by 2002 – up from 0.5 per cent in 1995. As investment banking becomes even more geographically diverse, will 'Wall Street', already a metaphor for the entire US financial system, have to re-invent itself as an electronic 'virtual' entity? Since then, towards the end of 2000, the euphoria surrounding the 'dotcoms' trading on the Nasdaq has subsided.

In answer to the second question, the expected rate of return on investment determines which sector of the property market (e.g. undeveloped land, industrial, commercial, housing), and which submarkets (e.g. office blocks, hotels, regional shopping centres, retirement villages, condominiums, hostels, low-cost renting, owner-occupation) attract funds at a given point in time. (Chapter 6 deals with these processes in greater depth.) Since building and health regulations all but eliminated the rack-renting and overcrowding associated with slum landlordism, the rental income from low-rent investment housing seldom matches yields from other forms of real estate. Hence most governments have to offer other incentives, such as tax breaks, to make investment in low-rent housing profitable.

PROPERTY INVESTMENT, FIXED CAPITAL AND THE BUILT ENVIRONMENT

Capitalist development has … to negotiate a knife-edge path between preserving the exchange values of past capital investments in the built environment and destroying the value of these investments in order to open up fresh room for accumulation.

(Harvey 1978, 123–4)

TABLE 3.9 *The three premier world financial centres: mid-1980s*

	London	New York	Tokyo
Foreign banks	399	254	76
Foreign stock exchange members	22	33	6
Share of international banking (1985, %)	24.9	15	9.1
Share of foreign exchange turnover (1985, %)	32.6	22.3	5.3
Stock market turnover (1985, £bn)	52.8	671.3	271.5
Stock market capitalisation (1985, £bn)	244.7	1302.2	648.7

Source: Leyshon and Thrift (1997); based on data from Price (1986, 16).
Reproduced by permission of Routledge

Quite simply, this was the hard choice facing many central city governments during the decade of high unemployment in the 1980s when, on the one hand, heritage buildings and precincts cried out for preservation while, on the other, redevelopment promised to create new jobs in construction, financial and business services, entertainment and tourism.

Susan Fainstein (1994, 24–57) presents case studies of two of the most outstanding examples of this 'knife-edge' redevelopment during the 1980s in her book *The City Builders*. The main catalyst for this investment was the ascent of the City of London and Wall Street to leading positions as global financial centres. Table 3.9 encapsulates the relative standings of London, New York and Tokyo just before the introduction of full-throttle financial deregulation in the UK in 1986 – the so-called Big Bang. This helped to refuel redevelopment in London's financial heart at the expense of investment in the domestic economy elsewhere in regional Britain. In the City, the Big Bang also led to spectacular overcapacity in the markets for cross-border loans, international securities and equities. By November 1987, Britain's overseas assets, at £114.4 billion, exceeded those of any other country, including Japan (King 1990b, 95). The story that Fainstein tells for Central London and the Docklands, and New York's Battery Park City and Times Square, can be -retold for many other nineteenth-century cities, like central Melbourne (Fig. 3.5).

While not all cities have participated in the financial networking of the global urban hierarchy, the added presence of international and regional offices has greatly increased the financial turnover and wealth generated in capital markets like the City

FIGURE 3.5 *Bruno Grollo, property developer, overlooking central Melbourne from the Rialto Building*
Source: *Sunday Age* (2000)

of London (King 1990b, 93–100). A special feature of the research on the geography of monetary transformation done by two University of Bristol geographers, Andrew Leyshon and Nigel Thrift (1997,

135–55), is the connection that they draw between the earnings of the City's workforce and the pattern of expenditure in south-eastern England on household consumption and real estate. They assess the City's salary bill in the mid-1980s for the 390,400 workers at about £5.4 billion, representing about £3.25 billion after tax and other deductions.

The Japanese experience in the 1980s provides a telling illustration of how the flow of capital from industry and finance into real estate can potentially undermine an entire economy. Japan's industrial miracle was followed by the partial opening-up to the outside world of its once closed financial system. Flush with funds, Japanese corporations began to purchase real estate at the same time as international firms began to open offices in Tokyo and other provincial cities. The demand for prime office space pushed up rents, and land prices followed. In the three-year period 1987–9, the value of Japanese property assets in the corporate sector rose by as much as the 1988 size of Japan's GDP (Fig. 3.6). Landed property in Japan in 1991 was at some 20 per cent of the world's wealth! According to Mera (1998, 181), at peak the land value of the Emperor's palace in the centre of Tokyo alone was estimated to be equivalent to the land value of the State of California, or all the land, houses and factories in Australia.

In turn, corporate Japan used this high-value land as collateral in the late 1980s to purchase real estate and undertake property development in the United States (Fig. 3.7), not to mention in cities throughout Southeast Asia (Fig. 3.8). Honolulu's Waikiki Beach and Los Angeles' Civic Centre were transformed as a result of this spending spree, while landmark properties like the Rockefeller Centre in downtown New York and Pebble Beach Golf Course on the Californian coast changed hands.

The property 'bubble' burst around 1990–1 (Fig. 3.6) and seven years later, with Japanese banks staggering under US$365 billion of bad debts, property prices fell to about a fifth to a quarter of the peak level in Japan. Such erratic fluctuations in investment patterns and asset values have a crippling effect on urban development. By 1990 in Tokyo, 4.85 million square metres of office space were under

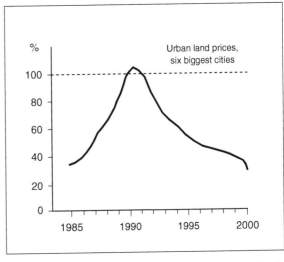

FIGURE 3.6 *The rise and fall of the urban property market in Japan's six largest cities, 1980–2000*
Source: Japan Real Estate Institute

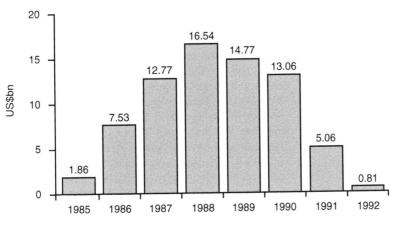

FIGURE 3.7 *Japanese investment in US real estate in the 1980s*
Source: Haila (1995)

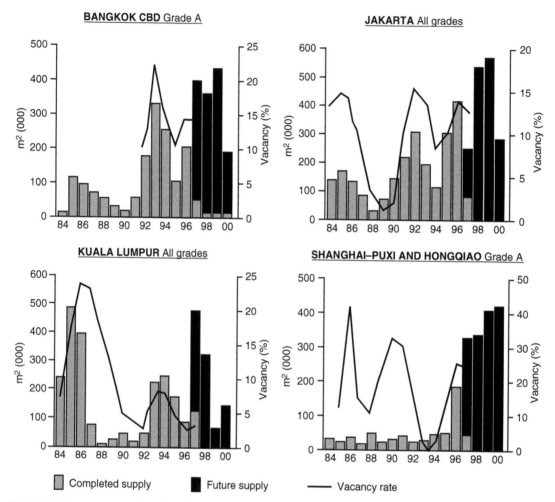

FIGURE 3.8 *Property market trends in four Asian cities, 1984–2000*
Source: JLW Research

construction. By 1993, the requirement had fallen to 2.32 million square metres (Mera 1998, 182). In the case of cities like Bangkok, Jakarta, Kuala Lumpur and Shanghai, building activity virtually came to a standstill when the property bubble was punctured in 1997–8 (Fig. 3.8).

Susan Fainstein (1994, 25) concludes that, 'Property development belonged to the eighties financial boom as cause, effect, and symbol. Profits on large projects, huge tax benefits from real-estate syndication in the US, and trading margins from mortgage securitisation formed the basis for vast fortunes.' While funds loaned for large-scale urban development projects that are now transforming the skyline of

cities and pushing them outwards at the edge circulate the globe like vapour trails, property developers are mostly the 'agents' who first conceived of, and then packaged up, these projects. 'As their wealth and visibility made them prominent actors in the cultural and social scene, the names of developers like Donald Trump, William Zeckendorf, Jr., and Mortimer Zuckerman in New York, or Stuart Lipton, Godfrey Bradman, and Trevor Osborne in London become widely publicised' (Fainstein 1994, 25). Japan's counterpart is a property developer, Minoru Mori, whose empire encompasses about 90 leased buildings in Tokyo, Osaka and several other Japanese cities.

An architect's view of the wealth created and extracted by property developers

In the hands of Haussman, who created Paris' Grand Avenues and Mitterrand with his Paris Grand projects, development always was and is a political issue – the purpose of building has as much to do with a vision of certain kinds of state as it does with the pragmatic concerns of civic life. But for the most part it is the private developer who has the last word in shaping the city, limited by what the market will bear and working with the current rather than against it.

In an era of corporate gigantism, when a massive investment is needed to break into any manufacturing market – from dog food to portable computers – property speculation is one of the last businesses that allows an individual with no resources, beyond a sense of his own worth, to amass huge wealth in a very short period and then lose it equally quickly. It is an edgy, maverick undertaking, requiring successful practitioners to make the difficult adjustment from street-corner huckster to pin-striped corporate responsibility. The naked realities of guile, bravado, aggression and ego, tactfully concealed in more mature, not to mention anaemic businesses, are still disconcertingly close to the surface in property.

In the early stages of the creation of a property empire, bluff, hyperbole and vainglorious publicity seeking are as much part of the developer's repertoire as financial skill. And by extension, the truth is that, while it is commercial development which shapes our cities, it is in the hands of those who have no interest in using their powers for the long-term future.

(Extracted from the annual Lloyd Rees Memorial Lecture, presented in 1997 by one of Australia's leading architects, John Denton)

Trying to make sense in Harvey's terms of the processes responsible for capital formation and the production of the built environment in what were formerly command economies is doomed to failure (Badcock 1986). The over-building that took place in Shanghai defies the market. According to an investment adviser with the Dutch bank ABN-Amro Hoare Govett Asia, 'In terms of overdeveloped real estate, Shanghai can't be beaten anywhere.' While Shanghai's total office space rose 10-fold between 1994 and 1997, rents fell by about half in the same time. Vacancy rates for Shanghai City averaged 40 per cent in 1998, and were as high as 70 per cent for prime office space in Pudong New Area on the east bank of the Huangpu River. With an eventual population of 2 million, Pudong is intended to take some of the strain off greater Shanghai's other 15 million residents. According to the Dutch investment adviser, 'By being unwilling to accept a market-driven deflation of property prices, the government is crashing the economy to save a sector.' This is to imply that the Chinese risk repeating the mistakes that prolonged Japan's slump by almost a decade. A Finnish researcher who studied the Chinese property system in the late 1990s asked herself, 'Why is Shanghai building a giant speculative property bubble?' (Haila 1999). She attributes it to the willingness of overseas Chinese to invest in China, 'fuelled by expectations of huge profits in China', the selling of the city, together with the edifice complex beloved of property developers, and state subsidies.

In the context of 'market socialism', the Chinese approach to urban development involves keeping central planning intact, though this does not necessarily guarantee effective coordination. The Pudong New Area Administration was under the charge of the Mayor of Shanghai, Xu Kuangdi. But as well as Shanghai's 760 state-owned and 310 collectively owned development companies, not to mention local developers, several overseas developers had a significant stake in Pudong by the mid-1990s. They included: Hong Kong's billionaire tycoons Walter Kwok, Li Kashing, Ronnie Chang and Cheng Yung-tu; Robert Kuok of Malaysia; Dhanin Chearavanont's CP Group of Thailand; and the Mori Brothers from Japan.

This reflects the growing interdependence of Asian urban development due not only to the interlocking portfolios of Japanese or Overseas Chinese investors, but also to the increasing influence exerted by institutions based in cities such as Hong Kong, Taipei, Manila, Nagoya, Washington DC, London and Paris. Kris Olds (1997), an urban geographer teaching at the National University of Singapore, observes that this influence extends as well to elite non-Chinese design professionals, on investment processes, urban planning and social policy reform processes in Chinese cities.

CITIES, REGIONAL CYCLES AND ECONOMIC WELFARE

As the course of economic restructuring unfolded during the closing decades of the twentieth century, some of the initially stark regional dualisms, such as the north–south divide in the United Kingdom, or the 'snowbelt–sunbelt' schism in the United States, assumed a more complicated patina. Restructuring is impacting unevenly, as ever, but also *inconsistently* upon different regions and cities within the UK and US space economies. What is also becoming more noticeable is the extent to which the flow of investment into and out of the urban property market at different phases of the regional cycle can amplify or detract from the economic welfare of cities.

Regions and cities like the rustbelt states and steel towns in the United States, which were especially hard-hit in the 1970s and 1980s due to their traditional industrial structure, were not as severely affected by the recession at the beginning of the 1990s as some of the states in the middle of the continent concentrating employment in energy and resources, or the coastal states. While the 1981–2 recession hollowed out the industrial heartland of the East North Central and East South Central statistical divisions, in the early 1990s its greatest impact was on the Atlantic and Pacific states. The investment boom, including the flood of Japanese finance into Californian cities, boosted jobs in finance, insurance, real estate (FIRE) and construction on the two coasts. The stock market crash of 1987, which precipitated the recession of the early 1990s, wiped out all of the job gains Manhattan had made since 1980 (Markusen and Gwiasda 1994, 181–2). Between 1988 and 1990, the New York region lost an aggregate of 37,700 jobs, or −0.9 per cent. But as well as losing nearly 40,000 jobs in manufacturing, the region lost 10,000 jobs in banking, 15,500 jobs in securities and 3300 jobs in business services. Since 1987, New York's pre-eminence in producer services has gradually been whittled away by the decentralisation of these jobs in almost all subsectors (Markusen and Gwiasda 1994, 180). Boston was also set back by this 'east coast' downturn: job numbers fell by 78,700 between 1988 and 1990, or −4.5 per cent of the total workforce.

Boston's vulnerability was linked to some extent to its regional strength in the defence and aerospace industries. As part of the 'Peace Dividend' following the end of the Cold War, national defence-related employment in private industry was cut back by 30 per cent from a peak of 3.7 million jobs in 1987. As a result of the concentration of jobs in FIRE and defence-related industries, California went through its worst economic downturn since the 1930s. In the first half of the 1990s, the state lost half a million jobs as military bases were closed or scaled down, defence contracts were reviewed and the aerospace industry downsized (Fig. 3.9). Hence Los Angeles also began to run into difficulties as the wave of defence cuts began to flow through its heavily defence-oriented economy in 1989. When Lockheed closed its Burbank 'hub', more than 14,000 reasonably well-paid blue-collar jobs in and around north Los Angeles were threatened. Yet in the second half of the decade, the job loss was not only stemmed but reversed by the birth of fast-growing young companies dubbed 'gazelles' in the multimedia and other information-rich sectors. The jobs lost in aircraft construction, for example, have been more than made up for by those created in motion pictures and cinema (Fig. 3.9). In 1995 alone, 40,000 new jobs were created in the motion pictures industry. Consequently, California is now home to over half of all America's multimedia firms, including printing, publishing, graphic design, movies, computer software and hardware, communications and advertising. This makes California the 'fast breeder' of the information-based economy.

The Boston region also plunged into recession in the early 1990s after a period of unprecedented prosperity that can partly be attributed to a real estate boom between 1984 and 1987 (Case 1992). Average house prices rose from US$82,600 in 1983 to US$182,200, or 121 per cent, by the end of 1987. As a consequence, Boston's home-owners were much better off and some borrowed against inflated home equity to consume more goods and services. In Massachusetts, employment levels in the construction and FIRE sectors climbed by 57 and 27 per cent between the beginning of 1984 and the end of 1987. But at the same time house price inflation curbed the growth of New England's labour supply through the 1980s because migrating workers were reluctant to move to a high-cost region. Case concludes that, while declines in defence spending and retrenchments in the high-tech sector all contributed to the severity of recession in the Boston region,

there is strong evidence that the real estate cycle amplified the business cycle on the way up and on the way down. Not only is the region giving back jobs that were added directly because of the real estate boom, but the boom did serious damage to the cost structure of the region, making it less attractive to both existing firms and potential new entrants to the region.

(Case 1992, 182)

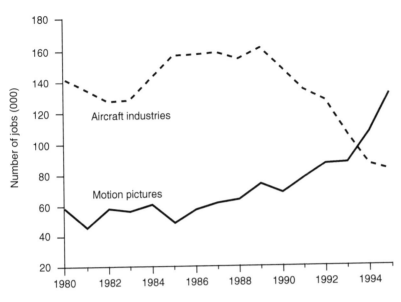

FIGURE 3.9 *The switch from jobs in aerospace and defence to multimedia and movie-making in the Los Angeles region during the 1990s*
Source: California Employment Development Agency

In the case of the UK, despite the relative rebound in the 1990s of some cities like Birmingham or Glasgow from the depredations of de-industrialisation and the emergence of others within otherwise struggling regions, the economy has reverted to the well-worn pattern of an ailing industrial north and a prosperous south dominated by Greater London. With manufacturing shrinking to only 20 per cent of the economic output of post-industrial Britain, the setting of interest rates is increasingly swayed by the ebullience of London's regional economy. According to the ONS, the population of London grew by 4.3 per cent over the seven years prior to 2000; the south, in 2000 accounted for well over 60 per cent of GDP; unemployment in the south-east was down to 3.8 per cent, compared with 5.9 per cent nationally and 7.1 per cent in the far north; and, London house prices had risen by 20 per cent in 1999–2000, threatening a break-out of mid-1980s proportions. Problematically, any increase in interest rates to dampen growth in the south threatens to check the fitful growth in the north. What is economically in London's best interests is no longer necessarily in the interests of the rest of the British economy.

The same regional imbalances and policy quandaries are present in the Australian context where spending for the Sydney Olympics injected enormous sums into an already inflated property market.

By the beginning of 2000 unemployment rates had fallen to 4.7 per cent across the Sydney region, and even lower in better-educated communities on the North Shore and around the Harbour. Yet in nearby Newcastle, now without its steel mill, and the Hunter region, the jobless rate exceeded 10 per cent, while further up the coast in northern New South Wales it approached 12 per cent. With the Reserve Bank foreshadowing a tightening of monetary policy, this prompted concern that 'the rest of the nation is being made to pay…for the booming Sydney economy when many regions are still struggling. The Reserve Bank needs to be careful to distinguish between what is happening in Sydney, and to a lesser extent in Melbourne, from elsewhere' (reported in Wellington's *Evening Post*, 21 January 2000).

Competion for new investment and jobs

In a world of instantaneous communications and globally footloose corporations there is now enormous pressure on city leaders and managers to attract new investment to their region rather than elsewhere. This demands a much more entrepreneurial approach to urban governance and management (Harvey 1989). The competition among cities to win new investment and jobs has taken three main forms. Firstly, urban governments have to be able successfully to market the superiority of their region as a potential production site or 'events' venue. The hosting of major international events is an avenue open to cities trying to attract investment, and at the same time heighten international awareness of their region. Secondly, a city has to be able to assemble a more attractive financial package of incentives than a rival bidder. Lastly, with urban 'growth machine' regimes leading the way in North America, cities are now turning to entertainment-based consumption – the downtown version of theme parks – and 'cultural industries' as an added means of revitalising central city economies.

'Place marketing'

'Place marketing', or the selling and promotion of a city's competitive advantage and distinctive attractions, has become the ultimate form of product differentiation in consumption-led societies. The challenge from a marketing perspective is to 'stand out from the crowd'. This involves image-making and then trading upon the special identity of place:

Partly because the traditional force of locational factors is eroded, partly because the *Zeitgeist* of the 1980s and 1990s is so heavily pro-competitive, cities tend to market themselves rather like competing consumer goods. And, according to traditional rules of marketing, this leads to the imperative to establish a market niche: city administrations find themselves impelled to establish some unique quality for their city, some magic ingredient that no other city can precisely match. Since the sources of the new economic growth are so various and finally perhaps so fickle, the possibilities are endless. But one central element is quality of life. It is no accident that, as never before, rankings of cities dominate the media ...

(Hall 1999, 70)

For example, the business journal *Fortune* now publishes an annual ranking of the best cities in which to conduct business in the United States, Europe, Asia and Latin America. In 2000, New York City headed the US league table followed by San Francisco, Chicago, Washington DC and San Jose. London led Frankfurt, Helsinki, Amsterdam and Dublin in Europe. Four of the top five cities in Asia – Hong Kong, Sydney, Singapore and Auckland – all had the strongest Internet connections. Taipei, which

did not make the final five, nudged out Tokyo on Internet connections. The top five in Latin America were ranked as follows: Buenos Aires, San Juan, Mexico City, São Paulo and Santiago (reported in *Fortune*, 27 November 2000, 106–13).

'Bidding wars'

In order to counter the effects of trade liberalisation and the proliferation of lower-cost regions in developing economies, cities in the North have sought to improve the local 'business climate' by lowering labour costs and granting financial incentives to prospective investors. In particular, urban development agencies have to be able to convince manufacturers that the costs of doing business in their city will be lower than elsewhere. Many states and cities in the USA fiercely market their locational advantages as well as offering 'start-up' assistance in the form of tax breaks, the waiving of rates, subsidised energy and water costs, training programmes to equip workers with the right skills and expertise, and ample, inexpensive land in industrial parks. Sunbelt cities in the United States, for example, have also been able to capitalise on the corporate reaction to the political climate in older cities like Mayor Daly's 'fiefdom' of Chicago.

According to an article in the *Financial Times* in October 1993, a point was reached in the mid-1990s where 'bidding wars' threatened the financial security of some state and urban authorities. In the mid-1990s, Alabama paid US$253 million upfront, plus tax breaks, which will be worth an additional US$230 million over 25 years, to lure a Mercedes-Benz plant to the state. Indianapolis won the contest for a United Airlines facility but in the process put the city deeply into the red. A number of states, led by the Carolinas, have invested heavily in worker training programmes that are delivered through a network of community colleges. This is part of a package that clinched the deal for corporations like Michelin, BASF, Adidas and Hoechst.

Now, with the transition to the knowledge economy, the imperative for governments at all levels is shifting to an emphasis on innovation and creativity. The British geographer-planner Peter Hall (2000) believes that creative milieus, concentrating an abundance of highly educated and creative minds, are places where innovation is most likely to flourish in the post-industrial era.

The 'events syndrome'

Because of the economic 'multipliers', cities now bid for the right to host major events, whether they be international trade exhibitions, conferences or festivals, or sporting events like the Olympics or Formula 1 Grand Prix. For example, organisers of the Hanover Expo 2000 predicted it would attract 40 million visitors, thereby rivalling Seville's Expo in 1992 (41 million visitors). As well as stimulating economic growth – construction of the Hanover Expo injected DM3.5 billion into the local economy – such an event requires improvements to urban infrastructure. In 2000, nearly 4800 events were programmed by the convention and exhibition market in the United States. They were attended by over 110 million visitors. By 2003, Chicago, Orlando and Las Vegas will all have over 200,000 square metres of exhibition space, and at least five other US cities will have 100,000 square metres (reported in the *Financial Times*, 12 May 2000, 2–4).

Table 3.10 gives a rough idea of what some of these events have been worth to Australian cities. It is notoriously difficult to come up with accurate estimates of the costs and benefits of staging these events. For example, it is impossible to put a final figure on the value of international television coverage of an event and the tourist visits that it subsequently generates. Be that as it may, this is a high-stakes, 'winner-takes-all' game that, like the 'bidding wars' between cities competing to win new investment and jobs, may leave city rate payers with crippling long-term debts. Host cities invariably have to build facilities and infrastructure, and provide accommodation to meet peak demands that may not be approached again in the foreseeable future. After the experience of Montreal and Barcelona, this is the concern of any city that now bids to host the modern Olympic Games.

Examples of the 'events' syndrome abound, but two will have to suffice. The always successful efforts of the owners of professional sports teams – baseball, ice hockey, gridiron, basketball – to funnel public money into private stadiums illustrates how corporate welfare works in North America. In the case of baseball, during the 1990s more than US$10 billion in underwriting was promised to team owners by those cities that secured a new stadium – San Diego, Pittsburgh, Denver, Cleveland, Seattle and Baltimore, to name a few. This generally takes the form of free land or lucrative parking concessions. 'Boosters' claim that these stadiums generate spill-over benefits, including the injection of jobs and spending in poor, run-down inner-city neighbourhoods. (A study by the Brookings Institute estimates that Baltimore's Camden Yards stadium, representing a US$200 million investment, nets only US$3 million in additional jobs and taxes a year.) Most of the new stadiums are funded by raising local taxes on retail sales, hotel rooms and rental cars, and through lotteries and municipal bond issues. Cities know they are at the mercy of franchise owners, who tend to be billionaires or large corporations, but still try to out-do each other to bring a professional team to town for the fans. In 1999, Hartford, Connecticut, 'stole' the New England Patriots NFL team from Boston by promising to underwrite a US$350 million stadium. Massachusetts refused to provide the franchisee with the subsidies he was seeking to stay in Boston.

The hosting of the America's Cup by Auckland in 2000 provides the other illustration of the costs and benefits of succumbing to the 'events' syndrome. The event cost NZ$300 million to stage, and transformed an under-utilised harbour basin in the process. Feasibility studies predicted that the regatta would generate NZ$600 million, create an extra 6800 jobs in the Auckland region and inject about NZ$1.3 billion into the national economy. The international exposure was valued at NZ$100 million, and the number of overseas visitors was expected to increase by about 5 per cent over two years. In fact, the volume of tourists from overseas was up by 100,000 in 1999 and they spent an extra NZ$87 million on accommodation. But the Auckland region benefited at the expense of other parts of New Zealand, where tourism was down. Boat-building and marine exports, which reached about NZ$250 million in 2000, also stand to win in the longer term.

'Urban playgrounds'

The possibilities surrounding the conversion of idle and undercapitalised central-city or waterfront land into 'urban playgrounds' gave a fillip to the business core of numerous cities in North America, Europe and Australasia. This is bringing residents back into the city as condo or apartment dwellers, creating jobs for locals, bolstering core area spending and reviving the declining rate base of central city

TABLE 3.10 *Economic benefits of hosting Hallmark events in Australia*

Event	Year	International visitors				International media	Total region/city impact (US$m.)
		Total number	Average length of stay (days)	Average daily expenses (US$)	Economic impact (US$m.)		
Sydney Gay and Lesbian Mardi Gras	1998	5190	21.5	347.25	41		98.8
Victoria: Bledisloe Cup	1997	8350	5.75		15		61
Ford Australia Open	1997	5512	5.75	323.48	11.6		82.6
World Cup Soccer qualifier	1997	1700			1.28		35
Melbourne International Flower Show	1997	*					2.36
Australian International Air Show and Aviation Expo	1997	806	7.3		1.25		63
Transurban Grand Prix	1996	3100			10.1	120 countries 500 m. viewers	96
Melbourne Comedy Festival	1995	*					17
Melbourne Festival	1995	*					13
South Australia: Adelaide Festival	1996	450	10.78	138.54 (daily)	2.02	40 int. media	13.1
Queensland: Indy Grand Prix	1997	4008	6.99	186.9			46.7
World Master Games	1995		18.2	212.85		16th most televised event in the world	50.6

* Not separately estimated.
Source: Marsh and Levy (1998, 17)

administrations. Starting with the redevelopment of Baltimore Harbour in the 1960s, the process 'is depositing an infrastructure of casinos, megaplex cinemas, themed restaurants, simulation theatres, interactive theme rides and virtual reality arcades which collectively promise to change the face of leisure in the post-modern metropolis' (Hannigan 1999, 1). The redevelopment of these precincts as 'entertainment destinations' for inner-city dwellers, suburbanites and tourists alike is tantamount to bringing Disney's theme park to town.

The development of waterfront precincts, followed by the 'urban entertainment destination', or UED, caught on first in North America as a kind of lifeline for the declining business centres of traditional cities. But because they tend to have more vibrant city centres that needed rebuilding after the Second World War, European cities have been more cautious about giving over the downtown area to themed entertainment – London's Millennium Dome notwithstanding. By contrast, cities in the Asia-Pacific region with an established or emerging middle class and eager to attract tourists (Tokyo, Hong Kong, Sydney, Melbourne, Brisbane, Singapore) all have, or are planning, their versions of UEDs. These projects, which are usually on a vast scale (see Chapter 4), bring together planning agencies, property developers, financiers, retailers and corporate entities in the leisure and entertainment business, e.g. hotel chains, casinos, sports foundations, movie distributors and multimedia houses.

'Cultural industries'

'Culture is now seen as the magic substitute for all the lost factories and warehouses' (Hall 2000, 640). While culture and the arts have always had a significant presence in large cities, the growth of multimedia and digital forms of creative expression are now linking the cultural industries of cities like London, Paris, Berlin and New York into global markets. 'Cultural tourism' can run the gamut from the attractions of cityscape, urban heritage and design, to great galleries and museums, to the 'living arts' including theatre, music and ballet, to cultural festivities, etc. In 1999, New York City opened eight new museums.

Proceeds from lotteries run by the Millennium Commission have helped to fund a series of 'cultural showcases' in cities throughout the UK hoping to achieve what the Guggenheim museum of art has done for the Spanish city of Bilbao. These millennium projects include London's Royal Opera House in Covent Garden, the Tate Gallery (both for the new Tate Modern in the old Thames Bankside power station in Southwark and a tranche of other galleries) and the British Museum's Great Court extension. There is also a third generation of cultural projects emerging. The Imperial War Museum North is being built opposite the Lowry Gallery on the Manchester Ship Canal. Newcastle is building its International Centre for Life. Rotherham has its Magna project, Glasgow its Science Centre, Belfast its Odyssey regeneration project. According to the *Observer* (7 May 2000, 21), 'All piously hope that these groupings will create cultural microclimates, achieving jointly a critical mass of visitors.' If the Millennium Dome is included, the value of these cultural projects in UK cities approached £2 billion in 2000.

The current fad among consultants is to urge civic leaders to project an image of 'innovation and creativity' by capitalising on their intellectual and cultural assets, and the arts. Charles Landry, an urban designer, has literally taken the 'creative cities' concept to the ends of the earth, selling the package to about 25 to 30 other cities before he finally got the call from the cities of Adelaide (1998) and Wellington (2000) – by which time it had lost all novelty value as a selling point.

CONCLUSION

What this extended treatment of the geography of wealth creation in cities tends to emphasise is that there is no urban development strategy that, alone, will guarantee the economic welfare of a city and its workforce in the long term. First, the strategy has to be appropriate to local circumstances and conditions. California's Silicon Valley assumed the status of a development paradigm in the 1970s and 1980s. In due course, 'technology parks' began to mushroom like 'breeder reactors' as part of local economic development strategy in cities as far apart as Manchester and Melbourne. But as Anna-Lee Saxenian wryly observed in the late 1980s, it does not automatically follow that, 'Like a soufflé which exceeds the size of the initial ingredients, a region endowed with the proper mix of institutional and economic resources will be the lucky recipient of rapid high-tech growth' (in Brotchie *et al.* 1995, 258).

Rather than seeking any longer to identify 'what causes firms to locate' in a particular place, attention has come to focus more on the dynamics of new industrial districts and the intricate and often indecipherable advantages offered to firms in the form of agglomeration economies. And what research reported in the mid-1990s began to confirm is just how essential scale, or city size, actually is to ongoing business performance and urban growth (Harrison *et al.* 1996).

FURTHER READING

Jane Jacobs' books (1969; 1984), *The Economy of Cities* and *Cities and the Wealth of Nations,* are very readable and provocative accounts of why some cities prosper while others stagnate.

Influential ideas relating to urban regions and industrial clusters are set out in *Competitive Advantage and the Wealth of Nations* (**Porter** 1990).

William Lever and **Ivan Turok** (1999) edited a volume of *Urban Studies* that contains debate and case studies on competitive cities.

Ann Markusen (1999) provides a very lucid critique of the debates that have raged around the notions of post-Fordism and Marshallian industrial districts.

In *Creative Cities,* **Peter Hall** (2000) argues that, with knowledge holding the key to wealth generation in post-industrial societies, cities that can attract and hold creative people will have a monopoly on competitive advantage in the future.

The contribution of urban property investment and development to wealth creation is implicit in **Susan Fainstein's** (1994) analysis of property, politics and planning in London and New York.

4

LOCATION OF ECONOMIC ACTIVITIES IN CITIES

INTRODUCTION

This chapter is devoted to making sense of the geographical needs of firms and the way land is allocated within cities in support of these locational requirements. In mixed economies this is accomplished via the property market, with government agencies intervening to make sure that outcomes are in the public interest. Forms of intervention in cities include strategic planning, land use zoning and building regulations. The intention is to concentrate on the spatial patterning of economic activities that forms in cities under capitalist market conditions. This means setting aside land assignment in pre-capitalist and medieval cities, or in socialist cities. The urban geographer James Vance provides good coverage of the practices adopted at the time in pre-capitalist and medieval cities (Vance 1971).

Within the communist system there is nothing equivalent to the pricing mechanism that allocates land to the highest bidder in the property market. Before the collapse of the command economies, industry, housing and collective services like schools, hospitals or parks were allocated administratively in socialist cities (Bater 1980; French and Hamilton 1979). But often this raised as many problems of misallocation as sometimes occur under market conditions in capitalist societies (Badcock 1986). With the demise of communism, cities in eastern Europe and the former Soviet Union are once more subject to some of the same city-building processes that shaped them before the Second World War (Andrusz et al. 1996).

This chapter, then, traces the changing spatial pattern of investment and jobs in manufacturing, retailing and wholesaling, the service industries and residential development. The distribution of these activities in European and North American cities, for example, has changed over time with developments in production, transport and communications technologies, the introduction of new regulatory regimes and shifts in patterns of consumption. Four main phases of urban development stand out, with readily identifiable arrangements of land use in cities. In this chapter I will expand upon each of these four eras in urban development and show how the costs and benefits of different locations in the city have periodically been revalued by firms as their business requirements have changed. Of course, as the locational requirements of firms change over time with technological innovation, this in turn brings forth changes in urban form. Perhaps the most significant modification to the structure of western cities since the mid-nineteenth century has been the progression from a free-standing urban place dominated by a single city centre (Chicago in the 1920s), to vast metropolitan regions dotted with subregional activity centres. Los Angeles, formerly '50 suburbs in search of a city', is now portrayed by the 'prophets of post-modern urbanism' as the harbinger of the future urban landscape in North America (Dear and Flusty 1998).

The four main phases of urban development can be summarised as follows.

* During the era of steam-powered cities this form of energy dictated that the machinery

of production and labour power must be in close proximity.

- 'First-phase' suburbanisation dates from when the electrification of the mass transit system allowed workers on modest incomes to live uptown, or even in a dormitory suburb well removed from their city jobs.
- 'Second-phase' suburbanisation – the decentralisation of manufacturing, shopping and housing – had to await the widespread adoption of trucking and car usage in conjunction with the construction of urban freeways in the 1950s and 1960s.
- The development of the microchip and telecommunications – the 'digital economy' – has given firms and their workforces, together with increasing numbers of retirees, even further locational freedom to break away from the continuous expanse of outer suburbia. Based on their analyses of the 1990 census round in the USA and Europe, Frey (1993) and Cheshire (1995) both suggest that the signs are emerging of a new phase of urban development. In the United States, this peri-metropolitan development has been described as the 'edge city phenomenon' and likened to 'stealth cities'.

The dispersion of economic activities beyond the edge of cities has also gone hand in hand with a degree of inner-city redevelopment and living in some older cities in North America, Europe and Australia. Beginning in the 1980s, Cheshire (1995, 1045) detected a 'significant degree of re-centralisation in many northern European cities, with nearly half of all core cities gaining population'. In these particular cases, re-centralisation of business activity is helping to spark a partial revival of the inner city that for some optimists is nothing less than an 'urban renaissance'.

During the course of the discussion in this chapter, reference will be made to some of the better-known location-allocation models that have been developed to aid our understanding of the arrangement of land uses within cities.

STEAM-POWERED CITIES

Apart from those forerunners of the nineteenth-century industrial city mentioned in Chapter 2,

by today's standards the earliest manufacturing towns were highly centralised and compact. They mostly sprang up where factors of production – land, labour and capital – could be brought together at the so-called least cost location. Canals and railway lines were built to freight raw materials into the nineteenth-century cotton and woollen mill towns where capital, steam power, labour and factory housing came together on the coalfields. The use of steam power in industry and the savings to be gained from minimising the dissipation of energy encouraged the construction of multifloor factories and storage capacity near rail terminals and the waterfront:

Steam worked most efficiently in big concentrated units, with the parts of the plant no more than a quarter of a mile from the power-centre: every spinning machine or loom had to tap power from the belts and shafts worked by the central steam engine. The more units in a given area, the more efficient was the source of power.

(Mumford 1961, 519)

As the degree of specialisation increased, industrialists turned their hands to steel and tool making, then light and heavy engineering and, ultimately, the design and production of industrial equipment itself. Great cities like Manchester and Liverpool, Leeds and Sheffield, and Pittsburgh and Dortmund-Essen, owe their prominence to these groups of industries. Because the size and spatial extent of these cities were ultimately limited by transport and building technologies, activities like manufacturing, warehousing, commerce and government competed for the prime sites near the core of Victorian cities.

But eventually the space requirements of expanding industries would force them to relocate. By the 1870s, engineering workshops had been established in the suburbs of Gorton and Newton Heath near Manchester, Springburn near Glasgow and Stratford in east London. In Sydney, the railway corridor serving western New South Wales, together with plentiful land along the course of the Parramatta River, helped the westward spread of manufacturing in the 1870s and 1880s. Clyde and Granville, which were also home to large railway workshops, were promoted by land developers as the 'Birmingham of Australia'.

In a classic paper on the spatial organisation of land use in the nineteenth-century city, two urban economists laid out a set of general principles that account for the location of much of the large-scale manufacturing found in mature nineteenth-century cities (Fales and Moses 1972). After considering the applicability of the von Thünen model, they cast that aside in favour of the ideas of Alfred Weber, who wrote the *Theory of the Location of Industries* in the 1920s. Drawing on Weber's search for the least cost point of production for a given combination of processing and transportation costs, they developed an empirical model for Chicago in the 1870s, which incorporates the material intensity of production functions, along with the respective costs of transporting workers and goods, and transmitting information (Fales and Moses 1972, 61–5).

By 1870, Chicago's industries included blast furnaces and foundries processing heavy materials as well as lighter industries making clothing and leather goods. The city's population grew from about 30,000 in 1850 to 306,000 in 1870, by which time its spatial form was pretty well fixed. Basic roading, water and other municipal services were in place. Three horsecar lines and a number of feeder routes provided a regular connection to the downtown area. Fales and Moses argue that, because much of Chicago was destroyed by the Great Fire of 1871, the firms that survived had relative freedom to make their locational decisions all over again. They did this in the context of a very active land market, the absence of zoning restrictions, and plentiful capital. Brick making, foundries, blast furnaces and meat packing were especially sensitive to the least cost locations, owing to the significant reduction in weight that accompanies processing. These weight-losing industries acquired sites either adjacent to marshalling yards, or along the canals and the frontage of Lake Michigan (where ice used in brewing could be cut during the winter).

Fales and Moses (1972, 68) decided that a materials orientation was more decisive than being close to the market in the nineteenth-century city for the following reasons.

- Scale economies in inter-regional rail and barge transport helped to undercut the costs of moving freight within cities. The impressive economies of scale in line-haul operations meant that cartage within the city might well exceed inter-regional transport costs by up to 25 to 30 times per ton mile.

- The movement of the city workforce was more efficient than intra-urban freight. Therefore, the cost of moving goods within cities was high relative to the cost of moving people. Industry largely oriented itself to the river and the various rail terminals, and households were located with reference to the horsecar lines that fed into major employment districts.

In Greater Manchester, over 100 textile mills, breweries, chemical works and foundries lined the Aston and Rochdale canals. The two-mile stretch of canal between Borderley and Aston in Birmingham packed 124 works along its banks, including soap, varnish and tar making. With their great appetite for water, distilleries, chemical plants and foundries favoured the canal system in the Port Dundas area to the north of central Glasgow. In Melbourne, right through the nineteenth century, the Yarra River was despoiled by the heavy users of water, whether it was for washing and cleansing materials like wool, skin and hides, for cooling in the breweries and the soap and candle works, or for the disposal of waste from the boiling-down works and slaughterhouses.

The emphasis that Fales and Moses placed upon a materials orientation still leaves one very mixed group of industries unaccounted for: the small-scale workshops and craft industries specialising in clothing, lace making, furniture, printing and publishing, jewellery, watchmaking, instrumentation and the repair of machinery capitalised upon a range of externalities available by virtue of their location in the inner areas of cities like London, Birmingham and Nottingham on one side of the Atlantic, or New York, Boston and Philadelphia on the other. These included the facilities provided by the 'room and power' companies, a local pooling of specialist skills and close linkages with both the suppliers and receivers of components and semi-finished goods. When it came to the short trips and repeated handling of piece goods, the horsedrawn cart had no equal in nineteenth-century cities.

Fales and Moses also suggested that the limitations of the telegraph encouraged the clustering of those firms dependent on quick access to information close to the city telegraph exchange. While messages could be transmitted between cities quite rapidly, the early switchboard created a bottleneck for information flows around the city. Banks, manufacturer's agents, the dealers on the floor of the stock exchange, their credit agencies and accounting services located as closely as possible to the mid-town telegraph terminal. Consequently, between 1812–15 and 1885 the number of subscribing members of Manchester's Royal Exchange rose from about 1500 to 7500. Because of the premium placed on access to market information and doing business 'face to face', these are the firms, then, that traditionally sought, and continue to seek, office space in the heart of the Central Business District (CBD).

So far as housing is concerned, many factory owners built barrack-like rows of terraces within short walking distance of their factories. The hidden benefits were reflected in lower operating costs and higher productivity. Many of Manchester's factory owners extended the working week to between 78 and 84 hours in the early 1860s. By the same token, any worker dependent on casual employment – like dockers and builder's labourers, or country girls looking to 'keep house' for the gentry – needed to be near the centre of the Victorian city's labour market. At the height of the building boom in 1865, London's casual labour force probably exceeded 680,000 workers. Central London was the best place to hear of work and the place most accessible to all quarters of the city, whether it was the 'rookeries' of Drury Lane for the Covent Garden market, Tower Hill for the docks, or Soho for the West End tailoring trades and domestic services.

The only way slum housing could continue to compete with industrial and commercial users at the edge of the CBD was by building at densities not previously contemplated. In New York in the 1850s and 1860s, developers and spec-builders demolished older converted dwellings to make way for the notorious 'dumb-bell' tenement, which was named because of its shape. By 1900 the number of tenements in Manhattan had increased to 42,700 and they housed 1.6 million people, chiefly recent immigrants, at densities in excess of 35 persons per building.

Profits from housing, except in cases of extreme 'rack-renting', seldom matched the yields from industrial and office premises. Decaying inner-area neighbourhoods throughout European cities were cleared to extend or widen railway lines and docks, lay trunk sewerage and water mains, redevelop stations and goods yards, erect huge warehouses and bond stores, and provide for commercial and government offices. Land values rose accordingly: between 1861 and 1881 the rateable value of the City of London rose from £1.33 million to £3.48 million, and again by a third to £4.86 million in 1901. In Stedman Jones's opinion, 'In its arbitrary and unplanned way, demolition and commercial transformation in nineteenth century London must have involved a greater displacement of population than the rebuilding of Paris under Haussmann' (Stedman Jones 1971, 159). Thus began the depopulation of the central areas of British cities. Every ward of central Liverpool was losing households by 1871. In Birmingham four of the central wards lost population between 1851 and 1871. The City of London's population fell from 143,387 to 51,439 between 1861 and 1881.

The early Victorian cities were still largely unpaved, unsewered and ill-lit and lacked safe drinking water. The pressure coming from reformers like Charles Booth and Octavia Hill for housing improvements and proper drainage helped to force the hand of Parliament. With the re-organisation of local government in 1832, some Victorian cities began to confront the backlog of urban problems with much greater determination. Yet even more progressive urban corporations like Leeds, Birmingham and Liverpool followed bursts of lavish civic spending with long spells of penny-pinching. By 1870 there were 49 municipally owned gas companies in England and Wales. Further, under the Public Health Act of 1875 urban authorities had the power to municipalise gas companies, along with other privately operated utilities like gaslight, waterworks, tram and railway companies. This reformism has since been referred to, somewhat caustically, as 'gas-and-water socialism'.

The 'housing problem' in Victorian cities triggered a series of legislative measures between 1851 and 1875 that were condemned to failure partly because they ran ahead of public opinion. It was thought that the overcrowding of inner-city tenements would be relieved by developing suburbs. On the whole it was much easier for urban municipalities to treat the 'housing problem' narrowly as a sanitary issue and to enforce building codes in the forlorn hope that this might prevent new slums from forming. The situation did not improve until the turn of the century, when the London County Council (LCC) began to acquire land in the outer suburbs for the construction of working men's homes. Within five years or so the LCC was supplying 7 per cent of all working-class housing in London. This marked the beginning of 'an enormous stride towards large-scale munici-pal socialism in the field of working class housing' (Wohl 1977, 234). But otherwise, escaping to the sub-urbs with their promise of space for a garden, fresh air and verdant fields within walking distance were a dream that only middle-class workers could real-istically aspire to in the late nineteenth century.

Suburban rail and streetcar promoters were in cahoots with politicians right through the nineteenth century, starting with the landed aristocracy who owned estates near cities, and drawing to a close with the granting of utility franchises by municipalities in the United States. The nadir was probably reached in Australia during the boom decade of the 1880s when suburban rail promotion fuelled land specula-tion on a grand scale in Sydney and Melbourne. By 1888, the number of allotments subdivided for subur-ban housing in Melbourne exceeded the needs of London's population at the time!

ELECTRIFICATION OF MASS TRANSIT AND 'FIRST-PHASE' SUBURBANISATION

Initially, steam trains played a relatively minor part in moving workers about the city. 'First-phase' sub-urbanisation was a period during which industry and commerce was still tied to the city centre. Even in London in the mid-1850s, 10,000 commuters at most arrived by rail. This figure was completely overshadowed by the quarter of a million passen-gers entering the City on a daily basis by foot or by horse-drawn streetcar (Kellett 1969, 365).

The electrification of mass transit pushed the commuting corridors further out into the country-side and allowed those fortunate workers who could afford the fares to escape from the over-crowded and polluted inner core. In fact the Cheap Trains Act was passed by Parliament in 1883 in an attempt to relieve some of the overcrowding in cen-tral London. The purpose was to force the railway companies to extend the geographical coverage of the 2d working men's fares. But these genuinely cheap fares only ever amounted to 7 per cent of London's daily rail traffic: out of the 410,000 com-muters from London's suburbs in 1901 only 27,569 qualified for the 2d fare; 105,000 riders paid a rate set to attract clerical workers rather than blue-collar workers (between 9d and 11d daily); and the

remaining 278,000 paid the ordinary fare. Whether it was suburban London, Chicago or Sydney, transit riders were still restricted to walking distance around the drop-off points along the rail or tram track. Downtown Chicago was linked with over 100 railway suburbs strung like beads along the main lines out of town. These were the first 'dormitory' suburbs for city workers making the daily commute to work.

The development of the dynamo in the 1870s and the rapid electrification of the horse-drawn street-car routes paved the way for extensive suburban development in North American cities like Boston (Warner 1962). During the late 1880s and early 1890s, electrification pushed the area of convenient trans-portation out to the edge of a six-mile radius. By 1895 over 90 per cent of the streetcar network had been electrified, while private companies operated 850 trolleybus systems in US cities on about 10,000 miles of fixed track.

Diesel buses broke that dependence upon fixed-track modes of transport for commuters, but it was not until trucking and car ownership were much more widespread that industry and commerce on any scale could afford to relocate to the suburbs. As Moses and Williamson (1967) explain, 'A pre-requisite for decentralization was the breaking of the transport tie to the core.' With the arrival of the motorised truck, the costs of moving freight about the city fell to a point where, for the first time, firms were able to capitalise on the potentially lower wages and rents payable in outlying locations. They sug-gest that the effect of this development upon the spatial structure of cities can roughly be divided into two phases. During 'first-phase' suburbanisation the truck replaced the horse-drawn wagon for moving goods around the city. Truck registrations in Chicago rose from 800 in 1910 to 23,000 in 1920. However, the motorised truck could not yet compete with the rail-ways when it came to moving freight between cities. Industries could move to the suburbs, but unless they located on a rail siding they had to arrange for the cartage of their materials and finished goods in and out of the main rail terminal for shipment to other cities.

This tie to the freight yards in the centres of cities was weakened during the second phase,

when improvements in the truck and in the interre-gional highway system meant this mode could be used for long-distance transport. The full impact of this change was probably not felt until the revival of a strong peacetime economy after World War II. The attractiveness of the satellite area in this period was increased by the automobile which allowed firms to draw labour from a broad area.

(Moses and Williamson 1967, 213–14)

The 'trade-off' model and rent gradients

It was about this time, in the 1960s, that urban economists developed a land use model to try and explain the processes and land use patterns most commonly encountered in US and British cities. The most general form of the model has firms and households trading off locational costs against accessibility to the city centre. Alonso, Wingo, Muth and Mills were among the leading exponents of urban modelling in the United States; while in the UK, an economist extended the residential predictions of the original 'trade-off' model (Evans 1973). Despite the development of some well-laid-out industrial estates – sometimes known as trading estates in the UK – and the appearance of the first regional shopping centres in the suburbs, the central city was still the dominant destination for all work trips in the metropolitan region of the 1950s and 1960s.

The 'trade-off' model suited these conditions when economic activity was concentrated in the urban core and the suburbs were not only economically subordinate but the dominant pattern of movement in cities was radial in orientation. In fact, the distribution of land uses mapped in Figure 4.1 can only be derived if a number of simplifying assumptions are imposed from the start. These are the *a priori* conditions that yield a 'single-centre only' solution. First, the model assumes a circular city in which all economic activity and jobs are initially located at the centre, and transport is possible in all directions. Secondly, all households are assumed to be identical in terms of tastes and in terms of earning the same incomes. That is, they are said to have identical utility functions. With each household maximising utility subject to constraints on income and time, incomes, transport costs and the price of goods are fixed and known throughout the urban area. It is the price of land, therefore, that must vary. This gives rise to a rent gradient with rents declining with distance from the city centre, consistent with the fall in

net income and rise in the costs of goods as accessibility declines.

Once the equilibrium distribution of households and the resulting rent gradients are established, the cost of labour – the wage rate gradient – can be determined. It measures the wage a firm has to pay at various locations in the suburbs, assuming other conditions are also met, in order to attract workers away from the central city labour market. The remaining factor price gradient for capital is invariant with respect to distance from the city centre. The model specifies which general category of economic activity – manufacturing, offices, retailing, storage or housing – will be allocated to a given zone around the city centre (Fig. 4.1). Firms contemplating relocation, for example, have to 'trade off' a combination of higher rents and lower transport costs available in the more accessible core city area, against lower rents, larger sites, though higher transport costs, in the less accessible outer suburbs. The further the distance from the city centre, which optimises access to the metropolitan market, the higher the transport component of location costs. According to Moses and Williamson (1967, 212–13), the model implies that factor costs tend to be lower in the suburbs, which accounts for their need to explain the original concentration of economic activity at the core of nineteenth-century cities in terms of transport costs (see above).

Plainly, the suitability of a 'single-centre' model was breaking down with the growing momentum of suburbanisation in the 1950s and 1960s, more or less before the eyes of the modellers:

all workers are assumed to receive the same gross wage. CBD workers are nevertheless in equilibrium wherever they live in the suburbs because land rents just offset transportation costs. But some workers are employed in the suburbs and they are obviously better off than CBD workers living in the same neighbourhood.

(Mills 1972, 237)

The rising tide of suburbanisation that accompanied post-war prosperity in the United States is neatly summarised in Table 4.1. The table documents the shift of residents and jobs from the city to the suburbs in 90 Standard Metropolitan Statistical Areas (SMSAs) between 1947 and 1963. These SMSAs kept the same central city and suburban county boundaries throughout this period. In 1960, they contained 37 per cent of the country's population.

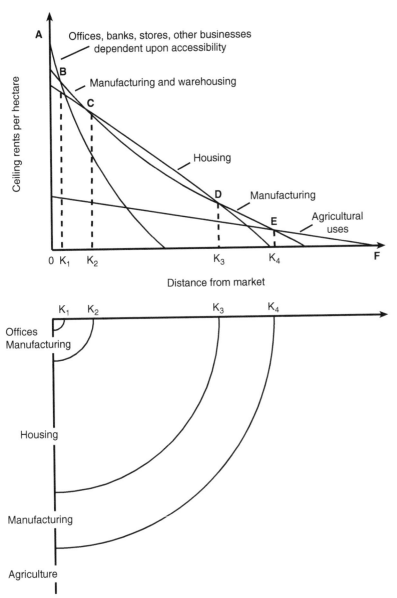

Ceiling rents per hectare

A

Offices, banks, stores, other businesses
dependent upon accessibility

B

Manufacturing and warehousing

C

Housing

D   Manufacturing

E

Agricultural
uses

0 K₁   K₂   K₃   K₄   F

Distance from market

K₁   K₂   K₃   K₄

Offices
Manufacturing

Housing

Manufacturing

Agriculture

Figure 4.1 *The bid-rent model of urban land uses in a city with a single centre*
Source: Nourse (1968); composite based on Figures 5–14 and 5–15, reproduced by permission of the McGraw-Hill Companies

While the population living in the suburbs of these 90 SMSAs climbed from 47 per cent to 60 per cent between 1950 and 1960, the corresponding share of metropolitan jobs between 1947 and 1963 rose from 32.7 per cent to 49.6 per cent. Table 4.1 also reveals that the fastest rates of suburban job growth took place in retailing (22.8 per cent), followed by whole-saling (20.0 per cent), services (18.4 per cent) and manufacturing (14.3 per cent).

TRUCKING, AUTOMOBILE DEPENDENCE AND 'SECOND-PHASE' SUBURBANISATION

The rising affluence that post-war prosperity brought, the mass production of automobiles and the construction of limited-access highways like the autobahn and autostrada in Europe, and tollways like

TABLE 4.1 *Suburbanisation of people and jobs in 90 US Standard Metropolitan Statistical Areas (SMSAs), 1947–1963*

Economic Activity	City		Suburban ring		City		Suburban ring	
	Number	%	Number	%	Number	%	Number	%
	1950				**1960**			
Population (millions)	26.7	53	23.5	47	26.6	40	39.5	60
	1947				**1963**			
Manufacturing jobs	3.8	60.5	2.5	39.5	3.3	46.2	3.8	53.8
Retailing jobs	2	71.5	0.8	28.5	1.7	48.7	1.8	51.3
Service jobs	0.6	79.5	0.2	20.5	0.8	61.1	0.5	38.9
Wholesaling jobs	0.9	85.2	0.2	14.8	0.9	65.2	0.5	34.8
Total jobs	7.4	67.3	3.6	32.7	6.7	50.4	6.6	49.6

Source: Mills (1972). Reproduced by permission of John Hopkins University Press

the Penn State Turnpike and NJ Parkway, all contributed to the outward extension of the metropolitan area. The federal highway programme in the USA was also of paramount importance to post-war suburbanisation. The entire federal transport budget of US$156 billion was diverted to roads between the years 1944 and 1961. Not surprisingly, every member of the motor industry, from vehicle builders to highway builders and oil companies, had a vested interest in paving America and getting commuters into cars. The Federal-Aid Highway Act of 1956 funded 90 per cent of the costs of building the 65,165 km Interstate Highways and Defence System. At its peak in the 1960s, highway construction consumed well over a third of all federal grants to state and local government. During the decade 1965–75 alone, over two-thirds of the long-term capital authorised by Washington for public works in the cities was earmarked for highway construction. This indirectly contributed to the running-down of urban public transport systems.

The urban expressway system constructed as part of the Interstate Highway programme acted as an umbilical cord by connecting dormitory suburbs with downtown office jobs. Just as importantly, the inter-city routes mandated by the 1956 Highway Act required that central city bypass routes be built into the urban highway network, especially in the congested Northeastern Seaboard conurbations. By the early 1970s, more than 80 of these circumferential

beltways had been constructed in the outer zones of US metropolitan areas under the interstate highway programme: 'Even during the 1970s, it was apparent that retailing, office employment, commercial-services provision and light manufacturing were concentrating around strategic locations on the metropolitan freeway network, especially where radial and circumferential expressways intersected' (Muller 1997, 46).

Decentralisation of manufacturing jobs

Manufacturing jobs began to follow the workforce to the suburbs in significant numbers from the beginning of the 1950s. The 1960–70 gain in suburban blue-collar employment in the United States was 29 per cent, against a 13 per cent loss for the central cities. In Sydney, Australia, by 2000 the manufacturing workforce based in the inner-city area had fallen to one-fifth of its size at the end of the Second World War. Behind the decentralisation of industry lie the competitive pressures that force firms to embrace technical substitution and strive for greater labour productivity. Production processes had long since been liberated by the electric generator from the central boiler plant where the steam power was generated, and from the rail siding. New forms of automated processing and goods handling rendered obsolete much former factory accommodation in the oldest inner manufacturing districts.

Plants with ageing equipment in old inner-city districts were faced with mounting congestion and pollution costs.

Yet the limited space available in core areas ruled out the possibility of building a modern plant on site. Firms sought out spacious sites in the suburbs where the assembly line could be organised to achieve maximum output and where materials and finished goods could be moved about the plant by forklift trucks. Free-standing, single-floor premises situated near a major freeway interchange, a container port or air freighting facilities avoided congestion costs and lowered freight charges. Trucking materials and goods door to door obviated the need for double-handling at break-in-bulk points like the docks or railhead. Besides, many industrial processes have become progressively cleaner and quieter, which has made them more acceptable to suburban communities.

But this is only part of the decentralisation of manufacturing 'story' in Europe and North America. Regional assistance programmes played a significant role in helping firms to leave the inner city and relocate in depressed areas in the UK. Currently one-third of the European Union's budget is spent addressing the problems of de-industrialising regions. Member countries provide matching support (reported in the *Independent*, 2 May 2000, 15). In the USA, southern counties recruited northern firms with packages that include tax breaks, community contributions to setting up costs and promises to discipline labour. Hence, tax revenue – state and federal as well as local – is being used to take jobs away from some communities and place them elsewhere. And most of these moves are from the centre of one city to the suburbs of another. In turn, the trend is fortified by the tendency for new jobs, in manufacturing (and services), to spring up where the moving jobs go, rather than where they are coming from.

This generalised picture of manufacturing decentralisation in large metropolitan areas is based on net changes in establishments, output and employment. Inevitably, the picture of factory closures, openings and relocations in the larger metropolitan region is much more complex. For example, the net loss of 172 manufacturing plants detected in a study of Clydeside between 1958 and 1968 conceals gross changes of 711 plant deaths and 529 plant births (Bull 1978). In addition, 607 establishments transferred their production during the ten-year period

to new locations within Clydeside. Excluding plant births, these adjustments represented over half of all the plants located in the Clydeside in 1958. At the time, this pattern of industrial restructuring was also consistent with other British and US centres of manufacturing. In Greater Manchester, 41 per cent of all the plants had relocated or closed between 1966 and 1972. In the two-year period 1967–9, 11.5 per cent, 7.6 per cent and 10.2 per cent respectively of a total 1966 stock of 39,128 plants in the New York SMSA relocated, closed or opened.

Neither the post-war industrial estates in the outer suburbs of the old manufacturing regions in North America or Europe, or their equivalents in the sunbelt cities, escaped unscathed. Yet the 1990s saw new rounds of investment, especially in advanced manufacturing technologies. Industrial parks in suburban counties around both large and small US metro areas captured the lion's share of plants using programmable automation (PA) technology. A national survey of 1363 metalwork-ing establishments concluded that these innovative clusters 'now have their own complements of business, educational, and technical services, and increasingly diverse mixes of industrial activity, even as companies so situated continue to have access to the social, political, and physical infrastructures of the densest core urban areas' (Harrison *et al.* 1996, 151–4).

As Australia's leading centre of manufacturing, Melbourne bore the full brunt of de-industrialisation in the 1970s and 1980s. The western suburbs of Melbourne struggled to attract investment and jobs, even during the 'good times'. In the early 1990s, the Keating government set up a new tier of regional bodies to help stimulate economic activity. For example, the Western Melbourne Regional Economic Development Organization (WREDO) is an investment facilitator between the private sector, state and commonwealth governments, and the six local government authorities in the west. Construction of a western ring road to bypass the congested city core and link the national highway system with Melbourne International Airport was a crucial first step in attracting new industry into these formerly unappealing regions. Since its completion in 1998 an extra A$7 billion has been committed to new manufacturing, business and housing development in the west. Land sales of A$800 million in the 18 months to June 1997 accounted for 60 per cent of land sold for industry in Victoria. Much of this

TABLE 4.2 *Office floor space and rents in selected cities 1990*

City	m² (m.)		US$/m²/per year	
	Prime CBD	Urban region	Prime location	Suburban location
Tokyo	4	45	1600	685
London	15	25	1200	400
Paris	13	35	735	335
Frankfurt	2.7	8	540	275
New York	29	—	522	na
Los Angeles	2.6	17.8	305	375
Brussels	4	6.5	205	175
Amsterdam	0.5	5.2	200	130

Source: Korteweg and Lie (1992). Reproduced by permission of Blackwell Publishers

new investment in Melbourne's western suburbs can be attributed to the expansion of foreign companies into the Asia-Pacific region, with 15 Japanese manufacturers opening plants in the five years to 2001, including a new Toyota plant at Altoona. Estimates place the job multiplier for the west at 17,000 construction jobs and around 2000 permanent jobs in the retail, service and industrial sectors. While this has not been sufficient completely to eliminate unemployment in these working-class suburbs, the sign of some new manufacturing jobs has given a real lift to community morale in Melbourne's west.

Office jobs in the suburbs

The decentralisation of office-based activities such as business and professional services has been ongoing in most advanced economies since the 1950s. This has multiplied with the growth of employment in service industries and associated office space. In the USA, the total amount of office space doubled between 1959 and 1979, and then doubled again between 1980 and 1990. By 1986, 57 per cent of the nation's office space was located outside the central city (Pivo 1990). While the effect of this has been to detract from the importance of the CBD in many urban regions, marked differences exist from city to city. The degree to which office floor space was still concentrated in cities like London, Paris, Brussels and Sydney, as opposed to Tokyo, New York and

Los Angeles at the beginning of the 1990s, is evident in Table 4.2. Northern New Jersey had attracted over 35 million square metres of office space by 1990. At the time this exceeded the combined volume of downtown office space in Chicago and Los Angeles.

Figure 4.2 illustrates these trends, and some of the differences that can occur, for six North American cities – Los Angeles, Houston, San Francisco, Seattle, Denver and Toronto. The dramatic surge in suburban office development that occurred during the 1980s is apparent, as is the extent to which downtown Los Angeles and Houston are overshadowed by 'cities in the suburbs'. On the other hand, Seattle and Toronto still possess reasonably dominant CBDs. Whereas very little of this office development in the suburbs of American cities has been managed as part of a regional growth strategy, in Europe it tends to reflect the relative success or failure of government efforts to disperse office employment to the suburbs, or even direct office jobs away from cities to job-poor regions.

In the UK, the dominance of central London in the regional office market (Table 4.2) persists, despite a concerted government effort in the 1960s and 1970s to divert office jobs away from London (Alexander 1979, 63–79). Between 1948 and 1963, planning permission was granted for the construction of some 5.4 million square metres within London's central area, representing a net increase of 4.6 million

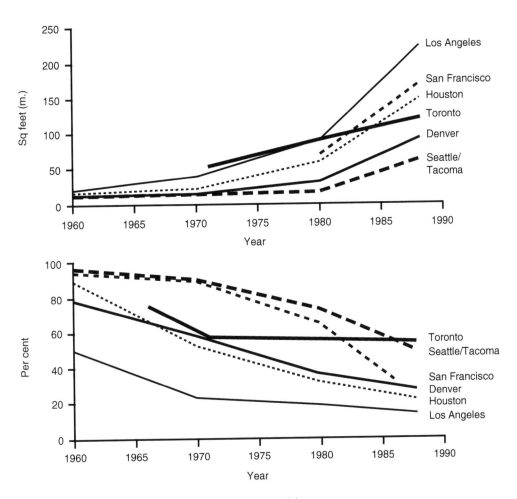

FIGURE 4.2 *Office decentralisation in six North American cities*
Source: Pivo (1990). Reproduced by permission of the American Planning Association

square metres, or 50 per cent, over space available in pre-war London. In the early 1960s, the Location of Offices Bureau was established to regulate office development and to encourage offices to relocate or start up outside central London. Applications for Office Development Permits (ODPs) to construct a total of 2.9 million square metres of office floor space were refused between 1965 and 1976. Estimates obtained in the mid-1970s of the number of office jobs diverted range from 170,000 to possibly 250,000 – or about one-quarter to one-third of central London's 756,000-strong office workforce in 1971. Only 30,000 of these were civil service jobs, or about 35 per cent of the 86,000 central government

jobs located in the capital at the time. For example, in the 1970s, 4000 clerical workers employed in the Motor Registry went to Swansea and 6000 DHSS workers went to Longbenton in the north-east. In the 1990s, 6000 Ministry of Defence jobs were relocated to Bristol's northern fringe suburbs alongside corporate giants like Hewlett Packard and Orange. Other European countries like France, Sweden and the Netherlands also developed a range of policies about the same time to try to redistribute office employment.

In the Netherlands, office activity is widely dispersed among the four main cities that form the conurbation known as Randstad – Amsterdam,

TABLE 4.3 *Locational factors rated highly by office firms based in major Dutch cities*

Location factors	Amsterdam %	Rotterdam %	The Hague %	Utrecht %
Characteristics of accommodation				
Size of building	64	64	67	77
Flexible floor space	58	56	52	66
Possibilities for expansion	39	42	40	59
Identity	63	69	53	67
Standing	87	87	79	77
External appearance	62	75	64	73
Rent/price	69	65	66	75
Cost of services and energy	44	41	53	75
Facilities for computers	50	56	57	72
Security	58	72	63	78
Characteristics of location				
Accessibility by car	94	87	87	83
Location near motorway	76	62	66	58
Parking facilities	95	90	87	85
Accessibility by public transport	69	71	76	82
Near a railway station	44	41	60	61
Prestigious environment	70	79	68	70
Visibility from motorway	30	10	11	18

Source: Korteweg and Lie (1992, 257). Reproduced by permission of Blackwell Publishers

Rotterdam, The Hague and Utrecht (Korteweg and Lie 1992). And within these cities, office jobs have been steadily moving to the suburbs since the 1970s. Between 1981 and 1987, the centre of Amsterdam lost 4600 office jobs and about 100,000 square metres of office floor space. With the trend continuing, the office workforce fell to 45,000 by the end of the 1990s, when the central areas of Amsterdam, Rotterdam, The Hague and Utrecht only accounted for about 19, 40, 36 and 30 per cent of the office floor space in their respective city regions. Rotterdam, The Hague and Utrecht have allowed some office redevelopment in their central areas. But parcels of land and canals often dating from medieval times limit the supply of land and buildings suitable for modern offices. In addition, developers wanting to demolish or convert city buildings run up against height and size restrictions, listed landmark buildings and designated preservation zones. This has driven firms with medium to high floor space requirements out to suburban office centres like South Amsterdam and Amstelveen, where they are also closer to Schipol Airport. By 2000, total office floor space in south Amsterdam exceeded 650,000 square metres. As Table 4.3 suggests, accessibility by car and ease of parking, followed by environmental amenity, are key considerations for corporations locating in these 'subcentres'. When it comes to the accommodation favoured, external qualities (standing, identity, external appearances) vie with rent and functional qualities (technical facilities, floor space flexibility, security) in importance (Table 4.3).

TELECOMMUNICATIONS AND 'DETACHED' PERI-METROPOLITAN DEVELOPMENT: EDGE CITIES

In the 1980s and 1990s, developments in telecommunications (satellite, cable TV, fibre-optics, the Internet), linked with the rapid take-up of high-speed microprocessors in business, gave further momentum to the decentralisation of economic activities. Firms in remote locations, including the suburbs, have access to national and global telecommunications networks on a par with those serving firms at the heart of 'world-cities'. They can be linked into the global economy without having to direct information flows via the CBD core. With this advantage, suburban firms are able to match their downtown counterparts in all but a few specialised areas of business and finance. One estimate put the suburban share of international telecommunications traffic in the Greater New York region as high as one-third in the early 1980s, and suggested that by 2000 it would exceed one-half (Muller 1997, 51).

These developments helped to sustain suburban growth in North America, Europe and Australasia – though at a slower rate in many cities. What changed in the 1980s is that US suburban growth was accompanied for the first time by a quite new form of urban development even further out beyond the suburbs. This 'detached' peri-metropolitan development is now popularly referred to as 'edge city'. A journalist working for the *Washington Post*, Joel Garreau (1991), coined the term 'edge city' to describe this 'third phase' decentralisation in metropolitan economies.

Because edge city development in the United States is taking place in counties that overlap with suburbs, it is often difficult to disentangle. But some of the fastest-growing cities in the United States – including Atlanta, Phoenix, Dallas, Houston, Denver, Las Vegas and Austin – are those with the best-developed edge cities. Between 1980 and 1990, the rate of population growth in the suburban ring of US cities more than doubled the rate for the central city. Two-thirds of net employment growth in America's 60 largest metropolitan areas was located in the suburbs (US Department of Housing and Urban Development 1995, 10). Over the same period, the suburbs captured 120 per cent of the net job growth in manufacturing, while central cities suffered absolute losses in manufacturing employment. Indeed, almost one-third of the 1.6 million jobs lost to the US manufacturing sector in the 1980s involved plants leaving central city locations in north-eastern or mid-western cities to set up mainly in the suburbs and outskirts of cities in the south and the west.

While a number of European cities, such as Bristol, Frankfurt-Rhine-Main and maybe Zurich, saw office and industrial parks established at their edges in the 1990s (Keil and Ronneberger 1994), the phenomenon does not bare strict comparison with edge city development in the United States. Rather, decentralisation in the UK has involved the movement of housing and jobs to the suburbs, and to outlying towns and rural areas. Figure 4.3 reveals the extent to which the regional pattern of investment favoured towns and rural areas over a fifteen-year period. While the 20 largest cities in Britain lost 500,000 jobs between 1981 and 1996, the rest of the country gained 1.7 million. In Britain, where the suburbanisation of jobs is part and parcel of a broader redistribution of economic activity, the process is commonly referred to as 'urban–rural shift'.

Edge city development in North America went hand in hand with the property boom in the 1980s. Developers capitalised on the opportunity to build giant, free-standing edge cities close to the major interchanges on highways skirting US cities. As built form, edge cities sit somewhere between the regional centres serving low-density suburbs and the high-density CBD (Fig. 4.4). They are also quite unlike the industrial parks or shopping malls built in the 1960s and 1970s. Developers had to adhere to strict land use codes that segregated industrial, commercial and service activities. Edge cities are characterised by 'mixed use development', which means that they tend to support a much broader mix of urban activities. This includes retailing, distribution, data processing, educational services, entertainment and leisure. Within the 'new economy' there are very few forms of 'knowledge producing' activity that are incompatible with consumption-based activity. The distinction between office jobs and jobs in advanced manufacturing, storage and the despatch of goods is increasingly blurred. And, unlike suburban shopping malls with their low-wage, retail jobs, edge cities also attract many of the high-wage, white-collar jobs of the 'new economy' (Fig. 4.4). What also sets edge city development apart from conventional post-war dormitory suburbs is that commuters typically live in 'master-planned' residential communities quite close to the office where they work.

The amount of floor space devoted to campus-like office blocks and commercial centres in some of these edge city nodes often rivals, or even overshadows, their downtown CBD in terms of retail sales and office space. By the late 1980s, for example, Costa Mesa's vast South Coast Plaza in Southern

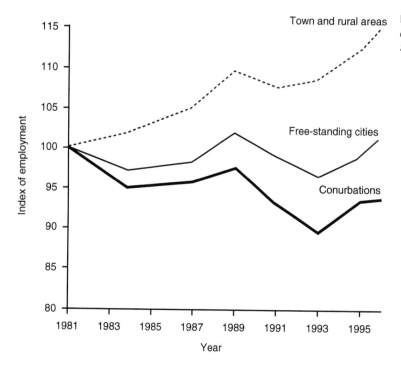

FIGURE 4.4 *Metropark, a major office park in central New Jersey adjacent to the junction of routes US-1, the New Jersey Turnpike, the Garden State Parkway and the NJ Transit Northeast Corridor Line*
Source: Author's collection (composite image)

California's Orange County had a higher retail turnover than central San Francisco. Edge cities are also now competing directly with the traditional downtown for highly specialised functions such as mortgage banking, corporate legal offices, accounting services, publishing, conference and exhibition venues and luxury hotels. By setting lower limits of 450,000 square metres and 55,000 square metres of

TABLE 4.4 *Employment in producer services, central city and suburban ring of the Greater New York Region, 1984–1996*

Services	Year	New York City (%)	Suburban ring (%)
Advertising	1984	84.8	15.2
	1992	76.8	23.2
	1996	72.5	27.5
Legal services	1984	60.2	39.8
	1992	56.3	43.7
	1996	53.7	46.3
Computer and data-processing services	1984	31.6	68.4
	1992	24.6	75.4
	1996	19.8	80.2
Engineering and management services	1984	49.9	50.1
	1992	41.2	58.8
	1996	37.9	62.1

Source: Muller (1997, 56); compiled from US Bureau of Census, County Business Patterns (annual publication). Reproduced by permission of Sage Publications

leasable office space and retail space respectively, Garreau (1991, 423–39) identified 123 edge cities and 78 emerging edge cities located beyond the built-up suburbs of US metropolitan areas. At the time, in the early 1990s, five edge cities were in the planning stage for the Phoenix area. He regards Tyson's Corner, outside Washington DC, as the archetypal edge city, but others that have achieved national recognition in the United States include Walnut Creek (east of San Francisco), King of Prussia (Philadelphia), Schaumburg (Chicago), Irvine (Los Angeles) and Bellevue (Seattle).

The proliferation of edge city development in the 1990s was aided by the continuing exodus of producer services from the commercial core of cities like New York (see Table 4.4). Urban geographer Peter Muller has tracked the moves made by the top 500 American corporations listed in the *Fortune* magazine annual survey to suburban and edge city locations. Starting with 47 corporations in 1965, this figure grew to 56 in 1969, 128 in 1974, 170 by 1978 and 233 (or 47 per cent) by 1994.

This trend is especially evident in the Greater Los Angeles region and can be partly attributed to the way the southern Californian economy has re-adjusted since the severe downturn of the late 1980s (Kotkin 2000). Los Angeles now bears little resemblance to the big-company city of 25 years

ago, which rested on the three 'pillars' of military contracts, movie making and land development, and was run by corporate titans like Lockheed, Disney, Security Pacific, First Interstate, Great Western Financial, Atlantic-Richfield, Times Mirror (publisher of the *Los Angeles Times*) and Carter Hawley Hale (a statewide retail chain). These giant corporations have either been acquired and/or have moved their headquarters from downtown Los Angeles. More and more, the sprawling mega-city resembles a 'pillarless economy' dominated by a plethora of dynamic and loosely linked firms in the software, information and digital entertainment sectors.

Once touted as 'capital of the Pacific Rim', Los Angeles now has fewer Fortune 500 corporations that San Francisco, let alone New York or Chicago, but has a greater proportion of firms employing fewer than 100 people. Civic Centre, which is the Los Angeles CBD, has been eclipsed by the Westside district in terms of total office space, and rental and occupancy rates. Even the San Fernando Valley rates three Fortune 500 companies, including Walt Disney, which is now probably the region's most important corporation.

In New York, where the Manhattan skyline has always 'taken the prize' as the twentieth century's iconic modernist built environment, the departure of headquarter offices has 'lately surged far ahead of the national pattern' (Muller 1997, 53):

As the location cost differential between the metropolitan core and ring equalized around 1970, a corporate suburbanization movement was launched that has still not run its course. With noneconomic factors taking over, headquarters have been guided by perceived improvements in the quality of workplace life, the geographic prestige of new suburban business milieus, and overall convenience for a workforce whose majority already resided in the suburbs.

In 2000, the suburbs were home to about 65 per cent of all the Fortune 500 head offices based in the Greater New York region. Most of them are clustered near a few strategic junctions on the regional transport system in south-west Connecticut, northern New Jersey (Fig. 4.4) and suburban New York State. AT&T, for example, quite effectively controls its world communications network from a headquarters at Basking Ridge, New Jersey, about 50 km out of Manhattan. Therefore, while Manhattan remains the chief supplier of corporate services, its position is steadily eroding as head offices leave New York City and tempt growing numbers of support providers in advertising, legal services, computer and data-processing services, and engineering and management services to move with them to peri-metropolitan locations (Table 4.4). Even New York City's highly concentrated 'finance, insurance and real estate' (FIRE) sector saw its share of regional employment fall from 63 per cent in 1984 to an estimated 52 per cent by 1996 (Muller 1997, 56). Furthermore, in the aftermath of 11 September 2000, it is still too soon to tell whether the attack on the World Trade Centre will play a part in the future pattern of office investment in the New York Region.

Likewise, although the City of London has a commanding position within the British economy, taken together the major cities have not recorded the expected rates of growth in the sectors making up the emerging knowledge economy, even in cultural industries and consumption. This loss of 'share' to the suburbs and rural towns led Turok and Edge (1999) to conclude that the notion of city-led, service-sector growth does not accord with the broader reality of 'urban–rural' shift in Britain. At the very least, it does suggest that the gravitation of services towards bigger cities is progressively breaking down with the levelling of cost differentials between the inner city and rural locations. Given the inroads

made by advances in telecommunications and trucking, smaller towns and rural localities throughout Britain are now able to exploit a range of competitive advantages. A survey of over 1000 manufacturing and service businesses undertaken in the late 1980s by Keeble and Tyler (1995, 94–5), established that success in business in rural locations is closely associated with a 'high-amenity living and working environment, greater labour force stability, quality and motivation, good management–labour relations, and lower premises, rates, and labour costs'.

Edge city development in Houston

The growth of Houston after the Second World War illustrates the main drivers and characteristics of peri-metropolitan development in the USA (Feagin and Beauregard 1989). Between 1940 and 1980 Houston grew from the 21st largest city in the country with a population of 400,000, to the fourth largest with a population of 1.6 million. While much of this population growth has been attributed to 'regional shift', fewer than half of the interstate migrants came from north-eastern or north-central regions. 'Thus, while suburbanization has been foreshortened by annexation, extra-metropolitan spatial shifts in population have been mainly intraregional and intrastate' – that is, within Texas (Feagin and Beauregard 1989, 163).

Houston's manufacturing sector has been dominated by three industry groups since 1947: petroleum products, petrochemicals and oil field machinery and tools. In the meantime, this has been significantly augmented in two other areas. In the early 1960s, the National Aeronautics and Space Administration was established in Houston with a workforce of 10,000. Then, more recently, the region has benefited from the growth of the health care sector. With more than 26,000 personnel and a major teaching hospital training 8600 students, the Texas Medical Centre is now one of the largest employees in the Houston area. Covering a 100-hectare site, the centre groups 29 hospitals, research institutions and health insurance funds.

Office-based business and professional services also have a prominent place in Houston's spatial economy. The phenomenal growth in the office submarket – 800 per cent between 1969 and 1986 – coincides closely with the cyclical rise and fall of oil prices. Between 1971 and 1980, 202 office buildings of 10,000 square metres or more were constructed, and then another 157 were added during the 1980s (Feagin and Beauregard 1989, 175). By the end of the decade there were 18 different office activity nodes in the Houston region. While about a quarter of the office floor space is in the CBD, two other nodes contain a further 35 per cent (the Galleria-West Loop at the intersection of I-610 and I-59, and West Houston). Likewise, most of the region's retailing is now dispersed outside the CBD, with over one-quarter of it contained in regional shopping centres of at least 50,000 square metres of floor space.

Figure 4.5 shows how these activities are located with respect to the regional highway system. Houston is well served by interstate highways. Ten spokes meet in a road that rings the CBD, and these spokes

are interconnected again about 8 km from the CBD by an inner circumferential, I-610. Further out is another circumferential (State Highway 6). This web-like highway system – more than 400 km in length – incorporates a series of major interchanges that are fairly evenly spaced throughout the Houston region. This is where large-scale activity nodes like Busch Gardens' tourist attractions, the giant Gulfgate Shopping Centre, the Astrodome and the Astroworld amusement park are located.

Multinational presence on the edge

In 'post-suburban' America, the arrival of a corporate newcomer in suburban or edge city locations,

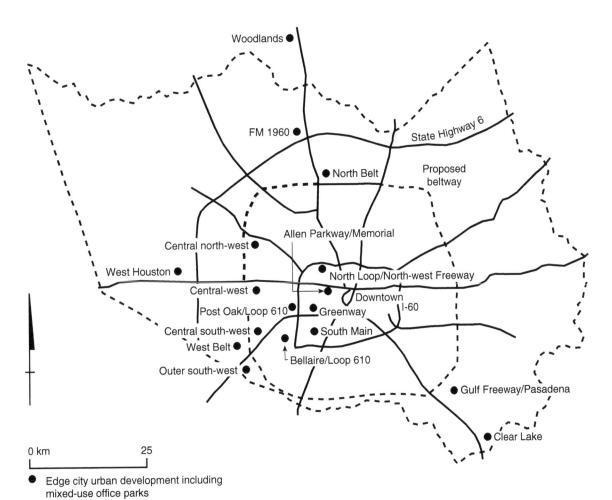

FIGURE 4.5 *Edge city development attracted to beltway locations beyond the suburbs of Houston*
Source: after Feagin and Beauregard (1989)

especially the head office of a multinational or a 'business migrant', carries with it the prospect of a workforce recruited from around the world. According to the Los Angeles Regional Technology Alliance, up to half the region's high-tech companies were founded by immigrants, particularly in heavily settled areas like the San Gabriel Valley. It is also the case that, when possible, recent immigrants – drawn predominantly from Latin America and Asia in the 1980s and 1990s – are settling closer to low-skilled job opportunities in the suburbs (Frey 1995).

The geographical and ethnic changes taking place in the business sector are also beginning to unsettle the American suburban monoculture (Chapter 7). In the Dallas–Fort Worth metroplex, about half of the firms with foreign owners are located in the suburban ring. Moreover, workers born overseas tend to cluster more tightly in the suburbs, where, among other things, they can maintain strong business connections. Of the 60,000 or more expatriates working for Japanese companies in the New York area, most live in the city's northern and western suburbs. Many of the overseas students studying in the United States, or Britain and Australia for that matter, are snapped up when they graduate. This accounts for the presence of a strong south Asian workforce in north-eastern New Jersey with expertise in the skills group that Robert Reich (1993) calls 'symbolic analysts'. Monterey Park, just east of Los Angeles, has a majority Asian population including immigrants from Taiwan, Hong Kong,

Southeast Asia and China. Almost an extension of Pacific Rim Asia, Monterey Park and nearby towns in the San Gabriel Valley have been transformed by capital from Asia finding its way into high-rise apartments, retail emporiums and a range of other enterprises.

Silicon Valley, to the south of the San Francisco Bay area, is also a locational destination for business migrants with capital and skills to invest in the region's economy. Silicon Valley is now home to over 7000 electronics and software companies, with about 10 starting up each week. An estimated 25 per cent of these start-ups are run by Chinese and Indian engineers, who are among the one in three engineers working in 'the Valley' that happen to be immigrants. But apart from their direct contribution to job and wealth creation, these immigrant entrepreneurs and their families form tight-knit communities scattered throughout the Greater San Jose region.

Absence of edge city development in Australia

The rates of population growth in the outer suburban and fringe zones of the Australian metropolitan areas were double the national average during the decade 1986–95 (Table 4.5). Peri-metropolitan development accounted for over half of the extra people added to the Australian population over the nine-year period 1986–95. Yet the contribution that edge city development, with its activity nodes and

TABLE 4.5 *Patterns of population and housing growth in urban Australia, 1986–1995*

Urban zone	Share of population 1995	Rates of population change (%)		Share of population change in Australia		Share of dwelling approvals 1988–96
		1986–91	1991–95	1986–91	1991–95	
Core	4.4	−0.1	−0.4	−0.4	−1.8	3.7
Inner	6.5	−0.1	−0.1	−0.6	−0.9	2.8
Middle	21.5	0.6	0.1	8.4	2.0	13.4
Outer	27.5	3.0	2.1	49.7	51.5	35.0
Fringe	9.9	2.9	2.3	17.0	20.3	15.2
Greater Metro Areas	69.9	1.6	1.1	74.1	71.2	70.1
Australia	100.0	1.5	1.1	100.0	100.0	100.0

Source: O'Connor and Stimson (1997)

master-planned residential communities, is making to the reshaping of US cities has not been matched to any extent in Australia.

Whilst there is some clustering of mixed-use development, including business and technology parks, in the outer suburbs of Sydney and Melbourne, whether the Australian economy is large enough to support genuine edge cities is a moot point (Freestone and Murphy 1998). In the mid-1990s, the 391,500 jobs in the main commercial activity nodes in the Sydney region were distributed between the CBD–North Sydney (286,000) and the subcentres of Parramatta (30,000), Chatswood (16,000), Liverpool (11,000) and Bankstown (10,000). There were also about 22,000 office jobs along the St Leonards section of the Pacific Highway, and another 16,500 in the mixed-use business and science park at North Ryde near Macquarie University. Only North Ryde, with 500,000 square metres of office/industrial floor space and 110,000 square metres of retailing, qualifies as an edge city in Garreau's terms, but is located in the middle suburbs on a ring road. The main activity clusters (and tenants) include pharmaceuticals (Beiersdorf), electronics (Sony), computers (Microsoft) and publishing (Prentice Hall).

The stage for American edge cities was set back in the 1960s and 1970s with disinvestment, the running-down of infrastructure, the escalation of crime rates and mounting racial tension in the inner city (Kirwan 1980). Although all city centres have lost ground to the suburbs in terms of office employment and retail turnover since the mid-twentieth century, for the most part, the inner areas of Australian cities were never abandoned, and continue to be revitalised. For example, in 1995 about one-third of Australia's total businesses in the 'finance and insurance services' and 'property and business services', respectively, were located in the central city. The corresponding figures for the 'fringe' zone, on the other hand, were 8 per cent and 9 per cent respectively (O'Connor and Stimson 1997, 45). All the time, manufacturing firms were rather more decentralised, with 25 per cent in the outer suburban zone and 9 per cent in fringe locations. 'On balance, the restructuring of Australian metropolitan economies has probably tended to accentuate the corporate primacy of the CBD' (Freestone and Murphy 1998, 292). The limited nature of edge city development around Australia's major cities can also be attributed to a lack of urban freeway-building on the North American

scale. And, despite the suburbanisation of employment and the construction of a few privately funded tollways in the 1990s, most city transport networks and traffic generation patterns are still overwhelmingly radial in their orientation.

Of equal importance in explaining the absence of edge cities in Australia is the much stronger regulatory tradition in Australian urban development. There is nothing comparable to the *laissez-faire* market ethos that gave rein to edge city development-by-stealth in cities like Houston, Dallas, Denver or LA's Orange County during the 1980s (Knox 1992). Urban land release and staging of corridor development is firmly managed by state government planning agencies. Also, by specifying where they would be located and earmarking land for 'district centres' as part of a metropolitan strategy, state governments have been able to curb speculative business development beyond the edge of the built-up area.

'New town' development at the edge of Asian cities

During the 1980s, the first signs emerged of private land assembly and investment at the edge of Asian cities like Bangkok, Taipei, Manila, Kuala Lumpur and Jakarta, which exceeded the piecemeal development of suburbs in the 1960s and 1970s. Some of Asia's wealthiest families moved beyond the development of one-off projects to planning, financing and developing fully integrated towns on the edge of major cities for the growing middle class. In Bangkok, the market is dominated by Bangkok Land, Tanayong, and Land and House. Even Ho Chi Minh City has a new town project – Saigon South – which is a joint venture between Taiwanese interests and the People's Committee.

These new developments are on a huge scale. While the entire area of the Capital City region of Jakarta covers 66,000 ha (Fig. 4.6), by the mid-1990s over 90,000 ha outside the zone had received government approval for urban development. With only 25,000 ha committed, it is estimated that the remainder of this land bank held by a few large Indonesian cartels will meet Jakarta's needs until 2020 (Dick and Rimmer 1998, 2313–6). Lippo Karawaci (2360 ha) and Lippo Cikarang (5500 ha) are being built by the Lippo Group (Fig. 4.6). By 1997, Lippo Karawaci had a CBD with multiple office towers, a 100,000 square metre shopping mall,

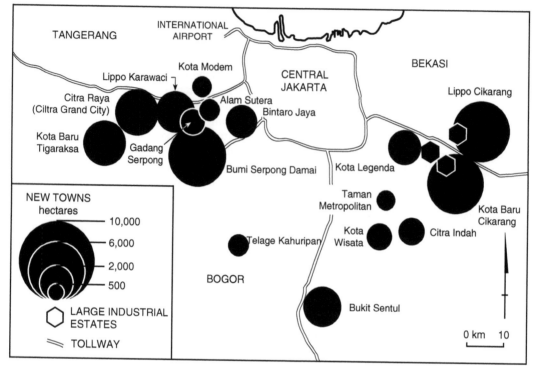

FIGURE 4.6 *New Towns either under development or planned to cope with Jakarta's future urban growth*
Source: Dick and Rimmer (1998). Reproduced by permission of Taylor & Francis Ltd

two condominium towers (52 and 42 storeys), a 328-bed international hospital, a private school and university, and the essential golf course, country club and five-star international hotel. The projected population for 2020 is one million. As Dick and Rimmer (1998, 2316) note, 'These projects are unambiguously "bundled" cities which contain all significant elements under the control of a single developer. This is clearly First World not Third World.'

REVITALISING DOWNTOWN OR 'URBAN RENAISSANCE'?

At the very time in the 1960s when many inner cities in advanced economies were in decline and the central city was losing its competitive edge in office and retail employment to regional centres in the suburbs, the political economist Stephen Hymer saw the possibilities of a downtown revival built upon the corporate needs of regional and global firms:

one would expect to find the highest offices of the multinational corporations concentrated in the world's major cities – New York, London, Paris, Bonn, Tokyo. These, along with Moscow and perhaps Peking, will be major centres of high-level strategic planning ... Since business is usually the core of the city, geographical specialization will come to reflect the hierarchy of corporate decision-making.

(Cohen *et al.* 1979, 64–5)

While the advent of telematics in the 1980s freed up routine data transcription and telemarketing to locate beyond the city centre, cities with high-quality, low-priced rapid transit systems (like Paris, Frankfurt, Zurich, Hong Kong, Singapore, Vienna, Berlin and maybe London) are still very competitive office locations. Here, firms are right at the heart of the most concentrated pool of clients, high-level skills and expertise, and suppliers in any metropolitan labour market. For these reasons, some central cities in North America and Europe have managed to retain a greater share of the global and regional head offices specialising in high-level traded services than car-dependent American cities like Los Angeles, Houston or even New York.

Cheshire (1995) reports a doubling in the number of urban cores experiencing growth among those cities in northern Europe larger than 330,000 – up from 22 per cent in the second half of the 1970s to 47 per cent at the end of the 1980s. This suggests that the conditions were in place to sustain at least a degree of redevelopment at the heart of cities suffering from inner-city decline. This is likely to continue in Europe with the advent of a single currency and the dismantling of borders. As investment opportunities in the capital city office sectors begin to dry up, property investors are beginning to look to central office locations in second-order cities like Lyons, Milan, Düsseldorf, Rotterdam, Munich, Barcelona and Birmingham as alternatives (*Australian*, 9 June 2000, 38).

While the conditions for some recentralisation might be helping to arrest the 'hollowing-out' of the urban core of troubled cities across North America and Europe, there is no single underlying factor at work. As world trade and global financial deals multiplied during the 1980s, the languishing fortunes of cities like London and New York were temporarily revived. 'Although other European and American cities added larger amounts of space in percentage terms at the peak of the development boom in the late eighties, none equalled the increment to London and New York in absolute numbers' (Fainstein 1994, 34). London added nearly 30 per cent to its office supply and New York more than 20 per cent during the 1980s. Nevertheless, while impressive in absolute terms, both London and New York are steadily losing their share of producer services (Markusen and Gwiasda 1994; Turok and Edge 1999).

In New York City's case, as with so many other ailing North American cities, though unlike central London, the initiative for promoting redevelopment activity came from a local growth coalition of business leaders and governmental officials (see Chapter 9). As well as participating in a range of projects like the Javits Convention Centre and Battery Park City, where public authorities like the Urban Development Corporation took the lead, private developers also took advantage of tax-subsidy programmes to redevelop whole city blocks. For example, the Grand Hyatt and Times Square Marriott hotels were among the first of Donald J. Trump's New York ventures.

Urban administrations throughout Europe also instituted bold regeneration schemes to make market conditions in their city centres more attractive for property investors in the 1980s and 1990s. The list includes cities like Berlin, Birmingham, Amsterdam, The Hague, Rotterdam, Barcelona, Munich and Leipzig. So far as London was concerned, it was 'the urgings of the national government, [incorporating] Margaret Thatcher's views that private investors operating in a free market would create local economic growth [that] opened up London's once highly regulated property development arena for speculative ventures' (Fainstein 1994, 37–8). Furthermore, London's competitive advantage in the British urban system extends beyond purely economic determinants to its political and cultural dominance of national affairs.

Yet, at the very centre of these cities, the hold of the prime sites has loosened somewhat with investment banks and stockbrokers commissioning a new generation of 'smart offices' – buildings with large, clear floors for trading, high floor-to-ceiling heights to accommodate a raised floor, and high-performance mechanical, electrical and telecommunications services. In New York City, during the 1980s building spree, financial services spread out from the Wall Street canyon at the bottom of the island, to the once pre-eminent World Financial Center on the Hudson River and to Midtown Manhattan (Fig. 4.7). In much the same way, commercial property developers are beginning to convert old downtown warehouse buildings into 'telecommunications hotels'. These buildings are ideal for 'web-hosting', co-location or 'server farms' because they offer the centralised locations close to fibre-optics, massive and redundant sources of power, large floor plates, heavy floors and high ceilings that telecommunications companies need. On Manhattan, the block-long 111 Eighth Avenue building in Chelsea, and 75 Broad Street, a 67,500 square metre building, provide examples. In NYC, Atlanta, Los Angeles, Boston, Chicago and Dallas, the opening of these telecommunications hotels and data centres is bringing new business and jobs to formerly marginal neighbourhoods.

Likewise, in the City of London, financial services have moved to the Broadgate office precinct, spread along the banks of the Thames to London Bridge

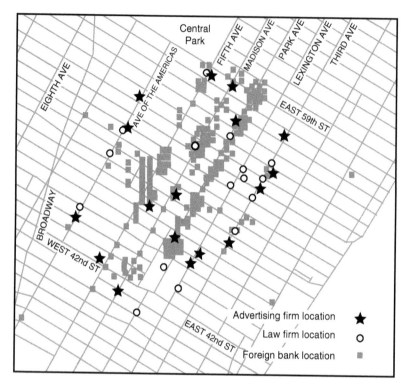

FIGURE 4.7 *The concentration of banks, business and financial services, law firms and advertising agencies in the mid-town district of Manhattan, New York*
Source: composite of figures in Moss (1991)

and Charing Cross, and out to the Docklands in East London. A feature article in the *Independent* (19 December 1990, 17) on London's office boom in the mid-1980s described Broadgate in the following terms:

Colossal air-conditioned and computerised trading floors slot into a steel frame masquerading behind heavy granite cladding. The building's construction and its lighter side – high quality art-works, bars and restaurants, and a public ice-rink at Christmas – cannot be faulted. The baroque design gives way to a daring-engineering-style *Bladerunner* block nearing completion on Worship Street.

With the exception of the building programme in Berlin, the redevelopment of the London Docklands dwarfs any of the other regeneration schemes in Europe. Despite the setback suffered in the late 1980s by the backers of the Canary Wharf office complex (Fig. 4.8), if anything, the shift of office activity from the City into the Docklands has gathered pace during the 1990s. By the beginning of 2000, 529,533 square metres of office floor space had been built and leased, and another 230,000 square metres was due for completion by first-quarter 2001, while a further 260,000 square metres was proposed in mid-2000.

The expansion of office employment in conjunction with the emergence of global cities has had a 'knock-on' effect in the residential sector. Between 1981 and 1987, the housing stock expanded by 110,000 and 50,000 dwelling units in Greater London and New York City respectively. This represented the first time in 25 years that NYC's housing stock had grown continuously for six consecutive years (Fainstein 1994, 35–6). Much of the residential property development in inner-city neighbourhoods close to downtown offices has taken the form of building conversions targeted at the gentrification submarket (see Chapter 6). In inner London boroughs like Camden, Kensington and Westminster, this has involved flat break-up and the inevitable displacement of tenants on a grand scale (Hamnett and Randolph 1988). While on the other side of the City, upmarket housing has appeared along with the office jobs created in the Docklands (Fig. 4.8).

As well as the residential conversions that have taken place in gentrifying neighbourhoods on Manhattan's Lower East Side and on the Brooklyn foreshore, many of the lofts in the garment-making district of Chelsea have been lost to the real-estate developers (Zukin 1995).

FIGURE 4.8 *The Canada Building, No. 1 Canary Wharf, under construction during the 1980s*
Source: London Docklands Development Corporation

San Francisco's Northeast Mission district

The whole area immediately east of San Francisco's CBD out to the Bay is bound to be completely transformed over the next 10 to 20 years by a combination of developments that are now in train. The redevelopment of the Northeast Mission district as part of this transformation reflects the locational requirements of the users that are now competing for space in this former industrial backwater (Fig. 4.9). San Francisco's Northeast Mission district was once a bustling neighbourhood of large manufacturing and warehousing operations, laced with active rail spurs (Cohen 1998). The only industrial uses that remain are cleaner light industries, wholesaling and small 'niche' manufacturers. Towards the end of the 1990s they were joined by an increasingly varied mix of commercial uses such as business services, restaurants and retail establishments, artistic and 'multimedia' enterprises, 'live/work' and loft residences.

With the extension of San Francisco's Muni light-rail into the neighbourhood south of Market, the 3000-acre Mission Bay redevelopment and the new Giants ballpark going up just to the north, the 5000 remaining blue-collar jobs were threatened. Since 1990, more than 1400 lofts have been built in the City of San Francisco, and another 1534 were in the pipeline awaiting approval in mid-1999. From 1990 to 1997, live/work lofts accounted for 13 per cent of housing development in the city. Conversions of old industrial buildings into live/work spaces and lofts were originally exempted from zoning restrictions in the 1980s as a way of promoting inexpensive housing for working artists in San Francisco. But, as with loft conversion in the neighbourhoods of

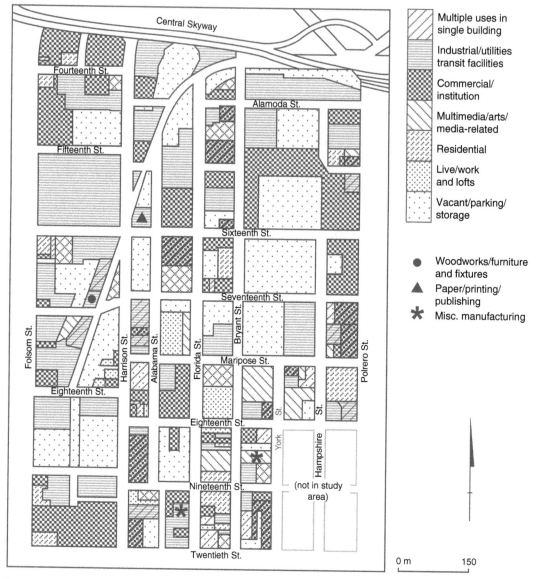

FIGURE 4.9 *Transformation of the NorthEast Mission District on the eastern edge of San Francisco's CBD*
Source: based on Cohen (1998)

lower Manhattan, this has steadily been overtaken by buyers working in software development, multimedia and knowledge production. Now these live/ work spaces are caustically referred to as 'lawyer lofts' (Cohen 1998, 11). Furthermore, the Planning Commission has pushed ahead with plans to revitalise the area, providing for the conversion or construction of about 8000 dwellings by 2020.

CONCLUSION

This chapter has considered how the location of economic activity in mainly western cities has changed along with developments in technology over the last 200 years or so. By taking a sweep from the beginnings of the nineteenth century onwards we begin to grasp how property markets, together with the gradual introduction of town planning and a regulatory framework, shape the economic geography of cities. In the earliest manufacturing centres of Europe and the New World, factory owners and, later, residential zoning determined where the workforce would live in cities.

The location of economic activities in cities is an urban labour market in another guise. Where firms locate in cities determines the distribution of job opportunities available to households in relation to where they live. In the next chapter, which deals with the work performed in cities, we look more closely at how labour markets are restructured and how that impacts upon urban households. Particular attention will be paid to spatial aspects of urban labour markets.

On the other hand, where households live in cities is largely determined by what they earn and the access that provides to housing in a given price range. As well as discussing the implications of the linkages between labour markets and housing markets, Chapter 6 goes on to describe residential patterns and household moving behaviour in cities. These two chapters lay the foundations for the social geography of cities with its demographic, gendered and cultural nuances. We will pick up on these aspects of urban life later on, in Chapter 7.

FURTHER READING

The study of how land uses re-formed in Chicago following the Great Fire of 1873 remains the best treatment of the general principles underlying the location of economic activities in cities (**Fales and Moses** 1972).

Chapters 4 and 5 of *Unfairly Structured Cities* (**Badcock** 1984) compress a lot of empirical material relating to the shaping of the economic geography of nineteenth- and twentieth-century cities in North America, Europe and Australia.

The account is extended in *The Restless Urban Landscape* (**Knox** 1993) and *Atop the Urban Hierarchy* (**Beauregard** 1989), both of which deal with aspects of economic and social restructuring in North American cities late in the twentieth century.

For further material dealing with the spread of economic activity beyond the suburbs to the edge of large American metropolitan areas, see **Muller** (1997) and **Garreau** (1991).

Postmetropolis (**Soja** 2000) also describes how the restructuring of the Los Angeles economy changed the geography of the region in the 1990s.

5

WORK PERFORMED IN CITIES

INTRODUCTION

The wide variation in the industrial structure of cities noted in previous chapters shapes the urban labour market, and ultimately the amount, type and quality of employment available for people wanting to work. However, there are some terms that need defining before we launch into a full-scale discussion of employment structure and labour markets in cities.

Commonly used terms

Firstly, economists refer to people in paid employment, including those who are seeking jobs but are temporarily out of work, as the 'economically active' population. They may work in the regulated, formal sector of the economy or, as in many Third World cities, in a multifaceted informal sector that often manages to avoid even minimal labour regulations. In Third World economies, the informal sector has played a vital role in absorbing the new migrants or refugees flooding into the cities from the countryside. Usually people in the age range 15–64 years qualify as economically active. Reference is also made to male and female 'workforce participation rates' to indicate the proportion of the eligible population that is working for wages. Housework goes unpaid and counts for nothing in the national accounts of countries adopting the UN System of National Accounts (Waring 1988), just as in some Third World cities the scourge of child labour still exists. It follows that the 'economically

inactive' are those people considered too young or too old to be in the paid workforce, or otherwise unable to work for wages. Mothers rearing young children fall into this category. Feminists argue that, along with growing and preparing food, running a household and educating children, this 'reproductive work' should be given its rightful place in the economic system. The ratio between the economically active and inactive components of the population is called the 'dependency ratio'.

Secondly, from time to time we will switch between 'classes of industry' and 'types of occupation' in discussing the employment structure of cities. Table 5.1 lists a whole range of industrial sectors in which people are employed. But to state that you work in the 'air transportation' industry, for example, gives no idea of the range of occupations that exists in that sector. It ranges from senior pilots and aeronautic engineers, through hostesses and flight controllers, to baggage handlers and aircraft cleaners. Clearly the qualifications and skills required, and hence the remuneration and conditions, of these people are vastly different.

Thirdly, economists identify three main forms of unemployment. First, 'structural' unemployment is caused by restructuring taking place within a sector of the economy, or within particular industries. The closure of coal pits and ship-building yards throughout western Europe, and the loss of jobs in the financial services sector following the introduction of computerised data processing and ATM banking, represent two forms of structurally induced unemployment. Typically, structural unemployment is longer term because of the falling demand for the skills acquired by workers in industries overtaken by new technology (Forrester 1999).

This differs from 'cyclical' unemployment caused by the lay-off of workers when demand for a company's products or services slumps during a cyclical downturn. The home-building industry, for instance, is very susceptible to both economic and property cycles. Finally, there is 'frictional' unemployment, which arises, say, when a big building project is completed and a contractor does not have another one coming on stream and has to lay off workers temporarily. Another important source of frictional unemployment is seasonal work like harvesting and fruit picking.

Fourthly, economists distinguish between primary and secondary labour markets. The 'primary labour market' supports core employment, comprising full-time, permanent jobs with good remuneration and working conditions, and maybe employer-funded pension and health insurance schemes. Increasingly, with the move to make employment conditions more flexible, there has been a growth of 'secondary labour markets' especially in countries like the United States, the UK, New Zealand and Australia. Census data show that Australian males still monopolise the permanent, full-time jobs in the primary labour market, with roughly half the female workforce relegated to part-time positions. More and more of the jobs being created are contract-based, casual or part-time. In the UK, for instance, between 1992/3 and 1995/6, only 9 per cent of the 750,000 jobs created were permanent and full-time.

Fifthly, the breakdown of industry and occupational groups by age, gender and ethnic differences contributes to the 'segmentation of urban labour markets'. For example, by the end of the nineteenth century in Britain about 1.5 million women – by far the largest group – worked as domestic servants. By 1991 the figure for women working as cleaners and housekeepers had fallen to 80,000, but women still remain grossly over-represented in clerical, sales and service occupations (like nursing and teaching), while men dominate in management and administration, trades and labouring. Moreover, even though women are breaking into male-dominated occupations like the professions, this 'gender division of labour' seems likely to persist, given the nature of many of the new jobs being created in the burgeoning service industries. By the mid-1990s, services accounted for some 70 per cent of the aggregate GDP and 65 per cent of

total employment for OECD countries (reported in the OECD *Observer*, December 1999). As the labour force participation rate of women continues to climb, some commentators are now calling this phenomenon the 'feminisation' of work.

Sixthly, labour markets in cities are constantly being changed by workers moving to new job opportunities. Labour mobility takes two main forms: 'intergenerational mobility' arises when sons and daughters occupy higher-status labour market positions than their parents occupied (or the converse); 'inter-regional mobility' involves the migration of workers between regions and cities, and between local labour markets within urban regions. Labour mobility is selective by nature. For example, Fielding (1992) has shown that the south-east of England operates as an 'escalator' region by recruiting workers with higher skills from the rest of England and Wales into professional and managerial positions in the region.

A seventh and final set of terms derives from the work that labour economists and economic geographers have done on the geography of urban labour markets. The daily commuting patterns of workers help to define 'labour sheds' and 'employment fields' in cities. An urban geographer by the name of Vance (1960) first recognised the need to discriminate between the residential catchments that employers at a given location in a city draw upon for their workers – the labour shed – and the scattered job opportunities in a city that individual workers access from their home base – the employment field. Hence the labour shed depicts the potential supply of labour within reach of an employment node – that is, from the employer's perspective – while the employment field depicts what is on offer and within commuting range of home – that is, from the worker's perspective.

The basic idea behind the labour shed is that catchment size is defined by the distances that potential workers can afford to travel to work. Clearly the pay rates of higher-status occupations enable commuters to recover the cost of travelling further to work, whereas there is a definite financial limit to the distance that unskilled workers can commute to poorly paid jobs. Beyond a certain point it makes no economic sense, which is the reality that confronts many young unemployed people and suburban housewives. In effect, they are spatially trapped with respect to distant jobs that they are otherwise qualified to compete for in a metropolitan catchment area.

As Morrison (2001) points out, the labour shed is a concept that was developed during an era of 'highly centralised employment that prevailed in much of the OECD prior to World War II rather than the highly decentralised patterns that we have seen evolve' in conjunction with suburbanisation and peri-metropolitan development. In some urban regions, therefore, restructuring has reduced the options for many women, youths and minorities who are hampered in re-entering the labour market by low levels of car ownership, or inadequate public transport (Cervero et al. 1999; Clark and Kuijpers-Linde 1994). The problem manifests itself geographically as one of 'spatial mismatch' between workers and jobs within the urban labour market. Aspects of spatial mismatch in cities are examined towards the end of this chapter.

Doreen Massey (1984) has shown with her work on the changing spatial division of labour that changes in local employment opportunities can only be fully understood by looking well beyond the workplace. Changes to labour markets in cities are the product of:

- broader economic forces (globalisation, new technologies, labour displacement, structural unemployment)
- political measures (deregulation, equal opportunities and child-care legislation, compulsory retirement, quotas on immigration)
- demographic trends (migration, ageing)
- changing social norms towards work (child-free households, working mothers, leisure)
- improvements to the urban transport system.

The aim of this chapter, then, is to begin to make sense of urban labour markets and the work performed in cities in the light of the profound economic, political and societal processes that are part and parcel of global change.

FORCES RESTRUCTURING EMPLOYMENT IN CITIES

The labour markets of cities around the world are caught up in an economic tsunami that is as tumultuous as the Industrial Revolution that overtook Europe in the middle of the nineteenth century.

In terms of its impact, this transformation of labour markets runs the gamut from the job shedding that Forrester (1999) rails against, to the new jobs in the digital economy about which economic journalists are prone to wax lyrical. Whilst the US economy is pioneering this present transformation in capitalism, economic niches that articulate with other parts of the global economy are also springing forth in the developing world. Cities like São Paulo, Buenos Aires, Taipei, Bangkok, Shanghai, Mexico City, Bombay and New Delhi all possess these niches with their recognisably internationalised segments of the labour market.

This is not necessarily to imply that these processes will lead to convergence on a global scale, although it sometimes seems this way given the hegemony that organisations like the World Bank and the IMF exercise in dictating policy settings to suppliant nations. In addressing unemployment, for instance, Anglo-Saxon governments have vigorously pursued a course of labour market deregulation coupled with more precise targeting of welfare beneficiaries over the last two decades. Many European governments, on the other hand, have opted for maintaining wage rates and decent social provision for those out of work.

In the series of points that follow I will briefly note how some of the key developments in the global economy, together with the policy responses of governments, have impacted upon the labour markets of cities in advanced economies since the early 1970s.

Global competition and productivity

The exploitation of cheaper labour in the developing economies by transnational corporations engaged in labour-intensive manufacturing and processing paved the way for a New International Division of Labour (NIDL) in the 1970s (Fröebel et al. 1980). The NIDL coincided with the lifting of oil prices by the Organization for Petroleum Exporting Countries (OPEC) in 1973–4 and again in 1979. As a result of the confluence of these two events, the traditional manufacturing sectors in Europe, North America and Japan lost their competitive edge and many firms moved to lower-cost regions in the developing economies to stay in business. Job-shedding did not cease with manufacturing in the 1970s and 1980s, but continued into the 1990s with significant jobs losses in public administration and finance, and now in telecommunications. With the onset of recession in 1991–2, the shake-out in banking and financial services reduced the City of London's workforce by an estimated 30,000 jobs (McDowell 1997). Over a ten-year period jobs in Australia's financial services sector rose from 302,000 in 1986 to a peak of 352,000 in 1991, and then fell to 321,000 in 1996 along with the introduction of electronic banking.

The short-term advantage that the USA, Japan and Singapore had at the dawn of the 'information age' is rapidly being eroded as the globalisation of services begins to quicken in pace. Indian cities, with an abundant supply of English-speaking graduates desperate for work, are rapidly becoming magnets for service jobs ranging from data processing to high-end engineering and software development. According to a report in the *Washington Post* (June 2000), McKinsey & Co predicted that India's remote-service industry could employ 700,000 workers by 2008. McKinsey, a leading international management consultant, operates its own research centre in New Delhi and a graphic arts centre in Madras. Both are staffed by computer experts and librarians who provide 24-hour back-up services to the company's global network of offices. From a New Delhi office, San Francisco's Betchel Group employs 400 engineers handling projects around the world. Ford employs accountants to work for its Asian outlets. Pfizer, the pharmaceutical giant, is using Indian laboratories to conduct trials on drugs. New Delhi is also the base for data and account processing operations for British Airways, Decision Support International, General Electric and American Express. Significantly, the export of these newly created, service-sector jobs occurred at a time when unemployment in the United States was less than 4 per cent, at the end of the 1990s.

Structural shift from manufacturing to services

'An important new development from the perspective of the urban economy is the growing demand for services by firms in all industries and the fact that cities are the preferred production sites for such services, whether at the global, national, or regional level' (Sassen 1997, 131). Between 1991 and 1996, the fastest job growth in the United States occurred in the business services sector (6.7 per cent per annum), with the creation of an additional 7 million

jobs, followed by leisure (6.2 per cent per annum), non-banking financial institutions (5.7 per cent per annum) and social services (5.6 per cent per annum). But the greatest number of overall employment opportunities was created within the state and local government sector (16.5 million jobs), health services (9.5 million jobs), and eating and drinking places (7.5 million jobs) (Table 5.1).

Significantly, by 1995 over 11 million people were employed in 'strategic business services' within OECD countries, or more than twice the number employed in the OECD's entire motor vehicle industry. These are the activities that are no longer considered to be the core business of a company, so they are outsourced to firms that have the specialist expertise. They fall into five main types: computer software and information-processing services; R&D and technical testing services; marketing services; business organisation and management consultancy; and human resource development and labour recruitment services. Firms like the Boston Consulting Group, Cap Gemini Sogeti, Pricewaterhouse, and the SAP and WPP Group have flourished by establishing an international presence alongside major manufacturing and trading clients.

The shift from manufacturing to services is also altering the pattern of earnings within OECD countries with the pay rates of non-professional service workers lagging behind the pay of skilled manufacturing workers. In the second half of the 1980s, the earnings of unskilled and skilled service workers in the USA were half as much and three-quarters as much, respectively, of skilled manual workers. Because of the unions' commitment to equity, higher minimum wages and the value that the respective societies place upon their public servants, the comparable figures were 89 per cent and 88 per cent for Germany and 89 per cent and 98 per cent for Sweden (Esping-Andersen 1990).

Table 5.1 illustrates some of these emerging disparities in the earnings of workers employed in different sectors of the US economy. For example, in 1993 average incomes in the primary metals industries, paper products, instruments, chemicals, and petroleum and coal ranged from US$42,000 to about US$70,000. In a variety of service sectors, however, average incomes started as low as US$12,000 in 'eating and drinking places' and US$15,000 in 'social services', where casual and part-time jobs are the norm, and rose to US$61,000 for legal services

TABLE 5.1 *Winners and losers by industry group, US workforce change 1991–1996*

Industrial group	Total jobs in 1996 first quarter (000)	Average annual growth rate (%)	Average annual earnings in 1993 (US$)
Business services	7009	6.7	22,499
Leisure	1505	6.2	21,018
Non-banking financial institutions	496	5.7	n.a.
Social services	2370	5.6	15,320
Brokerage	531	4.9	96,497
Local transit	439	4.7	20,496
Transportation services	430	4.4	31,617
Motion pictures	516	4.4	31,692
Agricultural services	599	4.2	n.a.
Museum and zoos	83	4.0	19,514
Auto repair and parking	1059	3.6	20,430
Furniture stores	950	3.4	21,208
Building materials stores	883	3.3	22,914
Health services	9463	3.3	34,200
Trucking and warehousing	1879	3.2	27,289
Engineering and management	2849	3.1	33,709
Special trade contractors	3334	3.1	26,443
Education	1982	3.0	20,088
Eating and drinking places	7419	2.8	11,920
Miscellaneous services	44	2.5	n.a.
Auto dealers and service stations	2234	2.2	25,433
Rubber and plastics	962	2.2	33,103
Air transportation	828	2.2	43,093
Lumber products	754	2.1	27,713
State and local government	16,584	1.5	28,859
Legal services	926	0.3	61,224
Stone, clay and glass	535	0.2	33,566
Food products	1674	0.1	32,369
Paper products	682	0.1	42,178
Printing and publishing	1532	−0.3	32,515
Primary metal industries	708	−0.7	47,020
Textile mills	642	−0.8	24,897
Chemicals	1026	−0.9	56,289
Apparel and accessory stores	1100	−1.1	13,971
Utility services	905	−1.2	55,722
Federal government	2781	−1.2	n.a.
Transportation equipment	1747	−1.2	n.a.
Banks and savings institutions	2022	−1.7	35,252
Metal mining	51	−2.3	56,964
Petroleum and coal	140	−2.4	67,996
Railroads	234	−2.7	55,707
Apparel and textiles	868	−2.8	19,225
Tobacco	41	−3.4	55,983

TABLE 5.1 *(continued)*

Industrial group	Total jobs in 1996 first quarter (000)	Average annual growth rate (%)	Average annual earnings in 1993 (US$)
Instruments	831	−3.4	45,795
Leather	99	−4.8	22,661
Oil and gas extraction	312	−5.0	36,011
Pipelines	14	−5.9	54,011
Coalmining	101	−6.4	62,044

Source: US Bureau of Labor Statistics (1996), data published in the *New York Times*, 21 July 1996, Section 3

and US$96,000 for brokerage. While average income for business services is an unimpressive US$22,500, it must be remembered that this industry group includes everything from accountants and consultants to data-entry personnel and stock clerks.

Deregulation, privatisation and subcontracting

The priority given by OECD governments to creating more flexible labour markets and reducing the size of the public sector has actively contributed to the changing employment structure of cities over the last two decades. There has been a significant shift by employers in favour of engaging workers on a casual or contract basis. Known as 'casualisation' (Sassen 1997, 137–40), this transformation within labour markets contains the seeds of the 'new urban poverty' (Marcuse 1996). It appears that, notwithstanding the considerable difference between the regulatory regimes of countries like the UK, Canada, Australia and New Zealand, on the one hand, and Europe's social democracies, on the other, there has been an almost universal trend within the OECD to casual and part-time work in services. For example, even though the unemployment rate in the Netherlands was as low as 3.2 per cent in mid-1999, well over half of all the jobs created between 1983 and 1997 were part-time and mostly in the service sector (reported in the OECD *Observer* 1999).

The privatisation and contracting-out of parts of the public economy have shifted many jobs in the service sector 'from being full-time, year-round regulated government jobs with fringe benefits to part-time or temporary jobs in subcontracting firms with

no fringe benefits and lacking the regulatory protection of the state' (Sassen 1997, 146). The scale of privatisation varies internationally, with New Zealand leading the way in the 1990s when measured as a proportion of GDP (14.5 per cent), followed by Britain (11 per cent), and the Australian Commonwealth and States (7–8 per cent). The labour markets hardest hit by the pruning of public-service jobs in the 1990s were capital cities like Washington DC, Ottawa, Canberra and Wellington. Tens of thousands of jobs were shed with the knock-on effects felt in all sectors of these capital cities' economies.

Immigration policy and immigrant labour

Because of labour shortages due to the ageing of their populations, in the 1960s governments in countries like France, Germany, Italy, the Netherlands and Japan began to recruit 'guest workers' or contract labourers from overseas to perform the less desirable work in industry and the cities. But, with mounting unemployment in the 1980s, many European Union countries began to restrict legal immigration. By the late 1990s, the foreign-born population represented the following proportions of the populations of the countries listed: Germany (9 per cent); France (6.3 per cent); Sweden (6 per cent); the Netherlands (4.4 per cent); Britain (3.6 per cent); United States (4.7 per cent); Australia (23.6 per cent).

Germany was still admitting 615,000 foreign-born immigrants in 1997. With an estimated shortage of 75,000 workers with programming skills by 2003, Chancellor Schröder planned to grant work visas to

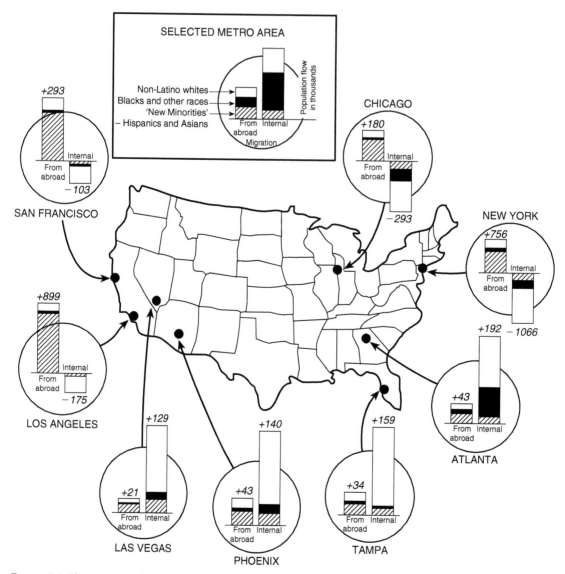

126 MAKING SENSE OF CITIES

FIGURE 5.1 *The movement of recently arrived ethnic groups between metropolitan areas in the United States, 1985–1990*
Source: Frey (1995). Reproduced by permission of Taylor & Francis Ltd

20,000 computer technicians from India and eastern Europe (reported in *Guardian Weekly*, 30 March–5 April 2000, 29). In the same year, 2000, the United States admitted about 800,000 immigrants. The main 'port of entry' cities for these new arrivals are New York, Los Angeles, San Francisco, Chicago and Miami (Frey 1995). Figure 5.1 charts the subsequent destinations of these immigrants from abroad, along with the internal flows of the main ethnic groups in the United States, between 1985 and 1990.

The presence of immigrants is helping to change the demography, the urban labour markets, and the ethnic and cultural diversity of large cities in these countries and others like them (see Chapter 7). The impact upon urban labour markets and the budget of cities is the object of considerable debate in the advanced economies. It centres on issues of job displacement, the drain on welfare and the role of immigrants in the informal economy of cities. While the informal sector has always been a significant employer of migrant labour in Third World cities, 'informalisation' is increasingly prevalent in First World cities receiving immigrants who cannot obtain jobs in the formal labour market. This expanding informal sector in places like the East End of London, the lofts of Manhattan and east-central Los Angeles frequently exploits (sometimes illegal) immigrants in sweatshops or via home-based 'outwork'. Over one in three immigrants landing in the United States after 1989 was poverty stricken. The poverty rate among Vietnamese, Cuban and Mexican arrivals during the first half of the 1990s averaged an estimated 80 per cent, 69 per cent and 49 per cent respectively.

A National Academy of Sciences study was commissioned in the mid-1990s to investigate the impact of immigration on the labour market. It concluded that the wages of native-born US workers without high-school diplomas fell by 5 per cent over a 15-year period (1980–95) because of the competition from immigrants (reported in the *Investor's Business Daily*, June 1997, B2). To be more specific, 44 per cent of the fall in wages for high-school 'dropouts' was attributed to the impact of immigration. On the issue of government outlays, the study reports that over a lifetime the average immigrant will consume about US$25,000 more in local and state services than they pay for in taxes. But at the federal level, they will pay US$105,000 more in taxes than the services consumed over a lifetime.

On the economic role of immigrants, Sassen (1997, 140–2) suggests that 'informalisation' owes its existence in First World cities not so much to immigrant survival strategies, as is the case in Third World cities, as to the growing dispersion of earnings in the mainstream economy and the way that this is impacting, in turn, upon prevailing consumption patterns. The highly paid are increasingly attracted to customised, rather than mass-produced goods and services, speciality items, expensive wardrobes and fine food. Small, full-service outlets relying on subcontractors and home-based suppliers have proliferated to provide an alternative to chainstores and supermarkets. But by far and away the greatest demand for the cheaper goods and services produced in the informal sector comes from poor households. Given their preparedness to work longer hours for lower wages, recent immigrants constitute a significant informal-sector presence in many cities – as both proprietors and hired labour.

Subcontractors and sweatshops have multiplied to meet this growing demand from both ends of the income scale. For example, according to Labor Department estimates, by the end of the 1990s there were between 25,000 and 30,000 small factories cutting and sewing garments in cities across the United States – 'many of them loft or basement operations that can disappear overnight and reappear elsewhere the next morning' (Reich 1998, 276). On his rounds as Secretary of Labor, Robert Reich got to meet some of the people in sweatshops uncovered by his staff. A 35-year-old Chinese woman, who lived in Brooklyn with her husband, parents and three children, and needed the work, rode the subway into Manhattan to a Seventh Avenue garment shop:

She worked sixty hours a week, earning $2.50 an hour with overtime pay. She says forty other women were crowded into the same small space on the top floor, with dim lighting and no fire exits. For three months before the shop suddenly closed, they weren't paid at all.

(Reich 1998, 277)

Changes to welfare and income-support systems

The technological displacement of manufacturing and service workers produced unemployment levels within advanced economies that had not been approached since the Great Depression. In mid-1999 there were still 35 million unemployed workers throughout the OECD, representing about 7 per cent of the total OECD labour force. On average, the jobless rate of the European members stood at 10 per cent compared with only 4 per cent for the United States (reported in *Guardian Weekly*, 17 July 1999). The magnitude of this gap reflects a fundamental difference in approach to welfare between these European countries and the United States. In the 1980s there was a systemic shift in emphasis from poverty reduction to cost reduction – though it was never taken as far by the social-democratic governments of Europe.

In the initial phase of unemployment, Luxembourg, the Netherlands, Switzerland, Sweden, Finland and the Czech Republic provide a package of unemployment benefits, family benefits and after-tax housing benefits to ensure that a married couple with two children continue to receive at least 80 per cent of their lost earnings (reported in the OECD *Observer*, December 1999, 60). In subsequent research designed to investigate whether this 'feather-bedding' of workers is behind Europe's higher unemployment rates, the OECD concluded that 'employment protection legislation strictness has little or no effect on overall unemployment'.

In 2000, France embarked on a radically different approach to lowering unemployment. One year after the introduction of a 35-hour working week for enterprises employing more than 20 workers, the unemployment rate had fallen, economic growth was steady and the workforce seemed happy with the extra holidays. The measures helped to make it the best job-creation year in France for a century (see http://www.frenchlaw.com/worktime.htm).

By contrast, a number of countries, including the United States, the UK, Australia, New Zealand and Canada, adopted a more punitive approach to cutting back unemployment and welfare dependency, which is now known as 'workfare'. However, the price for greater labour market flexibility, cheaper labour and getting more people 'off welfare' is the growing inequality in income and wealth that is now such a visible feature in cities where workfare has been adopted (Marcuse and van Kempen 2000). Following the passing of the Personal Responsibility and Work Opportunity Act 1996 in the United States, welfare rolls were been cut from 14 million to 9 million by forcing people back into badly paid jobs in a flexible labour market. For example, the city of San Francisco pays street sweepers working for 'the dole' a third of the union rates and they face the prospect of having their benefits docked for 30 days if they are 10 minutes late for their 6.30 a.m. start.

The United States has introduced some of the harshest tests for payment of social security benefits and income support. Male unemployment insurance lasts only for 26 weeks, so the choice is between jobs that many families can barely survive on or no income whatsoever. Tax rebates for low-income workers – the Earned Income Tax Credit – are meant to offset minimum wage rates, but young, single working mothers are seldom much better off once they have paid child-care and transit fares. By 2000, 20 states required mothers with children under a year old to work. On each occasion, the overhauling of the benefit system produced sizeable knock-on effects that rippled through local labour markets in every big city. Under President Clinton and Prime Minister Blair, voters traded off a high rate of unemployment for the 'new urban poverty' that is now endemic to many of the largest American and British cities (Marcuse 1996).

The buoyant US economy created an additional 3.7 million jobs for central city residents between 1992 and 1998. As a result, a number of central cities reported dramatic falls in their unemployment rates: Detroit, from 16.9 per cent to 7.2 per cent during the period; Atlanta, from 10 per cent to 5.6 per cent; Newark, from 16.6 per cent to 9.6 per cent; Hartford, from 12.6 per cent to 6.7 per cent; Santa Ana, from 11.8 per cent to 5.2 per cent. Yet unemployment rates still hovered around 10 per cent in cities like North Chicago, Flint, Miami, East St Louis, Atlantic City, Brownsville and Newark.

In 1998 the official poverty rate in the USA was 12.7 per cent, which was down from the peak of 15.3 per cent in 1993, but it still exceeded the rate in any year during the 1970s. Over the same period, poverty in the nation's central cities declined from 21.5 per cent to 18.5 per cent. But in one-third, or 170 central cities, poverty rates still exceeded 20 per cent. Cities with stubbornly entrenched poverty at the time included: Washington DC (21 per cent); New Orleans (34 per cent); St Louis (30 per cent); Philadelphia (24 per cent); Richmond (25 per cent); Newark (31 per cent); Hartford (35 per cent); Miami (43 per cent). So, despite the overall drop in poverty in the United States during the 1990s, it remains more rather than less concentrated in the big cities: 43 per cent of the poor currently live in inner-city neighbourhoods and 'are younger and more urban than they were a generation ago' (*The Economist*, 20 May 2000, 28).

Some of the unintended consequences of the changes to the US welfare system were apparent by the end of the 1990s in New York City's labour market. According to a social statistician reported in the *Weekend Australian* (16–17 August 1999, 19), 'This economy is as good as it gets, and we still have a quarter of the city's population living in poverty.' It is well known that the incidence is highest among African-Americans, Hispanics, women and children. But surprisingly, 'Poverty rates rose sharply among just the kind of families with children that were supposed to be safe: those that include two parents, a worker and a household head with more than a high school degree' (Bernstein 2000, B1). Figure 5.2 charts the rise in the proportions of working households that succumbed to poverty in NYC between 1987–9 and 1996–8. A study undertaken by the Fiscal Policy Institute estimated that the number

PERCENTAGE OF FAMILIES LIVING BELOW THE POVERTY LINE IN WHICH
THE HEAD OF THE HOUSEHOLD...

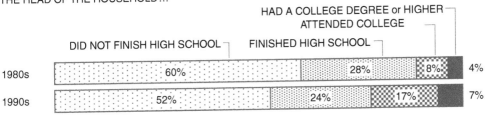

PERCENTAGE OF FAMILIES LIVING BELOW THE POVERTY LINE THAT HAD...

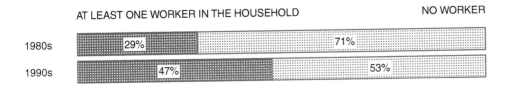

PERCENTAGE OF FAMILIES LIVING BELOW THE POVERTY LINE WHOSE
HOUSEHOLDS WERE HEADED BY...

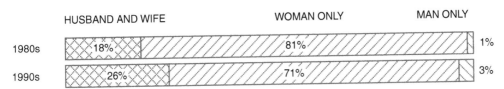

FIGURE 5.2 *The emergence of a new working poor in New York City during the 1980s and 1990s*
Source: Community Service Society of New York

of working poor families increased by 60 per cent in
New York City during the 1990s, which outstripped
the nationwide increase of 24 per cent.

Therefore a new class of 'working poor' has
emerged in those countries that have redesigned
their benefit systems around the principles of workfare. In the words of Viviane Forrester (1999),
'Neoliberalism has introduced a new economic paradigm. Increasingly, it offers the most vulnerable in
our society a quite new choice – poverty at work, or
poverty on the dole.' The nature of poverty in the
United States bears this out. While over 60 per cent
of poor families have at least one person in the
labour force some of the time, only 13 per cent of
poor adults managed to obtain full-time work in

1998 (*The Economist*, 20 May 2000, 27). Having a job
no longer guarantees escape from poverty. Reports
produced in the late 1990s by UNICEF, the children's arm of the UN, and the OECD both place the
proportion of children living in poor households in
the United States, Britain, Australia and New
Zealand at around 20 per cent. In Sweden, Norway
and Finland the rate is below 5 per cent.

FORMS OF LABOUR MARKET SEGMENTATION

Cross-cutting patterns of segmentation by occupational background, age, gender, ethnicity and space

have the potential to make for very complex labour markets in cities. Irene Bruegel's (2000) study of the restructuring of London's labour force between 1971 and 1991 reveals just how complex these dynamics can be when data on intergenerational social mobility are coupled with migration data. She concentrates her attention on a cohort of 10–25-year-old working-class 'children' growing up in the early 1970s in households where the father was a manual worker. With 36.5 per cent professional, managerial and intermediate white-collar jobs in 1991, London provides proportionally more high-status job opportunities than the country at large (30.5 per cent in 1991). Bruegel shows that both males and females from manual backgrounds growing up and staying in London did not match the occupational mobility of those moving to jobs in London over the 20-year period (1971–91). For example, by 1991 twice as many female in-migrants were in high-status jobs as London-born women. On the other hand, there was a visible 'creaming-off' effect among successful men from working-class backgrounds as they left London. By contrast, women with high-status jobs are less likely to leave London to start a family in the suburbs. Thus, singles were more numerous among people moving to jobs in London and married couples with children among those leaving.

Ageism and racism give structure to the hidden barriers that make it very hard, if not impossible, for some groups of workers to break into the job market. This is commonly the experience of middle-aged males who have been machine operators all their working lives; young, black, high-school drop-outs; skilled/qualified immigrants who are handicapped in terns of language use. The high incidence of poverty among minority groups provides a symptom of these barriers in the labour market. In 1995, 29.3 per cent of African-Americans and 30.3 per cent of Hispanics were classified as poor in the United States.

Racial discrimination appears to contribute to the relatively poor outcomes of 'London stayers' in Bruegel's study of intergenerational mobility:

While black people did better in London than elsewhere, London residence did not produce as great an advantage for black people as for whites. Given the continuing evidence of fairly systematic racial discrimination in the UK labour market, the racial composition of the group of London stayers in the cohort analysed here (approximately 12 per cent were black) goes some way to explaining the poorer outcomes of Londoners of long standing.

(Bruegel 2000, 89)

The way the urban labour market is dissected for analysis depends to some extent on the theoretical and conceptual orientation of the researcher. For the sake of simplicity, the discussion that follows assumes three main forms of labour market segmentation according to occupational status, gender and space.

Occupational division of labour

The division of labour according to educational standards and occupational qualifications and background – loosely speaking, occupational status – is without a doubt the principal form of labour market segmentation. Otherwise known as 'meritocracy', this is one of the organising structures upon which capitalist societies rest – unlike feudal or socialist societies.

Reich (1993) suggests that many of these people are 'symbolic analysts' in the post-industrial, 'digital' economy because most of their contribution to output revolves around manipulating information in a symbolic form. There are now over one million computer scientists, systems analysts and programmers in the USA, and, because top-echelon jobs in the knowledge economy command very high salaries, this is further segmenting labour markets in urban regions that are magnets for high-tech businesses.

With automation and robotics, machine tending in manufacturing and distributive activities, and data entry in public administration and finance, are now mainline (and very repetitive) tasks. Reich calls these tasks 'routine production services': 'The foot soldiers of the information economy are hordes of data processors stationed in "back offices" at computer terminals linked to world wide information banks' (Reich 1993, 175). And, while routine production work still accounts for about one-quarter of the jobs performed by Americans, 'Those who deal with metal are mostly white and male; those who deal with fabrics, circuit boards, or information are mostly black or Hispanic, and females; their supervisors, white males' (Reich 1993, 175–6).

However, the fastest-growing job sectors in a majority of OECD countries during the late twentieth century required only a modicum of training – shop assistants, nursing aids, hospital orderlies, security and prison guards, care givers in nursing homes, commercial and domestic cleaners, delivery van drivers, waiters and waitresses. Reich calls these 'in-person services' because they 'must be provided person-to-person, and thus are not sold worldwide' (Reich 1993, 176). In the United States, in-person services now comprise about one-third of all jobs. During the 1980s, well over 3 million jobs were created in fast-food outlets, bars and restaurants in the USA alone.

Hence the labour market for services is increasingly segmented with strong growth in high-paying jobs, but even greater growth in the number of poorly paid, often casual and temporary work at the bottom of the jobs ladder. As economies continue to restructure, therefore, the quality, as much as the numbers, of new jobs attracted by any city is of vital importance to the fortunes of its population. In the mid-1990s, for example, the 'Boomtown USA' phenomenon was exemplified by cities like: Las Vegas, Nevada; Austin, Texas; Boise, Idaho; Fayetteville, Arkansas; Provo, Utah; and Killeen-Temple, Texas. In particular, these cities were the magnets for growth in high-tech companies, leisure and recreational industries, and business services. All experienced annual rates of job growth in excess of 5 per cent to the first quarter of 1996 (reported in the *New York Times*, 21 July 1996, 3.10). However, the types of occupation supported by industries like electronics, computer software and medical technology, on the one hand, or resort and retirement towns, on the other, hardly bear comparison in terms of rates of pay and job security. High-tech jobs normally pay three to four times as much as jobs in tourism and aged care.

Gender division of labour

Labour markets are also segmented by gender. More formally, analysts speak about the 'gender division of labour'. The prime effects of a division of labour by gender show up in the roles traditionally attached to particular occupations by employers, and in the wages gap between men and women. Gender discrimination within the economy has its basis in yet another one of the organising structures – patriarchy – upon which capitalist societies have traditionally rested.

The steady rise in the workforce participation rates of women in the advanced economies can be linked to the shift in the demand for labour accompanying the enormous growth in service occupations. Women are now as likely to be employed in countries like Britain, the United States, Canada, Australia or Sweden, but they still do three-quarters of the work around the house. There are also countries like the Netherlands with somewhat lower rates of women in the labour force. The proportion of British males in the labour force fell from 88 per cent in 1984 to 84 per cent by 1999, while the corresponding figure for women rose from 66 per cent to 72 per cent. Data from the 1998 Australian Labour Force Survey indicate that 84 per cent of males and 66 per cent of females were in the labour force at the time.

At the beginning of the 1990s, more than half of all women in Britain and almost 60 per cent of the married women worked for wages, compared to just over a third of all women and a fifth of married women in 1951. 'The feminisation of the labour market is not, however, an undifferentiated process. Just as work is becoming increasingly differentiated as a whole in its conditions and rewards, women as waged workers are also becoming increasingly differentiated' (McDowell 1997, 17). Even though almost 3 million jobs were generated by the British economy for women between 1976 and 1990, they were disproportionately part-time and suffered from a decline in the total number of hours worked throughout the period. Likewise, while the scope for occupational mobility among well-educated women has opened up a little, the majority of employed women are still relegated to predictable occupations on the basis of stereotyping. In 1991, women in the US labour force comprised more than nine-tenths of all workers in 20 occupational categories, and more than one-quarter worked in administrative support positions. Nor are women making much progress in closing the pay gap that exists. In Britain, an Office of National Statistics report indicates that, in 1999, men's full-time earnings (£23,000) were still 42 per cent ahead of that of women working full-time (£16,000). This 30–40 per cent margin appears to hold across all occupational groups. In the case of part-time work, men are paid an hourly rate about one-third higher than that paid to women.

Of the 1.4 million new jobs created in Australia between 1980 and 1993, 60 per cent were part-time.

TABLE 5.2 *Persons employed full-time and part-time, Adelaide, 1981–1996*

Employees	Males (000)			Females (000)			Total (000)		
	1981	1996	% change	1981	1996	% change	1981	1996	% change
Part-time workers*	25.3	46.0	81.8	59.4	98.9	66.5	84.7	144.9	71.1
Full-time workers	206.0	188.2	−8.6	86.4	95.9	11.1	292.3	284.1	−2.8
(Hours worked not stated)	(8.2)	(4.3)		(8.0)	(3.6)		(16.2)	(7.9)	
Total	239.4	238.4	−0.4	153.7	198.5	29.1	393.2	436.9	11.1

Data are for Adelaide Statistical Division.
*'Part-time' is defined as fewer than 35 hours per week.
Based on ABS 1981 and 1996 censuses.
Source: Forster (1999). Reproduced by permission of Oxford University Press

Three-quarters of these part-time jobs were filled by mothers returning to the workplace after their children had entered school. Similarly, about 45 per cent of women in the UK workforce work part-time, at an hourly rate about one-third of that paid to men in similar jobs. Single mothers in the USA, who are disproportionately low-skilled, typically face a triple disadvantage: they have only one worker, then earn low wages and pay all the costs of raising their children (*The Economist*, 20 May 2000).

Table 5.2 illustrates how the shift to more flexible labour markets impacted upon a typical mid-sized city over a 15-year period at the end of the twentieth century. While the total workforce increased by just over 11 per cent between 1981 and 1996, job numbers for males shrank slightly but expanded by almost a third for females. Growth in part-time employment overwhelmed the creation of full-time work, which, in the case of males, actually declined. Two other points are worth noting: firstly, twice as many women as men were employed part-time in 1996; secondly, while the ratio between male full-time and part-time jobs is 4:1, for women it approaches 1:1 (Table 5.2). Since the job options of mothers with young children are subject to time constraints, it is not surprising that married women make up the majority of the part-time workforce.

Susan Hanson and Geraldine Pratt (1995) are two urban geographers based at Clark University in Worcester, Massachusetts. Worcester had a population of 430,000 in 1990. In the early 1990s they surveyed 698 respondents living in Worcester, of whom 526 were working at the time. At the time, occupational segregation showed no sign of losing its grip: female-dominated occupations were still filled predominantly by women (54 per cent) and male-dominated occupations by men (65 per cent). Table 5.3 makes explicit the degree of inequality that women still encounter in the labour market of a city like Worcester.

The negative effect of parenthood on the intergenerational occupation mobility of women who moved out of London rather than staying is clear from Bruegel's study:

While nearly a quarter of childless women from working-class backgrounds who lived in London throughout or who moved out of London had professional and managerial jobs in 1991, barely 10 per cent of mothers in employment had jobs at this level. The escalator effect of moving to London was particularly evident for childless women, with more than a third of these in-migrants from a manual background moving into professional and managerial jobs by 1991.

(Bruegel 2000, 87)

Spatial division of labour

The human geographer Doreen Massey (1984) suggests that there is a further form of labour market segmentation, which she calls the 'spatial division of labour'. In many older cities with Fordist regimes, economic restructuring replaced a predominantly

TABLE 5.3 *Labour market outcomes in Worcester, MA*

Earned benefits	Women	Men	Women in occupations		
			Female-dominated	Gender-integrated	Male-dominated
Average hourly wage (US$)	9.34	15.45	8.70	9.79	11.70
Median hourly wage (US$)	8.60	13.30	7.76	8.70	10.30
Percentage with health insurance benefits from own job	52.1	83.2	47.3	57.3	60.3
Percentage with retirement benefits from own job	46.9	63.7	42.5	53.2	46.7
Percentage with both health insurance and retirement benefits from own job	39.0	63.7	33.5	46.0	43.3
Average occupational prestige score	39.2	43.3	33.8	44.7	48.1
Percentage seeing possibility of promotion	43.4	55.1	36.8	49.6	56.7

Source: Hanson and Pratt (1995). Reproduced permission of Routledge

male, blue-collar workforce employed full-time on shift work in manufacturing, with female pink-collar employees working in service industries. This upturns the existing division of labour within cities by destroying certain occupations in one subregion and replacing them – or not – with others. In industrial centres across Europe, men who used to work on an assembly line or down the mines are having to adjust to childminding or being househusbands while their wives take up the paid jobs in local call centres, back offices, supermarket chains or nursing homes. In the north of England, for example, this is creating a new regional geography of employment that involves quite a different spatial division of labour.

Economic restructuring is also producing 'job poor' and 'job rich' areas in cities. Doreen Massey's work has made us aware that nationwide changes to the spatial division of labour can create very definite winners and losers with respect to jobs in local labour markets. However, this partly depends upon where workers can afford to live in relation to job opportunities and this is where the housing market comes into play (see Chapter 6). Cervero *et al.* (1999) are able to demonstrate these joint effects in their

study of the changing location of employment *vis-à-vis* housing opportunities in the San Francisco Bay Area. They compare the accessibility of jobs in 22 employment centres accounting for 47 per cent of all jobs in the region with the place of residence of workers in 100 randomly sampled census tracts in the Bay Area. According to their calculations (Table 5.4), the job accessibility of those workers in the highest-salaried positions increased the most during the 1980s. By contrast, blue-collar workers in manufacturing industries, who made up the bulk of the class defined as 'Other' in Table 5.4, saw their level of job accessibility slip the most during the 1980s. Changes to the spatial division of labour, therefore, are inherently unequal in their geographical impact. It is in this respect, then, that the property market – as a contributing source of spatial inequality in cities – is another one of the organising structures upon which all capitalist societies are based.

Three specific cases serve to show how urban labour markets are spatially segmented. Silicon Valley illustrates how the broad segmentation of the IT workforce also takes on its own unique spatial division of labour. The second general case relates to

TABLE 5.4 *Job accessibility indices by occupation, San Francisco Bay Area, 1980–1990*

Occupational class	Accessibility index[†]		Absolute change
	1980	1990	
Executive, professional and managerial	0.86	1.38	0.52
Sales, administration and clerical	0.83	0.57	−0.26
Services	−0.96	−0.82	0.14
Technical	−1.19	−1.04	0.15
Other	0.48	−0.09	−0.57

[†]Higher positive values indicate higher degrees of accessibility.
Source: Cervero *et al.* (1999). Reproduced by permission of Pion Ltd

the geographical concentration of chronic unemployment in inner cities on both sides of the Atlantic. The third instance where a segment of labour is excluded by geography from the primary labour market arises in the outer suburbs of some cities where female office workers are said to constitute 'captive workers' (Hanson and Pratt 1988).

Mismatch between jobs and housing in Silicon Valley

With a regional economy that supported over 7000 electronic and software companies, a gross domestic product of about US$49,000 per capita, and real wages growing at about five times the national average, Silicon Valley was a magnet for labour in the late 1990s. Silicon Valley is home to a sprawling population of about 2 million people residing in Santa Clara County and adjacent parts of San Mateo, Alameda and Santa Cruz. The 'Valley' has a classic segmented labour market with reportedly over 50 new millionaires a day during the heyday of dotcom start-ups, and average yearly incomes in software engineering exceeding the US$100,000 mark. If they are lucky, these people live in million-dollar homes in Palo Alto and Mountain View, not far from their work. These well-to-do suburbs have some of the best matches in the San Francisco Bay Area between residents' occupations and nearby job opportunities (Cervero *et al.* 1999, 1627). In fact, over 40 per cent of residents in executive, professional or managerial positions in the Palo Alto and Mountain View areas work in their local labour shed compared with an average of one in

three for the rest of the Bay Area. Figure 5.3 reveals that, during the 1980s, the accessibility of housing suited to the needs of these high-status workers slightly improved in the Palo Alto area.

The 70 per cent of the Silicon Valley workforce earning less than US$50,000 are typically assembly line workers in plants making computing equipment or fabricating computer chips and wafers. In 1990, 44 per cent of the Valley's workforce was female, and 38 per cent were from minorities. Up to 40 per cent of this workforce is employed part-time or on contracts, which means many do not earn enough to lift their families above the poverty line. Priced out of the costly regional housing market, their low wages force them to make long and costly daily commutes into and out of the Valley to work. Cervero *et al.* (1999, 1268) found that some of the greatest job opportunity mismatches in the San Francisco Bay Area occurred in San Mateo County's East Palo Alto/East Menlo Park, where African-American or Latino households are grossly over-represented. Again, Figure 5.3 indicates that the accessibility of housing suited to the needs of lower-status workers deteriorated the most in the San Jose area in the southern end of the Valley.

Localised unemployment in cities

Probably the most clear-cut case of 'spatial mismatch' arises when unemployed workers are left 'trapped' by the housing market in inner-city neighbourhoods where suitable job opportunities have all but dried up. Turok (1999, 909) regards this

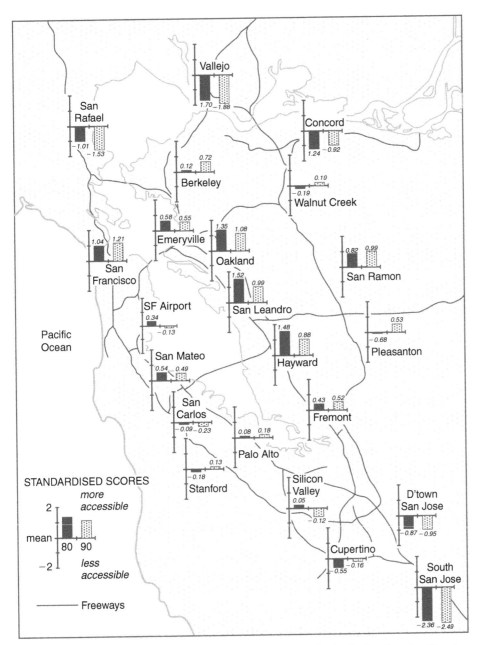

FIGURE 5.3 *Housing opportunities in relation to employment nodes in the San Francisco Bay Area, adjusted for occupational matching, 1980–1990*
Source: Cervero *et al.* (1999). Reproduced by permission of Pion Ltd

'jobs gap' in the inner city as 'the most important single issue facing Britain's cities'. While the same could be said for the United States as well, this version of 'spatial mismatch' tends to be reversed in the outer zone of Australian cities where the supply of jobs is insufficient to meet local demand. Outer suburbs with low-cost housing in Australia tend to be job deficit areas (Forster 1999).

In both Britain and the United States, poor households with low skills levels tend to be anchored in declining inner-city neighbourhoods further and further removed geographically from the suburbs where the job growth is occurring. This is a highly specific instance of the spatial division of labour in cities. In these areas, unemployment rates and the incidence of 'discouraged' workers invariably exceed the metropolitan averages by a large margin. In the late 1990s, real unemployment rates in the conurbation cores and coalfields of Britain reached between 25 and 35 per cent. According to Turok (1999), the increase in inactivity – mainly among the least skilled males aged 50 and over – is highest in those cities where job loss has been greatest. So for cities like Liverpool, Glasgow and Manchester, where the official estimates in January 1997 were 20.7 per cent, 16.7 per cent and 19.6 per cent respectively, independent research put the real level of male unemployment at 37.4 per cent for Liverpool, 35.3 per cent for Glasgow and 33.8 per cent for Manchester (Turok 1999, 906).

The level of real unemployment is even higher in some of the pockets of council housing at the centre of British cities. For example, a comparative study of Sheffield's inner-city and suburban labour markets by Paul Lawless (1995) reveals that the rates of male and female unemployment on the Thorpe and Kelvin council estates (2 km from central Sheffield) were both five times higher than in the middle-class, owner-occupied suburbs of Greenhill and Bradway (8 km from central Sheffield). Residents of Kelvin and Thorpe identified the shortage of suitable jobs in the local area and the wider labour market, ahead of their limited skills, as the main barrier to employment (Lawless 1995, 1108).

Kain (1968) was the first person seriously to suggest that the root cause of chronic unemployment and poverty in the inner areas of US cities might be the increasing physical isolation of these neighbourhoods from the job growth in the suburbs. For instance, the centres of Ohio's seven biggest cities added only 636 net new jobs between 1994 and 1997, compared with 186,000 jobs in the suburbs. Public transport rarely connects the two: 'In a lot of places, the bus literally stops at the border', according to Bruce Katz, an urban specialist at the Brookings Institute (reported in *The Economist*, 20 May 2000).

Since Kain's original diagnosis, a lot of divergent evidence has been produced on the 'spatial mismatch'

phenomenon. In reviewing this evidence, Holzer (1991) concludes that accessibility does matter. One study (Ihlanfeldt and Sjoquist 1990) found that accessibility to jobs was capable of explaining between 30 and 50 per cent of the difference in the employment rates of black and white teenagers. But, for others, the spatial mismatch hypothesis draws attention from the deeper racial divide in US society and the discrimination that minorities often run up against in the labour market. In an influential study of black households in Chicago, Elwood (1986) found similar rates of unemployment at similar levels of education regardless of whether they lived on the Southside, away from job opportunities, or west of the city near the booming corridor of Interstate 88. 'From this he rather pithily concluded that the chief reason for persistent unemployment among blacks is "race, not space" ' (Cervero *et al.* 1999, 1269). Immergluck (1998) also found that the most important determinants of the local area employment rate in Chicago neighbourhoods are high-school education and the proportions of local residents that are black and male. That is, the more women in the local area, the higher the employment rate.

A further look at the job accessibility of low-wage workers living in Boston's inner-city neighbourhoods indicates that they already have a clear advantage in terms of accessibility (Shen 1998). Some central cities in the USA have benefited from office growth and an increase in clerical and sales occupations that have mostly been filled by females. According to Shen (1998, 357), the key to job accessibility for low-wage workers living in the inner city is whether they have use of a car or must rely on public transport: 'Although low-wage workers living in inner-city neighbourhoods benefit from the central location of their residence, many suffer from low spatial mobility because they do not own any motor vehicle.'

McLafferty and Preston (1996) investigated the effects of distance on the job opportunities available to African-American, Latino and white women and men in the New York metropolitan area. Their results showed that African-American men and women living in the centre of the region have poorer spatial access to jobs than their white counterparts. Latinos suffer some mismatch in the central city, but less so than whites. The lower numbers of minority workers living in the suburbs do not appear to have significantly longer work trips than white workers, although African-American women and Latinos are more reliant on mass transit and this can lengthen commuting time.

The added difficulty that African-American and Latino youths without formal qualifications or skills face in a deregulated, low-wage market, is their very limited earning potential. On this basis, Immergluck (1998, 21) questions whether job-creation schemes in unemployment 'hotspots' can substantially improve local rates of labour force participation. Part of the problem is that, unless they are reserved for local workers without jobs, workers from outside the area with a stronger employment record get offered the jobs. As a matter of course, retail chains and banks recruit citywide, induct the trainees and then despatch them off to local stores and offices. Morrison (2001) concludes that, so long as there are not enough jobs to go round in the metropolitan labour market, and so long as labour remains mobile, it makes more sense to upgrade the skills of out-of-work residents in a depressed area than to fund local job-creation schemes.

Alternatively, programmes have been devised in the United States that relocate unemployed workers living in project housing in run-down neighbourhoods so they can improve their job prospects (Briggs 1997). One of the best known was developed by the US Department of Housing and Urban Development in the mid-1990s. Called 'Moving to Opportunity', the programme paid the moving expenses of public tenants and housed them closer to industrial and office parks in the suburbs (Rosenbaum 1995). In a similar vein, several federally supported pilot projects called 'Bridges to Work' were trialled in Baltimore, Milwaukee and other US cities in the late 1990s. They aimed at linking inner-city residents with suburban jobs by expanding door-to-door reverse-commute van services.

Gender gaps in work-trip length and suburban jobs: 'captive workers'?

A growing body of research, especially in the United States, attests to the existence of a sizeable gap in the respective distances travelled by men and women to their place of employment in American cities (Hanson and Pratt 1995). Across all metropolitan areas, there was a 3.5-minute difference in journey to work time between males and females in 1990 – 25.4 minutes for men compared with 21.9 minutes

for women, or a difference of one-sixth (Wyly 1998). In New York, Chicago and San Francisco, trip times for women are between 4 and 7 per cent shorter than for men, but in Philadelphia, Detroit and Boston they are at least 15 per cent shorter. The largest disparities are found on the outskirts of regional commuting fields around New York, Los Angeles and San Francisco. Long Island's Nassau–Suffolk MSA has one of the highest gender gaps in commuting time (12.4 minutes) owing to the clear gender and spatial division in the local labour market. High-income males commute down the island by rail into Manhattan at twice the ridership rate of females (12.7 per cent vs 5.9 per cent), while women are much more likely to find local employment. In this archetypal situation, the average male commute is 50 per cent longer than that for women (Wyly 1998).

Feminist geographers like Hanson and Pratt (1988) claim that this gender gap in the length of the journey to work in the USA is symptomatic of the degree to which college-educated housewives living in the suburbs are spatially entrapped in 'suburban pink-collar ghettos'. They hypothesise that, in this segment of the labour market, these suburban women are forced by the household division of labour and their child-rearing responsibilities to accept pink-collar jobs in nearby business parks for which they are overqualified. For example, two of the fastest-growing areas in the San Francisco Bay region, San Ramon and Pleasanton, were major recipients of back-office relocations from downtown San Francisco and Oakland during the 1980s (Fig. 5.3). They are located about 60 km to the east of downtown San Francisco and are home to two very large master-planned business parks. 'Among the many factors attracting firms to these outlying locations were their available pools of college-educated residents and in particular, back-office clerical and secretarial labour (especially middle-class married women, many of whom at the time were just beginning to enter the labour force)' (Cervero et al. 1999, 1272). As such, this constitutes a reversal of the 'spatial mismatch' problem faced by many non-professional women living in the inner city. Unable to find employment for which they are qualified nearby, they often have to commute long distances to low-wage service jobs or housework in the suburbs.

Kim England (1993) sets out to evaluate the 'spatial entrapment' hypothesis by surveying 200 women living in the suburbs of Columbus, Ohio.

TABLE 5.5 *Time spent travelling to work by women and men in four Australian cities, 1991*

	Minutes travel time (%)					Variable work place	Total
	≤15	16–30	31–45	46–60	>60		
Women without children:							
Aged 15–24 years, not married	40.4	31.4	13.6	8.3	3.9	2.4	100
Aged 15–34 years, :married	37.6	32.7	14.3	7.5	4.9	2.9	100
Women with children							
Married	50.3	29.5	8.2	5.1	1.7	5.1	100
Lone mother	52.4	28.0	8.8	6.5	*0.3	*4.1	100
Women:							
Aged 60 years and over, living alone	42.5	*28.4	*14.6	*	*3.9	*10.6	100
All women	45.2	30.4	11.5	6.2	2.4	4.4	100
All men	32.1	31.3	13.6	7.5	3.4	12.1	100

*Excludes persons living in non-private dwellings.
Compiled from Housing and Locational Preference Surveys; The National Housing Strategy: Housing and Locational Choice Surveys.
Source: National Housing Strategy (1992). Reproduced by permission

They all worked in clerical and administrative support positions. England also interviewed the personnel managers of ten relatively large back offices in Columbus. She established that the distances travelled to work by 'single' and married women in clerical jobs, and their husbands, were as follows: 9.1 miles; 9.6 miles; 11.6 miles. Apart from the danger of reading too much into relatively trite differences in travel time, England (1993, 236–9) questions the blanket assumption about the spatial entrapment of pink-collar workers in the suburbs.

This assumption was often confounded by the reasons pink-collar workers gave in negotiating multiple roles, and the factors advanced for the relocation decisions of employers. Many women explained that their short commute was not the outcome of their efforts to combine their various roles into a finite time period. For others, it was no longer a requirement, as the children had long since left home. Then there were two sets of women, one of whom had both young children and long commutes, while the others' travel time had shortened as the children

grew older, owing to a change of job (England 1993, 234–5). Furthermore, 'Only three of the ten personnel managers agreed that the demand for a spatially entrapped female clerical labour supply was one of the more important reasons in their firm's relocation decision' (England 1993, 228). The ensuing debate with Hanson and Pratt (1994) emphasises how context-specific the findings from a single case study like that of Columbus or Worcester is likely to be.

It is certainly the case that more women in Australian cities also work closer to home than men, and that women with children have the shortest work trips of all (Table 5.5). But it cannot automatically be assumed that this gender gap in work-trip length comes at the price of 'spatially entrapped' female labour. Analysis by Forster (1999) confirms that each major job centre in the suburbs draws upon a series of nested labour catchments: women work closer to home than men; part-time workers work closer to home than full-time workers; blue-collar workers originate from different suburbs from white-collar workers. Because many Australian

TABLE 5.6 *Gini coefficients for some OECD countries, 1970s–1990s*

Country	1970s	Year	Mid-1980s	Year	1990s	Year
Australia	0.291	1976	0.312	1984	0.306	1994
Canada	0.283	1975	0.289	1985	0.284	1994
France	0.296	1979	0.298	1984	0.291	1990
Germany	–	–	0.265	1984	0.282	1994
Netherlands	0.230	1977	0.234	1985	0.253	1994
Norway	–	–	0.234	1986	0.256	1995
Sweden	0.232	1975	0.216	1983	0.230	1994
United Kingdom	0.280	1981	0.330	1987	0.330	1996
United States	0.313	1974	0.340	1984	0.344	1995
New Zealand	0.269	1982	0.264	1986	0.331	1996

Based on Household Economic Survey, OECD.
Source: Statistics New Zealand (1998a)

suburbs house workers in the low- to middle-status occupations, there is a better occupational match with those pink-collar jobs that have moved to suburban centres.

Rising wage inequalities: income polarisation

The most outstanding legacy of labour market restructuring in countries like the United States, the UK, Australia and New Zealand is the growing gap between the incomes of job winners and job losers. Table 5.6 confirms that the Gini coefficients deteriorated most for these OECD countries over the course of two decades between the mid-1970s and mid-1990s. The Gini coefficient measures income inequality, with any movement towards one reflecting greater inequality. While the data are not exactly comparable, the graphs presented in Figure 5.4 record the extent to which the relativities between the top and bottom incomes in the United States and the UK have worsened since the early 1970s. A similar statistical picture emerges from Australian research. Modelling done on the changing distribution of earnings in Australia indicates that, 'From 1982 to 1996–7, the average income of the most affluent 10 per cent of Australians increased by almost A\$200 a week. This was three to six times more than for those at the middle and bottom of the income distribution' (Anne Harding, Director of the National Centre for Social and Economic Modelling, reported in the *Weekend Australian*, 24–25 June 2000, 19).

To take the case of the US labour market, where job creation has outstripped other OECD economies,

From 1989 to 1997, total employment increased by 11.3 million to 129.6 million, not seasonally adjusted. Over the entire period, job growth surged by 18.8 per cent in the highest earnings group, twice the increase recorded in the lowest earnings group. Employment in the middle-earnings category was unchanged over the period.
(US Bureau of Labor Statistics 1998)

At the same time, workers with higher-level skills and qualifications received progressively higher rates of pay as the demand for expertise and work experience grew (Danziger and Reed 1999). Between 1979 and 1998, real wages increased by 8 per cent for male college graduates in the USA while falling 18 per cent for male high-school graduates. This moved the weekly wage gap between male college and high-school graduates from 29 per cent to 68 per cent over the period. For women, the weekly wage gap between college and high-school graduates increased from 43 per cent in 1979 to 79 per cent in 1998. This growing wage gap is also apparent in the earnings of women and men at the top (>175 per cent of median earnings) and bottom (<50 per cent of median earnings) of the income distribution in Australia. Between 1982 and 1996–7, the full-time earnings of the top female income class increased by 22 per cent while those for the bottom female income

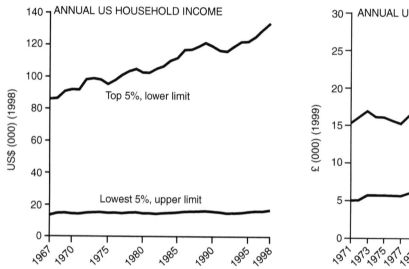

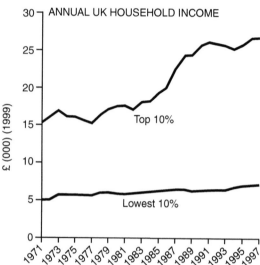

FIGURE 5.4 *The widening income gaps in the United States and the United Kingdom, 1970–1998*
Source: US Census Bureau; Office for National Statistics

class fell by 4 per cent (adjusted for inflation). For males in the top income class full-time earnings increased by 12 per cent, and for males in the bottom income class they rose by 6 per cent (reported in the *Weekend Australian*, 24–25 June 2000, 20).

To summarise, the various causes of the rising value of skill in the United States and other economies like it include most of the structural changes covered above:

Labour-saving technological changes have simultaneously increased the demand for skilled workers who can run sophisticated equipment and reduced the demand for less-skilled workers, many of whom have been displaced by automation. Global competition has increased worldwide demand for the goods and services produced by skilled workers in high-tech industries and financial services. Lower-skilled workers increasingly compete with low-wage production workers in developing countries. Immigration has increased the size of the low-wage workforce and competition for low-skilled jobs. Institutional changes, such as the decline in the real value of the minimum wage and shrinking unionization rates, also moved the economy in the direction of higher earnings inequality.
(Danziger and Reed 1999, 16)

They conclude that, 'Because growing income inequality arises primarily from long-term structural changes in the labor market that are unlikely to be reversed, US income inequality is likely to remain high in the coming decade.' Rifkin (2000)

also agrees that the productivity gains due to the surge in IT take-up by US companies reflect a long-term trend to technological displacement and increasing disparities in wage rates.

The '30/30/40 society'

Taken together, these processes are actively restructuring the labour force in advanced economies like the UK and producing new forms of labour market segmentation that Will Hutton, editor of the *Observer*, calls the '30/30/40 society'. The basic re-alignments that took place in the UK labour force in the mid-1990s are set out in Figure 5.5. At the bottom are disadvantaged workers and their families, who are most vulnerable to persistent poverty as they get recycled through the UK's much more flexible labour market. Most of the work they 'do find is part-time, casualised or insecure, so that their lives consist of unemployment interspersed with periods of insecure semi-employment' (Hutton 1995, 13). The 'Newly insecure' are defined not so much by their incomes – though three-quarters of working women in the UK are defined as low-paid – as by their relation to the labour market. Nevertheless, the fact that many part-time workers come from two-income households does not protect them against arbitrary dismissal without severance pay. Since 15 per cent of part-time workers had held

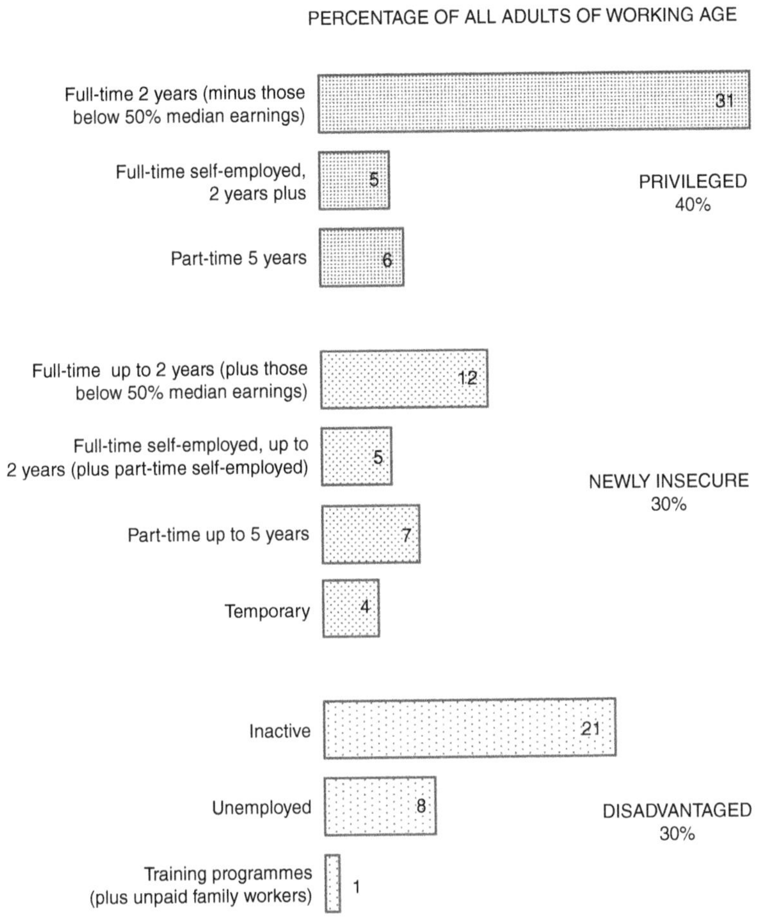

PERCENTAGE OF ALL ADULTS OF WORKING AGE

Full-time 2 years (minus those below 50% median earnings)   31

Full-time self-employed, 2 years plus   5

Part-time 5 years   6

PRIVILEGED
40%

Full-time up to 2 years (plus those below 50% median earnings)   12

Full-time self-employed, up to 2 years (plus part-time self-employed)   5

Part-time up to 5 years   7

Temporary   4

NEWLY INSECURE
30%

Inactive   21

Unemployed   8

Training programmes (plus unpaid family workers)   1

DISADVANTAGED
30%

FIGURE 5.5 *Changing class structure in the UK: towards the '30/30/40 society'*
Source: Hutton (1995). Reproduced by permission of the *Guardian* & *Observer* Syndication Department

their jobs for more than five years, Hutton puts them in with the 'Privileged' segment of the workforce, whose market power has increased in the UK since 1979.

As well as stressing that there is considerable segmentation within each of these categories, Hutton also points out that the 40-hour week, full-time job is under constant threat, so even the status of privileged workers is provisional in the UK's flexible labour market. While it is not so evident in Britain, labour economists in both the United States and Australia have detected an appreciable 'hollowing' of middle-status occupations during the decades between 1970 and 1990 (Gregory 1993).

The structure of Hutton's '30/30/40 society' is a bequest of the recent transformation of capitalist economies and their labour markets. Whilst it was devised to account for changes to UK society, it also describes the changes taking place in Australia sufficiently well for Latham to adopt it in his account of the impact of globalisation upon Australian society (Latham 1998).

The UK's new social class scheme

In 2001, the UK's Office for National Statistics introduced a new 'socio-economic classification' in recognition of the sea change that has taken place in the labour markets of advanced economies since the 1960s. For one thing, the classification it replaced needed updating to reflect the withering-away of the old manual working class. The new scheme depends mainly on a person's position in the labour market. It divides up people according to the nature of their employment 'contract'. Those at the bottom make a short-term exchange of cash for labour. Those at the top normally have more secure, longer-term contracts with a salary package.

This new scheme has seven classes:

I   higher professional and managerial
II lower professional and managerial
III intermediate (secretaries, police personnel, etc.)
IV self-employed and small employers
V   lower technical (e.g. skilled manual workers)
VI lower-level services and sales personnel like shop assistants
VII routine occupations (mainly semi-skilled or unskilled manual workers).

When it was tested, the scheme proved to be a good predictor of income – even though it is not based upon income – and to a lesser extent of health. As might be expected, the self-employed in Class IV are just as healthy as those in Classes I and II (reported in *The Economist*, 3 June 2000, 67).

CONCLUSION

In this chapter, I have dealt with the broad processes behind the restructuring of labour markets, and the availability of work and income in cities. These changes in the level, distribution and certainty of income, and in the location and educational/skill requirements for employment, are all likely to impact within the housing system. 'Labour market outcomes affect housing choices and the more "flexible" labour

market requires an adaptation in housing systems' (Maclennan and Pryce 1996, 1856). Significantly, it is the interplay of labour market adjustments with urban land and housing markets (Chapter 6) that ultimately helps to shape the social geography of cities (Chapter 7).

FURTHER READING

The Work of Nations (**Reich** 1993) represents an early account of how occupations are changing the nature of work and the implications for urban-industrial societies in the twenty-first century.

Rifkin (2000) extends the prognosis in a provocative book that foresees a decline in the global workforce and an eventual end to work as we know it. For a polemical treatment of the consequences of economic restructuring and how it is transforming working lives in Europe, see **Forrester** (1999).

Spatial Division of Labour (**Massey** 1984) develops a theoretical perspective showing how global economic restructuring transforms the local division of labour in cities and regions.

A very comprehensive review of urban labour markets can be found in an *Urban Studies* paper by **Morrison** (2001).

In a pioneering study, geographers **Hanson and Pratt** (1995) explore the complexities arising from the interrelationships of gender, work and space.

Linda McDowell's (1997) survey of workplace relations in London's financial district further broadens the scope of gender studies in urban settings.

The implications of an emerging 'working poor' in cities undergoing globalisation are taken up by **Peter Marcuse** (1996) in a monograph called *The New Urban Poverty*.

6

HOUSING MARKETS AND RESIDENTIAL LOCATION IN CITIES

INTRODUCTION

This chapter spells out how land and housing markets contribute to where households either choose or are constrained to live in cities. But households and the local communities they form can be very diverse in their demographic and social make-up (Chapter 7). These differences have a significant bearing on the sources and amount of income the household can afford to spend on housing. In simple terms, a household's 'demand function for housing' is strongly influenced by the position of wage earners in the labour market (Chapter 5). This, more than anything else, governs housing access and affordability.

With more 'flexible' and segmented labour markets, the distinction between households with no market income and those with two or more incomes is a growing source of inequity in housing markets. This is partly because the forms of housing assistance available to non-earning households like aged pensioners, single parents or disability pensioners vary a lot in generosity from country to country. For example, the state takes far greater responsibility for housing provision in Europe than in New World societies like the United States, Canada, Australia or New Zealand (Fig. 6.1). In these countries, the state's overriding goal has been to extend home ownership. The institutions created by governments to achieve this also play a part in regulating access and affordability. Alternatively, those households that cannot afford to buy or prefer to rent are also subject to access and affordability hurdles in the public or private rental sector.

In this chapter we will compare housing provision in different societies and how it impacts on the residential fabric and social geography of cities. This constitutes the 'supply side' of the housing system. The housing supply in cities is loosely segmented into a number of housing submarkets that vary

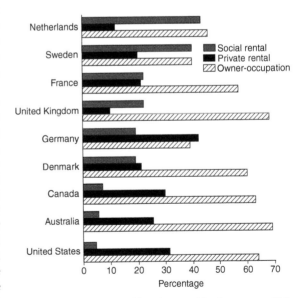

FIGURE 6.1 *Breakdown of housing provision by tenure within selected OECD countries in the mid-1990s*
Source: compiled from government statistics for the member states

according to:

- the era in which they were developed (e.g. pre-First World War, inter-war, post-Second World War)
- tenure (e.g. owner-occupation or home ownership, private rental, 'social' or public rental)
- dwelling type and density (e.g. high-rise flats or condominiums, tenement blocks or dormitories, medium-density townhouses, terraced housing, semi-detached villas, detached bungalows)
- locality (e.g. inner-city neighbourhood, council housing estate, master-planned community).

Figure 6.2 illustrates just a few of the wide range of housing submarkets that are found in most cities. Because big property developers and home-builders mostly build housing for a defined income range, recently developed suburbs are more likely to be homogeneous in terms of social class, as well as in their demography and family structure. On the other hand, a public housing presence introduces alternative types of housing stock, tenure patterns and community organisation into a city's neighbourhoods and suburbs. Not surprisingly, Singapore and Hong Kong, where government has been the chief housing provider, are entirely different cities from those found in societies like Australia, New Zealand, the United States, Canada and the UK, where private housing provision dominates.

The processes whereby individual households match their housing needs with a dwelling prompt a series of questions about residential preferences and moving behaviour. Who moves, why do they move and where do they move? It is the formation and break-up of individual households and their subsequent housing and residential location decisions that direct household flows within the housing system of cities. How these processes gradually change the geography of housing in cities will emerge from a handful of case studies.

Property markets play a casting role in the sorting of households within the city and in the dynamics of neighbourhood change. Models of neighbourhood change originally envisaged a sequence of formation, maturation, followed by slow decline and abandonment, and then redevelopment. But, as we will see, since the late 1960s a process called 'gentrification' has been revitalising whole neighbourhoods and suburbs in selected parts of the inner city. Eventually this new source of demand bids up property prices

and rents in the area, and can lead to the displacement of long-standing tenants. In big cities with severe housing shortages, like New York or London, tenants forced out of single-room occupancies (SROs) or rooming houses may become homeless.

Hence housing markets bring together households needing to be housed with the governmental agencies and firms in the housing industry that are in the business of meeting those needs. It is the operation of housing markets that gives form in this way to the residential layout and social differentiation of cities (Chapter 7).

The last part of this chapter looks at how the widening income gap is translated by housing market processes into growing social polarisation in British, US and Australian cities. Washington DC provides a graphic illustration of socially divided cities complicated by racial differences. One implication is that place of residence can be quite disadvantaging. Geographers call this 'locational disadvantage'.

HOUSEHOLD COMPOSITION AND HOUSING DEMAND

The rate at which new households are being formed – the household formation rate – determines the overall level of demand for housing in cities. While population growth is a fundamental driver of housing demand, in many OECD countries household formation rates have taken over as the key determinant of future housing needs. This is to say that the number of households is growing more quickly than the underlying population growth in countries with slowing rates of natural increase or limited immigration. A key reason for this is that the size of households has been falling imperceptibly in countries like the UK, Australia and New Zealand for many decades. In New Zealand, average household size has been in decline since peaking at 5.17 persons in 1886. The period of most rapid decline occurred between 1966 and 1991, when occupancy rates fell by 21 per cent, to reach 2.77 by 1996 (Statistics New Zealand 1998b, 45). In Britain, average household size dipped from 2.47 persons in 1991 to 2.40 persons in 1996 and is predicted to fall to 2.15 by 2021 (DETR 2000). Household size also declined significantly between 1960 and 2000 in Australia.

In turn, the decline in household size can be traced to changes in household composition. Within

FIGURE 6.2 *A cross-section of contrasting housing submarkets (a) Early twentieth-century tenements, Lower East Side, Manhattan NY; (b) Gentrified lofts in Kensington, London's West End; (c) Terraced Georgian apartments fronting a private square in Mayfair, London; (d) Street deck council flats, Walkley, inner Sheffield; (e) Public housing, Evry New Town, south-eastern Paris; (f) Suburban family homes, Blacktown, outer-west Sydney*

Source: author's collection

the last generation, households comprising a person living alone, childless couples, single-parent families and 'blended' families have begun to rival the traditional family household as the preferred living arrangement in many western societies. For example, according to the Australian Bureau of Statistics, the proportion of one- or two-person households in Australia rose from 31.5 per cent in 1954 to 55.2 per cent in 1996. By 1996, 11 per cent of adults were living alone in Australia. The number of one-person households is forecast to rise from 22 to 26 per cent of all households in the UK between 1991 and 2021 (DETR 2000). This greater household diversity is due partly to increasing longevity and partly to the effects of changing social norms, including rising divorce rates, deferral or the avoidance of child bearing, and opting out of marriage.

These developments are combining with the effects of workforce restructuring (Chapter 5) to widen the 'household income gap'. Only about 25 per cent of family households depend upon the earnings of one breadwinner any longer in North America, Europe or Australasia. More than half now have two or more paid workers, while the balance comprise households where no one at all is earning a wage. Among low-income households, there are a growing number headed by solo parents, usually a female, and dependent upon a supporting parent's benefit. By contrast, many high-income households comprise two wage earners who, quite often, receive equally high salaries. An Australian Bureau of Statistics survey of household income and expenditure undertaken in 1998 revealed that more than two-thirds of the top quintile (one-fifth) of income units in Australia had two earners. Conversely, 90 per cent of the bottom quintile, representing 1.8 million income units, were one-person households containing single people opting to live alone, widows or widowers, and lone-parent families (reported in *Weekend Australian*, 4–5 July 1998). This disparity between households without a wage earner, and those with two or more wage earners (dual- or two-income households), is of lasting importance in determining access to housing. For example, a Joseph Rowntree Foundation study of six British conurbations undertaken in the late 1980s revealed that, whilst 80–85 per cent of owner-occupied households had at least one earner, and half had two, the equivalent proportions for the council and housing association tenants were 40 per cent and 10 per cent

respectively. 'These patterns were clearly connected to changing labour market fortunes' (Maclennan and Pryce 1996, 1858). Another implication of the restructuring of labour markets and the associated loss of job security is that home ownership may no longer be sustainable at the high levels attained in countries like the UK, Australia and New Zealand towards the end of the twentieth century (Ford and Wilcox 1998; Yates 2000).

Income can also be highly variable at different stages in the working life of wage earners. For example, when married women leave the workforce to have children, earnings are severely depleted for a time. Retirement also brings a drastic loss of income (asset rich/income poor). Vulnerable households are perpetually caught up in an erratic pattern of retrenchment and rehiring throughout their working lives. Therefore, how much household members jointly earn, and how these earnings are pooled, have a big bearing on access and housing costs in different housing submarkets.

Housing provision

Figure 6.1 compares the structure of housing tenure in a selection of OECD countries in the mid-1990s. The three main forms of tenure are owner-occupied housing, private rental housing and social rental housing. The hallmark of the social rental sector is that rents are set at a level just to cover operating costs. It embraces a broad group of 'non-profit' or 'cost-rental' agencies: local authorities and housing associations in the UK; municipal housing companies and cooperative housing in Europe; state housing ministries and community housing in Australia; and publicly subsidised housing in the United States. Figure 6.1 highlights the contrast in the approach to housing provision taken by European countries like The Netherlands, Sweden, Austria and Switzerland with their strong social rental sectors, and the group of English-speaking countries known as 'property-owning democracies' (the UK, the United States, Canada, Australia, New Zealand). Significantly, although many people in this second group of countries regard high levels of home ownership and high standards of living as synonymous, they are no longer the most affluent member states of the OECD. This is to suggest that there is nothing preordained about the housing systems that societies develop.

Although not graphed, the city-states of Singapore and Hong Kong embarked upon very large-scale public housing programmes in the 1950s and 1960s. Between 1965 and 1995 Singapore's Housing and Development Board (HDB) built 630,000 dwelling units to house most of the population, which now exceeds 3 million. In the space of that same generation, average earnings in Singapore caught up with and then passed those of Australians and New Zealanders. But, whereas 80 per cent of Singaporeans rented from the HDB in the late 1960s, by the end of the 1990s the home ownership level had climbed to 90 per cent. The 'Home-ownership for the People Scheme' provided for the mass conversion of rising real incomes into mortgages backed by the state. Like Hong Kong, beyond the CBD the skyline is dominated by high-rise towers, either on large New Town housing estates or on privately developed condominium sites along the foreshore (Fig. 6.3). Hence a fundamentally different housing system has been responsible for producing Singapore's housing geography: 'A remarkable feature of the housing stock is the extent to which it is provided, regulated and managed by the government, while more than 90 per cent is owner-occupied' (van Grunsven 2000, 107). What is also remarkable about Singapore's housing programme is that this level of home ownership is never likely to be achieved in the UK, the United States, Canada, Australia or New Zealand.

Structures of housing provision

The housing system that has evolved in these home-owning societies is the product of quite another 'structure of housing provision'. This way of comparing the housing systems of different countries was originally developed by a British political economist with an interest in housing markets. It occurred to Michael Ball (1983, 17) that, 'Instead … of simply being one way in which housing may be consumed, owner-occupation has become associated with a particular way in which housing is provided; and with all the forms of land ownership, building, finance and market exchange that exist there.'

In English-speaking countries the state has actively sponsored home ownership ahead of alternative tenures; however, in the case of the UK, this has been comparatively recent. Although the benefits of home ownership are undeniable, the state's preferential treatment of the tenure in 'property-owning democracies' has come at the expense of those households who miss out. Firstly, property ownership as a source of wealth is closed off to tenants (Badcock and Beer 2000). Secondly, few tenants enjoy security of tenure. Thirdly, tenants in both the private and public rental sectors are increasingly besieged, as governments in countries like the UK, Australia and New Zealand scale back housing assistance to 'safety net' proportions.

Since the Second World War, home ownership has been accorded the status of a privileged circuit

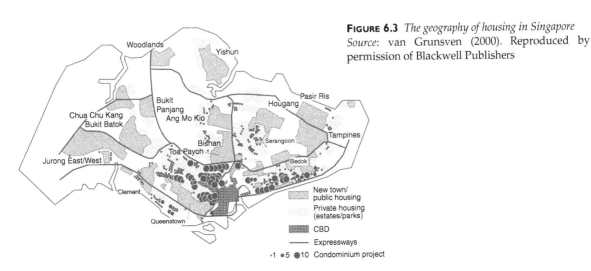

FIGURE 6.3 *The geography of housing in Singapore*
Source: van Grunsven (2000). Reproduced by permission of Blackwell Publishers

of capital within the Australian economy. Owner-occupation rose from 52 per cent in 1947 to plateau at 71 per cent in 1966. During this 'golden age' of home ownership in Australia, up to 90 per cent of households from the 1950s and 1960s owned at some stage in their housing career. This was a direct outcome of the housing policies pursued by the federal and state governments and the institutional arrangements created for providing housing. 'Viewed in the broadest sense, Australian housing policy ensures that home ownership is the dominant tenure by providing very little support for private renters, only a marginal public housing sector, and substantial tax incentives for home owners' (Bourassa *et al.* 1995, 83).

By contrast, home ownership in the UK has only reached 68 per cent in the late 1990s. Even as late as 1945, the UK was still a nation of private renters. Nearly 60 per cent of households had a private land-lord, while local authorities (LAs) housed only 10 per cent. Only one in three households owned a home. Prime Minister Attlee and Housing Minister Bevan made 'Homes Fit for Heroes' a priority of the welfare state, and, by the late 1970s, LA housing stood at about one in three dwellings. With post-war prosperity, home ownership steadily expanded (56 per cent in 1979), while private renting (13 per cent in 1979) contracted. By the beginning of the 1980s, therefore, housing in the UK was divided between two completely opposing tenures – owner-occupiers and council tenants. In a city region like London, the changes in tenure structure had created a housing 'doughnut' comprising an outer ring of home-owners (54 per cent owner-occupied) and an inner core of council tenants (39 per cent LA) in the east and private tenants (42 per cent private rental) in the west.

The Conservatives came to power in 1979 led by a Prime Minister determined to break up the mono-lithic council housing bureaucracies and sell off their dwelling stock to sitting tenants or housing associations. They introduced a series of Housing Acts that gave tenants the 'right to buy'. By the beginning of 2000, council house sales totalled 2,128,000, with large-scale transfers from the LA sector to housing associations numbering about 376,000. As a consequence, the level of owner-occupation in Britain rose from 56 per cent in 1979 to 68 per cent in 2000 (DETR 2000). As well as once more changing the tenure structure of housing

within British cities, the privatisation of council housing worked to alter voting patterns in the 1980s. Many 'cloth-cap' Labour voters and tenants switched their political allegiance to the Tories when they exercised their right to buy.

During the course of the 1980s in the United States the home ownership level fell from 66 per cent at the beginning of the decade to below 64 per cent at its end. In President Clinton's second term a home ownership drive was mounted by the US Department of Housing and Urban Development (HUD) to lift owner-occupation back to its earlier plateau. By a stroke of luck it coincided with the extended run of economic growth in the late 1990s. By the first quarter of 2000, just over 67 per cent of Americans owned or were buying their homes. Part of this success story is due to the higher rates of take-up among minority and inner-city households. While they cannot match the owner-occupation rate for whites (73.4 per cent in 2000), the rates for African-Americans (47.8 per cent) and Hispanics (51.2 per cent) reached an all-time high in 2000 (reported in the *New York Times*, 27 April 2000, G8).

Tax breaks form an important part of the struc-tures of housing provision in each of these property-owning democracies. Governments determine the scope for income tax relief and create schemes that help households buy. Although Canada and Australia do not allow buyers to set mortgage inter-est repayments against income, they exempt an owner-occupied residence from capital gains taxes. On the other hand, US home buyers can deduct the interest on mortgages worth up to US$1 million from their taxable income, but there is provision for clawing some of this back by levying capital gains on houses at the point of resale. In practice, though, ordinary home-owners in the USA pay little if they plough the proceeds back into a house of at least the same value when they move. Known as roll-over relief, this eliminates a possible break upon labour mobility in the US economy. By now, tax breaks for home buyers are an almost unassailable part of the electoral compact between parties on both sides of politics in the United States, the UK, Australia and New Zealand. Nevertheless, the Blair government voted to remove tax relief on home mortgages in its 1999–2000 budget.

Significantly, the tax expenditure forgone by these governments in pursuit of the votes of com-fortably off home-owners vastly exceeds outlays on public and other forms of non-profit housing. Described as middle-class 'welfare', this forgone revenue helps to underwrite what Galbraith (1992) calls the 'culture of contentment'. Yet, while governments turn a blind eye to middle-class welfare, the subsidies for constructing and modernising public housing in many OECD countries have steadily fell in real terms during the 1990s. This has occurred along with the introduction of market-related rents in the public rental sector, and a shift to housing-linked income support for tenants in the private rental sector. Known as Housing Benefit in the UK, the Private Rental Assistance Scheme in Australia and the Accommodation Supplement in New Zealand, these measures were meant to increase the supply of affordable accommodation, open up choice in the housing market, and close the gap between the housing costs of public tenants and private tenants. But, whatever the notional appeal of these 'reforms', there is no accounting for volatile interest rates or housing markets, not to mention the pressure that fluctuating unemployment rates bring to bear on the low-rent housing sector. In this way, international variations in forms of housing provision help to explain the different experiences of housing tenures and why they have developed in the various ways they have. A comparative study of housing systems in Europe and the United States shows just how much social progress depends on the provision of decent housing (Ball *et al.* 1988).

RESIDENTIAL PREFERENCES AND MOVING BEHAVIOUR: WHO, WHY AND WHERE?

'The residential choices of individual households in aggregate define the social areas of the city' (Ley 1983) – except that these choices are always subject to the constraint of income. Some familiarity with the processes that lie behind the decisions and actions taken by households shifting house – which could be from a rural homestead to a city townhouse or from a city flat to a suburban bungalow – is a prerequisite for beginning to understand how communities are formed in cities. While not all city neighbourhoods and suburbs are socially cohesive communities, they are normally the identifiable, named subareas that households gravitate to in

deciding where to live in a city. Therefore the social geography of cities is really a social mosaic produced by the myriad household moves being made at any one time.

Numerous studies point to the conclusion that there is much that is axiomatic about the residential preferences and moving behaviour of households in similar life circumstances. Professor Brian Robson, an urban social geographer at the UK's Manchester University, has modelled the decision to move based upon these generalisations. Another way to comprehend the processes is to ask, 'Who moves, why do they move and where do they move to?' Residential preferences relate to the assessment that a household makes about the services and environmental amenity available in the surrounding neighbourhood as well as to the dwelling. Nevertheless, most people place dwelling attributes ahead of neighbourhood amenity and environment when they choose a house. In general, movers tend to favour a better home in a less well-rated neighbourhood than the other way round, although sometimes access to schooling for children or to health services for the elderly may be an overriding priority.

Robson's model of residential mobility

Figure 6.4 is a flow chart setting out the decision path of a household contemplating a move. In the process of negotiating the decision path, the household will exercise varying degrees of choice, or encounter constraints, depending upon their life circumstances. In this sense, their move may be voluntary, or, at worst, forced upon them. For example, data from a British survey of homeless people in 1998 indicate that 27 per cent lost their home because friends or relatives could no longer accommodate them, 24 per cent because of relationship breakdown and 6 per cent because of mortgage arrears (DETR 2000). In a 1991 survey of Sydney and Melbourne households, 414,000 people in 147,000 households reported being forced to move involuntarily, or over one-half because of their inability to meet rent or mortgage repayments (National Housing Strategy 1991).

Figure 6.4 charts the decision path of a household reacting to sources of stress produced either by changing housing needs or aspirations, or by dissatisfaction with the present dwelling or neighbourhood. If the problem cannot be solved, the household

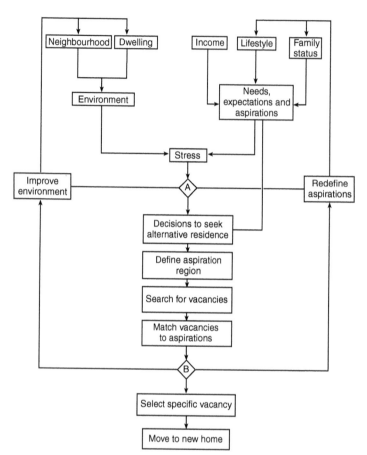

FIGURE 6.4 *The decision-making pathway of relocating households*
Source: Robson (1975). Reproduced by permission of Oxford University Press

begins to look for alternative housing, which involves matching needs to available housing. Among other things, the availability of housing depends on the vacancy rate in the local housing market. The balance between supply and demand is about right when 7 in every 100 dwellings are vacant. But when the vacancy rate falls below 7 per cent, as it does with a tightening of demand, it becomes harder and harder to find suitable accommodation at the right price.

For example, it may make little sense for an out-of-work council tenant in the north of England to forfeit secure and affordable housing for an insecure and poorly paid job in an expensive rental market like London; or for an unemployed single parent living in an inner-city project housing block in a US city to take up a job in the suburbs and incur crippling commuting costs in the process. Home-owners are sometimes trapped within a depressed housing market because they cannot find a buyer

for their house. Likewise, it is increasingly difficult for home-owners seeking to transfer from cities with below-average house prices like Brisbane, Adelaide and Perth, to a city like Sydney with grossly inflated house prices. This is also true for London, or for New York, or for Auckland in relation to other less expensive cities in their respective countries. These households have to weigh the sum of their earnings in the labour market against their housing costs within the property market. The outcome of this equation depends a great deal on the economic outlook and welfare of the respective cities (Chapter 3).

The staff of housing bureaucracies also control access to housing by acting as 'gatekeepers'. A long-standing complaint is that the process of determining a household's eligibility or priority on a public housing waiting list can be highly subjective. However, many of these objections have been removed by systematic reforms to the housing allocation process.

TABLE 6.1 *Overview of underprivileged households and their housing problems*

Household issues	Overcrowding	Substandard equipment, noise, other disturbances	Excessive rents in relation to income	Discrimination by landlords and other residents
Elderly	Seldom	Seldom	Sometimes	Seldom
Handicapped and ill	Seldom	Seldom	Sometimes	Sometimes
Single parents	Seldom	Seldom	Sometimes	Sometimes
Young families	Seldom	Sometimes	Very often	Seldom
Large indigenous families	Sometimes	Seldom	Sometimes	Sometimes
Large immigrant families	Very often	Often	Often	Very often
Other immigrants	Often	Very often	Sometimes	Very often
Unemployed	Seldom	Often	Often	Often
Students	Sometimes	Sometimes	Sometimes	Seldom
Other youth	Sometimes	Often	Often	Very often
Released prisoners	Seldom	Very often	Often	Very often
Drug addicts	Seldom	Very often	Often	Very often
AIDS	Seldom	Sometimes	Sometimes	Very often

Source: OECD (1996)

In New Zealand, for instance, it is argued that, because the Accommodation Supplement paid to tenants is calculated on the basis of income and housing costs, it is no longer subject to interpretation, as the Accommodation Benefit once was. None the less, underprivileged households are constantly confronted with a range of housing problems, which social housing providers are intentionally set up to deal with (Table 6.1).

The percentage of the population changing residence each year in a city provides a measure of housing turnover. The United States, Canada, Australia and New Zealand, for example, record the highest rates of housing turnover found in western societies. About 10–12 per cent of households change their address annually in the cities of these countries. This reflects a combination of labour mobility factors and the comparative absence of large social housing sectors offering permanent long-term accommodation to tenants. There is a wealth of material available for addressing the main factors that motivate households when they move house within the city. The illustrations used below are mostly drawn from a large-scale survey of Sydney and Melbourne households undertaken in 1991 in conjunction with a national review of housing.

Who moves?

Table 6.2 identifies types of household that had recently moved house in 1991 by employment status, age, income and country of birth, and the changes in tenure that eventuated. It confirms that home-buyers are much more likely than tenants – moving within or between rental submarkets – to be middle-aged couples with or without children, employed full-time and earning over A$46,000 between them. Contrast that with the status and background of tenants renting privately and moving into public rental, or transferring within public rental. Respectively: 29.3 per cent and 36.5 per cent are one-parent families; 18.1 per cent and 27.0 per cent are people living alone; 55.7 per cent and 62.5 per cent are not in the labour force; 57.4 per cent and 42.5 per cent are overseas-born; and household incomes are only A$13,566 and A$12,108. This means that, by the beginning of the 1990s in Australia, over 80 per cent of households living in public housing were dependent on social security benefits as their source of income.

More and more, council housing in the UK, along with public housing in the United States, Australia and New Zealand, is regarded as the 'tenure of last resort'. A doubling of the rate of turnover in council

TABLE 6.2 *Recent movers: household type and selected characteristics by tenure path, 1991*

Household type	Private rent to home purchase (%)	Private rent to private rent (%)	Private rent to public rent (%)	Public rent to public rent (%)
Couple only	23.6	20.0	12.9	6.8
Two-parent family	57.2	28.2	31.9	23.0
One-parent family	4.8	8.2	29.3	36.5
Relatives sharing	1.1	3.6	1.7	1.4
Group household	3.3	23.6	5.2	1.4
Lone person	9.0	15.0	18.1	27.0
Total %	100.0	100.0	100.0	100.0
N	610	1,014	116	74
Selected characteristics:				
% Employed full-time	85.6	67.6	19.1	15.3
% Not in the labour force	5.0	8.8	55.7	62.5
% Overseas-born	40.0	42.4	57.4	42.5
Age (mean years)	37	33	44	45
Household income				
(A$ annual income)	46,778	30,396	13,566	12,108

Source: National Housing Strategy (1991). Reproduced by permission

housing in the UK between 1980 and 2000 is one possible symptom of this. Despite a 35 per cent reduction in the overall stock of council houses in the 1980s, the number of former council tenants moving into owner-occupation or private renting rose from 33,000 to 76,000 between 1991 and 1996 (reported in a Joseph Rowntree Foundation press release, 12 October 1998). More than half those moving had lived in their council houses for less than five years. Nearly a third had moved out within two years. While many LAs increased rents ahead of inflation during the 1990s, it's clear that these tenants never saw council housing as a long-term option, or were quickly disillusioned.

Why do they move?

The main reasons for moving house are listed in Table 6.3. The weighting given by first home-buyers 'To purchase own home' (73 per cent) supports survey findings from most of the property-owning democracies where the desire to own is undiminished. Almost one in three changeover buyers (27 per cent), on the other hand, move to improve on the size or quality of their existing residence. 'Dissatisfaction with area' and 'To move to smaller dwelling' also loom much larger in the motivation of changeover buyers. Reasons are much more mixed for private tenants with a reasonably even spread between voluntary ('To increase size or quality', 'Life cycle/family influences', 'Employment related') and involuntary ('No choice', 'To reduce housing costs') factors.

Where do they move?

Are there any easily identifiable spatial regularities to the migration flows of households in cities? During the relatively affluent 1960s, Adams (1969) noticed that many households that were previously renting were leaving inner-city neighbourhoods in US cities to move to a home of their own in the suburbs. This gave a pronounced directional bias and orientation to the movement flows of movers – from the inner city to the middle and outer suburbs, and along the main arterials and rapid transit lines. His hunch was that this sectoral pattern of relocation is largely influenced by the 'mental maps' that take

TABLE 6.3 *Main reasons for moving, Sydney and Melbourne 1991*

Reason	First home-buyer (%)	Change-over buyer (%)	Private renters (%)	All households (%)
To purchase own home	73.0	18.2	0.2	19.2
To increase size or quality	3.9	27.1	16.1	16.2
Life cycle/family influences	11.8	8.5	17.8	13.8
Employment related	3.3	7.0	16.0	10.6
No choice	1.7	1.4	14.6	8.7
To reduce housing costs	1.4	1.9	10.9	7.3
Family contacts	0.8	6.5	5.0	4.8
Dissatisfaction with area	1.5	7.6	4.7	6.1
To move to smaller dwelling	0.2	6.2	1.2	2.4
To increase investment	0.6	4.1	0.3	1.5
Other reasons	1.6	9.8	11.3	7.8
Don't know/not stated	0.2	1.7	1.9	1.7
Total	100.0	100.0	100.0	100.0
Estimated no. of households	168,800	247,200	375,100	900,200

Note: 'All households' includes public and other renters, which are not shown separately in the table.
Source: National Housing Strategy (1991). Reproduced by permission

shape in the minds of city commuters; and that generally they confine their search pattern to three to five suburbs along this sector. Subsequent moves by changeover buyers are likely to be more localised so that established patterns of socialising, schooling, shopping and sports are not destroyed. A 1993 survey of housing demand in the Sydney region revealed a strong desire to live in a 'familiar' area, such as one where household members grew up or currently live. For example, 67 per cent of eastern suburbs households and 64 per cent of surveyed households in central-west Sydney had previously lived in the same area. Over 75 per cent of those thinking about moving said they would like to remain within their current area (Reid 1993, 35). Subsequent research on movement behaviour and patterns in many other cities have confirmed the general soundness of Adam's assumptions about household moving behaviour.

Who, why and where? A Melbourne case study

A study of the settlement of one of Melbourne's outer western suburbs encapsulates most of the regularities observed by other researchers. A sociologist,

Lynne Richards (1990), called this new housing subdivision 'Green Fields' to protect the anonymity of the residents. She sought answers to the following questions: who came to Green Fields; why did they choose Green Fields; where else did they search before settling for Green Fields and where did they come from? Somewhat pointedly, she called her book *Nobody's Home* because when she attempted to interview during the daytime she found most mothers at work. Richards surveyed 318 householders in 1978, and returned in 1981 to conduct 150 follow-up interviews with those who could be contacted.

The data revealed less of the social homogeneity that is often claimed for new suburbs. The clearest segmentation was on the basis of age and stage of family cycle. By 1981, 65 per cent of the respondents were under 36 years of age and first-time buyers. The remaining third, who already owned, were changeover buyers. As the younger women dropped out of paid employment to have children, the difference between single- and dual-income households assumed greater importance as a social divider on the housing estate. This was the time of maximum financial stress, with over 70 per cent of first-time buyers reporting serious financial worries.

TABLE 6.4 *Housing career model*

Life cycle stage		Housing career	
Household type	**Dominant age (years)**	**Housing type(s)**	**Housing tenure**
Young single	17–25	Room or flat, shared flat or house	Rented
Childless couple	21–24	Flat or unit, small house (2–3 bed)	Rented or owned
New family	25–29	Small house (2–3 bed)	Owned
Mature family	30–40	Larger house (3–4 bed)	Owned
Separated/divorced family	25–40	Small house (2–3 bed) or flat	Rented or owned
Blended family	30–50	Large house	Owned
Older family	40–50	Largest house (4-bed)	Owned
Empty nesters	50–60	Unit/small house	Owned
Senior aged	70 plus	Institution/granny flat	Owned or rented

Source: Maher and Burke (1991). Reproduced by permission of Pearson Education

Why Green Fields? Richards says that in view of feminist assumptions about the differing effects of the suburban environment on men and women she expected them to be 'seeking and seeing different features'. But 'Most people, women and men, had come for the house', and, just like suburbanites in the American studies of two decades ago, 'not for the social environment'. The 'environment' came into play after the house and an affordable location had been chosen.

Where did home-buyers 'search' and where did they come from? Green Fields home-buyers restricted their search field by familiarity and by vague ideas about an acceptable or desirable setting, played off against the specifics of an affordable location. Overwhelmingly, the new residents came from older, manufacturing suburbs closer to the city centre within the same corridor. 'Many did not look beyond an urban wedge familiar to them' (Richards 1990, 11). Several had picked this from childhood as the next stage out: 'the normal search procedure had been to explore all new housing in *their* wedge.'

MATCHING HOUSING NEEDS AND HOUSING STOCK

While the underlying geography of big cities is much more complex than implied, one of the major

decisions confronting households is whether or not to live at higher densities with restricted housing space, or in a purpose-built family home on its own allotment/block/section in the suburbs. Table 6.4 suggests in a very generalised way that certain house types and tenures are more suitable as needs change during the course of an individual's or family's housing career. When small area census data are examined, there is a significant amount of correspondence between measures of household type, stage in the life cycle, dwelling type and housing tenure. Table 6.5 summarises some of the key differences that arise between the central city, the suburban ring and the non-metropolitan edge in American metropolitan regions. For example, the demarcation between married couples with children in the suburban ring and solo parents and non-family households in the central city stands out very clearly. Married couples with children are twice as likely to live in the suburbs. Similarly, in the Sydney survey (Reid 1993), 55 per cent of single-person households and 28 per cent of childless couples lived in multi-unit housing, compared with only 14 per cent of couples with dependent children. Reid concludes that, in view of the continuing trend to smaller, non-traditional households in Australia and New Zealand, it can be expected that the demand for multi-unit housing in cities like Sydney and Auckland will go on growing in the future.

TABLE 6.5 *Household composition and location, US SMSAs, 1987*

Household type	Central city (%)	Suburban ring (%)	Non-metropolitan (%)
Family household	29.3	45.2	23.3
Married couple	26.3	49.4	24.3
With children under 18	25.0	50.9	24.1
With children under 6 only	28.7	49.7	21.6
Female householder	43.5	37.0	19.0
With children under 18	45.6	35.0	19.3
With children under 6 only	50.0	30.2	19.8
Male householder	36.7	44.4	18.0
Non-family household	41.9	36.9	21.2
Female householder	41.3	36.8	21.9
Male householder	42.7	37.0	20.3

Based on data from US Dept of Commerce.
Source: England (1993)

Although now dated, Grigsby's (1963) flow diagram tracing the aggregate movement of relocating households into, within and from the Philadelphia metropolitan area in 1955–6 is still hard to beat as a model of how household demand interacts with dwelling supply in cities (Fig. 6.5). The upper section of the diagram provides a count of the relative numbers of in-migrants, newly formed households, former renters and former owners moving into either rental units or owner-occupied units. For example, 79,000 renters continue renting, while 64,000 move into their own homes for the first time. A total of 70 per cent of those already owning buy another home (i.e. changeover buyers), leaving 30 per cent to return to renting. The lower section of the diagram indicates how the owner-occupied submarket is segmented between new and established dwellings. It shows that, while a majority of Philadelphia's changeover buyers bought new dwellings in the mid-1950s, the first home bought by tenants was more likely to be an established or 'used' dwelling. This pattern is consistent with an assumption commonly held by housing economists that dwellings 'filter down' the social scale while upwardly mobile households ascend stepwise in the housing market from renting to owning a house that has been passed down, before eventually buying a new home in an 'up-and-coming' suburb. Indeed, this pathway is not too dissimilar from that taken by most of the households that bought into Green Fields estate on the outer edge of Melbourne (Richards 1990).

Housing allocation and property markets

Residential location models that are a variant of the urban rent or 'trade-off' model described in Chapter 4 have been developed by urban economists like Alan Evans (1973) broadly to represent the residential density gradient, variations in the age and condition of the housing stock, and the distribution of income in North American and European cities. These models are treated in more depth elsewhere (Badcock 1984, 17–24), but the same essential assumption is made about the readiness of households with differing incomes to 'trade off' access to central-city jobs against more spacious housing in the suburbs.

In its simplest form the model predicts that the rent-density functions, and the land value and population density functions, will be 'negatively exponential' in shape. This means that population densities and land values are highest in the CBD and slope downwards with greater distance from the city centre. Then, by assuming that all households have income elasticities greater than 1 – this is to say they will consume disproportionately more housing relative to each additional dollar earned – household income can be shown to rise with distance from

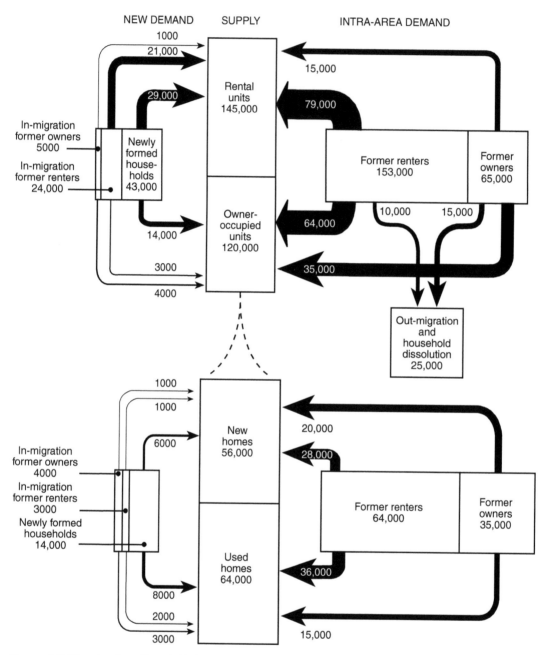

FIGURE 6.5 *The migration of households between housing tenures and submarkets in the Greater Philadelphia region, mid-1950s*
Source: Grigsby (1963)

the city centre. The model shows that the lowest income groups are able to meet the high rents charged in inner-city locations only by occupying the highest-density and often oldest housing stock. The highest-income groups, on the other hand, are prepared to commute to large suburban lots. As Mills (1972, 71) noted at the time, 'This is a remarkably realistic result, and it mirrors closely the predominant pattern in US urban areas.' It is otherwise known as the 'donut effect' in the United States because the poorest households happen to be black while the richest households happen to be white. In 1990, for example, more than half of the blacks in the metropolitan areas of Chicago, Washington DC, Philadelphia, Detroit, St Louis, Baltimore, Cleveland, Memphis and Buffalo lived in inner-city neighbourhoods that were at least 90 per cent black (US Department of Housing and Urban Development 1995, 12).

But how do you account for those wealthy households living in ritzy apartments on the Upper East Side of Manhattan, or in the West End of London, or central Paris, or the eastern suburbs of Sydney? Or the yuppies – young urban professionals – who have resettled formerly down-at-heel inner-city neighbourhoods in numerous North American, European and Australasian cities? Or the abundance of poor households on council or public housing estates in the outer suburbs of some British and almost all Australasian cities? In practice, these models can get you only so far. The role of the State Housing Commissions cannot be underestimated in the case of low-income groups on the periphery, which is the dominant Australian pattern. Quite simply, the land required for public housing in the 1960s and 1970s was much cheaper on the outskirts of sprawling cities like Sydney, Melbourne, Adelaide, Brisbane and Perth.

Gentrification, tenurial transformation and displacement

The urban geographer Neil Smith (1979) places much more emphasis on the structure of provision in accounting for the gentrification of inner-city neighbourhoods than on the lifestyle preferences and incomes of households. In the original version he argues that the formation of a 'rent gap' in inner-city neighbourhoods was a necessary prelude to the housing re-investment that has since transformed parts of cities around the globe from Amsterdam to Adelaide, and from Vancouver to Wellington. The 'rent gap' is the difference between the existing yields generated by buildings and land in their run-down state, and the potential yield once the site is redeveloped. It was this undercapitalised land in neighbourhoods close to the central-area labour market that offered an investment opportunity to

developers, professional home renovators and individual home-buyers willing to take a risk.

Gentrification is a two-pronged process, which is re-igniting residential property markets and heightening social polarisation in the inner areas of many North American, European and Australasian cities (van Weesep and Musterd 1991; Ley 1996; Smith 1996). One side of the coin involves the 'invasion' of an area by high-status households, who upgrade the housing stock and raise property values along the way. The other side of the coin involves the gradual displacement of the incumbent residents who may be economically weakened by the decline in traditional employment that used to be plentiful in the inner city, or by encroaching old age. Landlords eventually sell up and tenants are forced to move out of the neighbourhood, while long-time owners are 'winkled out' by offers that they can't refuse from well-heeled renovators or developers wanting to build upmarket townhousing.

If gentrification is allowed to run its full course, the process is capable of transforming the tenure structure of whole neighbourhoods from renting to owner-occupation (Hamnett and Randolph 1988). As a neighbourhood 'takes off' and competition among eager buyers gets more frenzied, house prices and rents begin their upward spiral. And once the renovation and new 'infill' townhousing is completed, the value of that work is then capitalised into property values. Often, in historic areas, the newcomers lobby city hall to designate urban conservation areas in order to protect their investment. Much less often, the existing inhabitants mobilise to defend their neighbourhood from 'middle-class takeover'. In this way, gentrification brings about social change at the neighbourhood level and invariably entails competition, if not conflict, between middle-class owner-occupiers exercising their market power against comparatively powerless tenants.

One of the best studies of tenurial transformation is Hamnett and Randolph's (1988) analysis of 'flat break-up' in London's West End between 1966 and 1981. They show how government home improvement grants were used to purchase, renovate and then sell tenanted Georgian, Victorian and Edwardian terraces and apartment buildings. In inner London, the number of unfurnished flats fell from 110,240 to 54,166 – a drop of 51 per cent. At the same time, the number of owner-occupied units rose by 217 per cent from 11,240 to 35,645. Much of this was due to the sale of flats by property trusts for conversion to owner-occupied units. According to Hamnett and Randolph (1988), a growing 'value gap' between the tenanted investment value and the vacant possession value of privately rented blocks offered property developers an attractive investment opportunity. As a result, extensive tracts of low-rent accommodation were lost within the heart of London – at first in boroughs like Westminster, Chelsea and Kensington, then on the eastern side of the City in and around the South Bank and the Docklands.

Since the formulation of the 'rent gap hypothesis', other researchers have questioned Smith's rather one-sided account of gentrification. Despite the driving force of investors seeking high yields in these initially undercapitalised housing submarkets, Hamnett (1991) and Ley (1996) insist that we also need to know, 'What's producing the gentrifiers who want to return to these neighbourhoods to live?' There is clearly a connection between economic restructuring, which is concentrating jobs at both ends of the socio-economic and pay scales (Chapter 5), the growth in professional and managerial jobs in 'global city' labour markets like London, New York and Sydney, and the gentrification of traditional working-class neighbourhoods not far from the office core. On this basis, Sassen (1991) argues that the 'production of gentrifiers' is explicable in terms of socio-economic polarisation within global city labour markets.

But in a city like London, which has not been a beacon for poor immigrants to the same extent as New York or Los Angeles, the process of socio-economic change is closer to one of 'professionalisation' (Hamnett 1994). This has resulted in relatively more high-end jobs being created in Central London, which helps to explain the presence in areas undergoing gentrification of university-educated, and therefore very well-paid, city workers. These are typically the young professionals who filled the glamorous jobs created in the financial and business houses in the City of London, or mid-town Manhattan, among the exuberance and excesses of the 1980s. Such households are popularly characterised as 'yuppies', 'dinkies' (double-income, no kids) or, more grandly (and by academics, of course), as the 'new middle class' (Ley 1996; Butler 1997). Although applied to London in the first instance, the professionalisation hypothesis offers a much more plausible rationale for the differing contexts in which gentrification is occurring in North American, European and Australian cities (van Weesep and Musterd 1991).

Damaris Rose (1988), a Canadian urban geographer, makes a claim for the importance of the changing role of professional women and the preference of those working in the city for an apartment close by. She also points out that single parents, and other women living alone or sharing, also find that the inner-city social environment is more supportive of their lifestyle than the family-oriented suburbs. The greater number of young women being appointed to the professions and management makes for more pressurised and highly paid individuals, or dual-career households in which two professionals share parenting and want inner-city housing close to work. Studies of London gentrifiers suggested to Lyons (1996) that young and single professional women – 'singles' – probably played a greater role in transforming inner London's housing market throughout the 1980s than households with two high-status workers. Where female partners were contributing a second wage to help with the purchase of a home in inner London, they were much more likely to have a less demanding part-time job than a high-status professional career. Lyons (1996, 341) concludes that 'feminisation of professional employment has as yet had only a marginal impact on gentrification, whereas casualisation of professional employment has been an important influence'. According to Warde (1991, 223), 'it is the ways in which women adapt to new patterns of employment that provide the most plausible explanation of the origins of the [gentrification] process'.

A further perspective emphasises the underlying importance of consumption to the lifestyle choices of the 'new middle class' and where they can best access fashion goods and leisure activities (Warde 1991). The inner city offers high-paid workers with disposable income, greater scope for eating out, visiting galleries and museums, taking in music and theatre, or attending festival events. In her study of loft living in New York City, Sharon Zukin (1982) elaborates on the role of 'cultural capital' in transforming inner-city neighbourhoods. Small businesses and craft industries occupied the lofts from the middle of the nineteenth century. Then, during the 1950s and 1960s, these businesses increasingly lost their competitive edge and had to close down, leaving a lot of disused multifloor buildings and warehouses in the lower part of Manhattan. In the early 1960s, struggling artists, craft workers, students from NYC university and those inclined to a bohemian existence began to remodel the lofts. As employment in the financial district and the lower part of Manhattan expanded in the 1980s, some of those workers began to appreciate the attractions of loft living. By 1977, the average rent for a loft (US$400) was bringing as much as a one- to three-bedroom apartment (US$350–450). Such a loft was the setting for the movie *Fatal Attraction*, which brought together one of the partners in an architectural practice, played by Michael Douglas, and his obsessed lover, played by Glenn Close.

Until 1975 the legality of occupying lofts in New York City was ill defined, and because of this banks were generally reluctant to lend funds for conversion and renovation. With more professional developers entering the market, the City of New York amended a section of its Administrative Code to offer a combination of long-term tax abatement and tax exemption to developers or owners undertaking the residential conversion of large commercial and manufacturing buildings. A series of amendments to the planning code during the period 1961–76 removed much of the uncertainty for developers and the investors that backed them, and eased the conversion of lofts (Table 6.6).

The displacement effects of gentrification are reflected in the ongoing struggle of long-time residents to 'save the neighbourhood' as much as in well-documented research. In the UK, following the Milner–Holland Commission of Inquiry, and in the Netherlands, tenants are generally afforded some security of tenure by the respective Rent Acts. But in the United States, tenants can be served with a 60- to 90-day notice to quit. As a result of the political outcry over condominium conversions in the 1970s, the US Congress directed the Department of Housing and Urban Development to investigate the scale, causes and impact of displacement over a six-month period (Le Gates and Hartman 1986). Some 20 years later, the survey remains one of the most thorough of its kind. Households in 12 major cities with high levels of conversion activity were investigated. In the three-year period 1977–80, 58 per cent of the original households had been displaced. Of the incoming residents, 70 per cent were new owners. An estimate for New York City puts the numbers of

tenants displaced during the decade before the real surge in financial and business services (1970–80) at about 250,000. But not all were casualties of gentrification because landlords intending to demolish or even abandon buildings to New York City have first to remove the occupants.

HOMELESSNESS

When pressures build up in local housing markets, this eventually leads to a loss of affordable accommodation. This was the fate of much of the cheapest housing in inner-city neighbourhoods targeted for revitalisation and gentrification on both sides of the Atlantic during the 1980s. On Manhattan's Upper West Side, for example, block after block of apartment buildings given over to single room occupancy (SRO) were reclaimed by the renovators for *Seinfeld*-like yuppies. The problem during the 1980s and 1990s was compounded in cities like Los Angeles because the collapse of affordable housing coincided with the economic marginalisation of poorly skilled workers and cutbacks in welfare programmes like Aid to Families with Dependent Children (AFDC), and the Food Stamp and Medi-Cal benefits (Wolch 1996). At the same time as the federal funding of public housing construction dried up and additions to multifamily housing stalled in LA County, tracts of cheap housing were being demolished throughout the west side (San Pedro) and around the central city (especially the Central American immigrant district of West Lake).

Los Angeles house prices, which were on a par with the national average in 1974, had leaped ahead 55 per cent by 1985. Regional vacancy rates deteriorated throughout the 1980s and pushed up rents by 50 per cent in real terms during the decade. But, as usual, the tightening rental market impacted selectively: low-rent stock numbers shrank from 35 per cent of the total rental stock in 1974 to 16 per cent in 1985; high-rent stock numbers grew from 14 per cent to 45 per cent during the same timeframe. As a consequence of this combination of structural unemployment, unchecked immigration, shrinking welfare services and a collapse in housing affordability, 'Los Angeles became the homeless capital of the United States in the 1980s' (Wolch 1996, 390). In 1990–91 an estimated 125,600 to 204,000 people were homeless in Los Angeles County at some point

TABLE 6.6 *Amendments to the planning codes that eased the conversion of lofts in New York City, 1961–1976*

Zoning resolutions	Building codes	Other laws
1961 Re-zoning of New York City		
		1962 End of rent stabilisation on commercial property in New York City
	1964 Article 7-B (New York State): legalisation of visual artists' residential use of lofts	
		1965 Landmarks Preservation Law passed (New York City Administrative Code)
	1968 Amendment to Article 7-B: expansion of artists' category to include performing and creative artists	
1971 Creation of artists' district within SoHo† manufacturing zone	1971 Amendment to Article 7-B: eased building restrictions for lofts in converted buildings; simplified artists' certification procedure	
		1973 Designation of SoHo as historic district (Landmarks Preservation Law)
		1975 Revision of J-51 program to provide tax benefits for large-scale residential conversion (New York City Administrative Code)
1976 Expansion of artists' district in SoHo†; legalisation of artists' residential conversions in NoHo‡ Creation of TriBeCa*, legalising residential conversion for artists and non-artists		

† Area south of Houston St.
‡ Area north of Houston St.
*Triangle beneath Canal St.
Source: Zukin (1982). Reproduced by permission of Johns Hopkins University Press

during the year, and between 38,420 and 68,670 people were homeless on any given night.

Although this coincided with a recession in the early 1990s, homelessness has become a permanent feature of urban life in all big cities regardless of overall affluence. A three-year survey of homeless people using shelters and soup kitchens in US cities, and carried out by the Urban Institute for HUD, shows that an estimated 470,000 used a shelter on an average night in February 1996 (reported in the *New York Times*, 8 December 1999, 1). Two-thirds were suffering from chronic or infectious diseases, not counting AIDS; 55 per cent lacked health insurance; 39 per cent exhibited signs of mental illness; 27 per cent reported a childhood history of foster care or living in an institution. Nevertheless, people using shelters constitute only about a quarter of those who are sporadically made homeless by economic circumstances. Over 40 per cent of those interviewed by the Urban Institute's Social Service Research Unit said what they needed more than anything else was a job.

THE CYCLE OF NEIGHBOURHOOD CHANGE

The dynamic forces that bring together supply and demand in housing markets underpin the formation and progressive evolution of neighbourhoods and suburbs. This cycle of neighbourhood, or suburban change, which is going on all the time, steadily changes the geography of housing and the built environment as well as the social geography of communities in cities (Chapter 7). Until the onset of revitalisation and gentrification in the late 1960s, conventional wisdom held that urban localities proceeded through a fairly orderly sequence of change from the original subdivision and settlement of a neighbourhood or locality, to consolidation and maturity, prior to entering a downward phase of disinvestment, thinning out, slum formation and finally abandonment. This left an area open to re-zoning, perhaps for a 'thruway', or ripe for clearance and renewal by a public housing agency.

However, Bourne's (1981) model in Table 6.7 modifies the traditional cycle of neighbourhood change to take account of the interruption of the sequence of decline by rehabilitation and gentrification. The chart indicates how the social composition of communities is expected to correspond with investment and construction of housing at differing stages of the neighbourhood cycle. New growth areas are likely to be dominated by single-family dwellings housing young, high-status and quite mobile families. Project developers have made a speciality out of large-scale subdivision, which makes for very homogeneous communities at first. Obviously

Bourne drew this up in the late 1970s with mainly Canadian and American communities in mind. For one thing, not all suburbs in Australia or New Zealand have been built for upper-middle-class families; for another, many of the fastest-growing suburban areas in the United States, both during the 1990s and in the foreseeable future, cater for retirees rather than young families.

HOUSING MARKETS AND INCOME POLARISATION IN CITIES

Property cycles and house price gaps

Booming property markets reward home-owners by increasing asset values and their net wealth, and penalise tenants by increasing their rents and undermining security of tenure. Since the Second World War, Britain has experienced three major booms (1970–3, 1978–9 and 1983–8) and Australia two (1972–4 and 1986–9). However, a slump in house prices, which tends to be triggered by the tightening of credit in the form of more expensive home loans, can catch out recent, or otherwise overextended, home-buyers. This is what happened in Britain in the early 1990s. In London and the south-east house prices fell by 30 per cent in cash terms and even more in real terms, so that some owners were hit by the 'negative equity' phenomenon. This meant that, after the slump in house prices, they were left to pay off mortgages that exceeded the written-down value of their asset. Almost half a million households lost their homes to bank repossession between 1990 and 1998 (Hamnett 1999c, 15).

Although most OECD countries appear to have tamed excessive price and asset inflation, and there are no more demographically driven booms on the immediate horizon in these countries, household formation rates are rising along with the growth in one- and two-person households. Also, stock market earnings since the late 1990s threaten to trigger a price breakout in some cities. Data on real estate trends from London and Manhattan during 2000 indicated that, if anything, housing markets in global cities have continued their upward spiral. For example, the average price for a Manhattan apartment in the first quarter of 2000 soared to US$733,860, which was 31 per cent higher than the same quarter in 1999 (reported in the *Australian*,

TABLE 6.7 *Neighbourhood and housing life cycles*

Sequence	Investment and construction		Social attributes				Other characteristics
	Dwelling type (predominant additions) and tenure	Levels of investment and construction	Population density	Household and family mobility structure	Class, social status, income	Turnover, migration	
Suburbanisation (new growth) 'homogeneity'	Single-family (low-density multiple); owner-occupied	High	Low (but increasing)	Young families; small, large households	High (increasing)	High net immigration; high mobility; turnover	Initial development stage; cluster development; large-scale projects
Infilling (on vacant land)	Multifamily; rental	Low; decreasing	Medium (increasing slowly or stable)	Ageing families; older children more mixing	High (stable)	Low net immigration; low mobility turnover	First transition stage – less homogeneity in age, class, housing; first apts.
Downgrading (stability and decline)	Conversions of existing dwellings to multifamily; rental	Very low	Medium (increasing slowly); population total down	Older families; fewer children	Medium (declining)	Low net outmigration; high turnover	Long period of depreciation and stagnation; some non-residential succession
Thinning out	Non-residential; construction – demolition of existing units	Low; increasing	Declining (net densities may be increasing)	Older families; few children; non-family households	Declining	Higher net outmigration; high turnover	Selective non-residential succession
Renewal	(a) Public housing; rental	High	Increasing (net)	Young families; many children	Declining	High net immigration; high turnover	The second transition stage; may take either of two forms depending on conditions
	(b) Luxury high-rise apt. and townhouse	High	Increasing	Mixed	Increasing	Medium	
Rehabilitation and gentrification	Conversions	Medium	Decreasing (net)	Few children	Increasing	Low	

Source: **Bourne (1981)**

22 May 2000, 37). In some parts of London – N2, SE14, E16 – house prices rose by as much as 45 per cent in the year ending March 2000. The price of a fairly average 'semi' in a fairly average London borough had climbed to £200,000. (A 'semi' is a semi-detached dwelling, usually with the living area downstairs and the sleeping area upstairs.) In gentrified boroughs like Westminster, and Kensington and Chelsea, the average weekly rent for a two-bedroom flat was £399 and £429 respectively (Hamnett 1999b).

Between 1995 and 2000 (Table 6.8) rising house prices in Australia did not approach earlier 'boom-time' conditions. Only in Sydney and Melbourne have house prices recorded significant real gains while those in the other cities are illusory. In Brisbane, Adelaide and Hobart, house prices in the cheapest suburbs actually fell in real terms right through the course of the longest economic upturn since the 1960s. None the less, the gap between the value of homes in the most exclusive and the poorest suburbs has continued to widen (Table 6.8). What tends to happen in 'bullish' housing markets is that sought-after homes in prestigious suburbs fetch higher prices at auction. Also, 'bargain hunters' descend on property hotspots in desirable postcode areas 'where incomes are high, supply is short and competition is fierce' (Hamnett 1999b, 2).

Housing wealth

The contribution of home-ownership to the distribution of wealth in the property-owning democracies has been hotly debated for some time (Forrest and Williams 1990; Saunders 1990). Are capital gains and losses within urban property markets repaying owner-buyers according to how much they earn? Or does home-ownership help to spread wealth more evenly across the household income distribution in cities? It is not a trifling matter, since most of the private wealth that households accumulate over the lifetime is held in this form (Hamnett 1999c). Until the 1990s, for example, when share market trading caught on, housing assets accounted for almost 65 per cent of personal wealth in Australia.

It seems that, when researchers delve into lifetime ownership, the pattern of capital gains and losses accruing to households who have owned consecutive homes provides quite a good match with lifetime earnings (Badcock and Beer 2000). This is also the conclusion reached by Andrew Burbidge (2000), who analysed the housing histories of over 2500 families who bought and sold houses in four Sydney suburbs and four Melbourne suburbs in the late 1980s:

When measured in dollar terms, analysis of the Melbourne and Sydney data showed that the distribution of net housing benefits had a substantial class bias in the years following deregulation of financial markets [in 1986], providing significantly smaller benefits to low- and middle-income homebuyers, blue- and lower white-collar occupation groups, those dependent on social security pensions or benefits and recent migrants. As Badcock found in Adelaide (1994, 266), the benefits of homeownership appear to increase absolute differences of income and wealth, helping to create a more unequal society.

(Burbidge 2000, 277)

Geography of income polarisation in cities

Processes in the housing market join up with the outcomes of labour market restructuring to change the map of income distribution in cities. Gregory and Hunter (1995) demonstrate how employment change affected the distribution of income at the Collector's District (CD) level in Australian cities between 1976 and 1991. Although CD and community boundaries do not necessarily coincide, they are sufficiently small areas to measure fine-grain socio-economic differentiation in cities. The 9483 CDs that they used had to have the same boundaries in 1976 and 1991. At the time of the 1991 census, they covered about eight million people living in cities over 100,000. Figure 6.6 charts the percentage change in the expected hourly wage of workers against the percentage change in average household income. It shows the two graphs diverging through the socio-economic scale with expected wages and household incomes falling in low-status CDs, and rising sharply towards the highest-status CDs. Significantly, the higher the socio-economic status (SES) of these urban localities, the greater the amount by which household incomes exceed expected wage gains (Fig. 6.6). Gregory and Hunter (1995) estimate that average household income in the lowest 5 per cent of SES 'localities' fell by about 23 per cent, or A$7,500, at 1995 prices, while rising by 23 per cent, or around A$12,500, for the top 5 per cent of SES 'localities'.

TABLE 6.8 *Growing gap in 'rich' and 'poor' suburb house prices (A$), Australian cities, 1995–2000*

Median house price in June 2000	Dearest suburbs	% change since 1995	Cheapest suburbs	% change since 1995
Sydney A$298,000	Mosman A$1,071,000	90	Campbelltown A$164,000	31
	Woollahra A$895,000	38	Wollondilly A$179,000	45
	Hunters Hill A$818,000	115	Penrith A$180,000	36
	Waverley A$669,000	83	Blacktown A$185,000	48
Melbourne A$245,000	Toorak A$1,040,000	80	Melton A$102,000	20
	Brighton A$630,000	78	Dandenong A$106,000	18
	Canterbury A$619,000	84	Healesville A$115,000	21
	Armadale A$596,000	58	Broadmeadows A$117,000	58
Perth A$150,000	Peppermint Grove A$1,128,000	39	Medina A$60,000	22
	Dalkeith A$721,000	45	Calista A$69,000	4.5
	City Beach A$521,000	52	Langford A$72,000	−2.7
	Cottesloe A$521,000	55	Parmelia A$75,000	4.1
Brisbane A$195,000	Hamilton A$499,000	58	Inala A$69,000	−10
	Ascot A$483,000	29	Redbank Plains A$79,000	−18
	Chelmer A$404,000	44	East Ipswich A$83,000	−5
	Brookfield A$347,000	−8	Deagon A$85,000	−27
Adelaide A$130,000	Adelaide A$363,000	120	Elizabeth A$64,000	−3
	Walkerville A$319,000	38	Salisbury A$89,000	8.5
	Burnside A$290,000	33	Noarlunga A$94,000	8
	Kensington-Norwood A$263,000	56	Munno Para A$98,000	29

Source: Collected from state valuation offices and reported in the *Weekend Australian*, 24–25 June 2000, 20–2

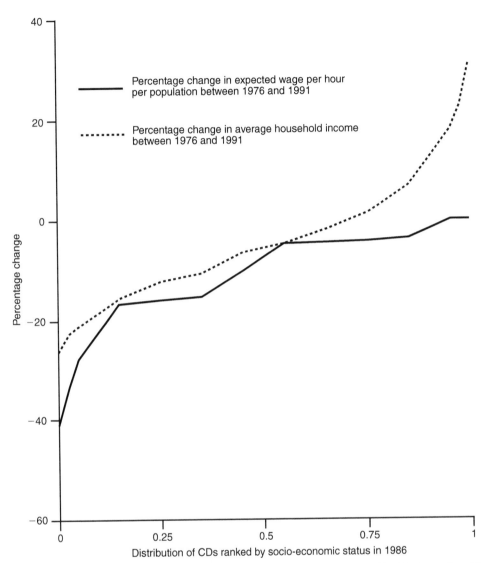

Figure 6.6 *Effect of employment change on income distribution at neighbourhood level in Australian cities, 1976–1991*
Source: Gregory and Hunter (1995)

At the risk of overdrawing these connections, it is apparent that the fortunes of households (and suburbs) in city after city are bound up with their participation in internationally exposed sectors of the economy (Chapter 3):

High incomes bring with them access to high cost housing areas and also a high propensity for outsourcing household consumption. In this fashion, high skill locations enjoy a virtuous cycle of high income, high consumption and high employment. By contrast, low skill areas are

locked into a cycle of low consumption and low employment. Successful individuals usually move out of these areas while new, poorly skilled residents move in, producing a churning effect.

(Latham 1996, 23)

This broad picture of 'winners' and 'losers' at the community level in Australian cities is supported by data from the Australian Taxation Office. The analysis is based on the reported taxable incomes of individual wage earners, since the social security

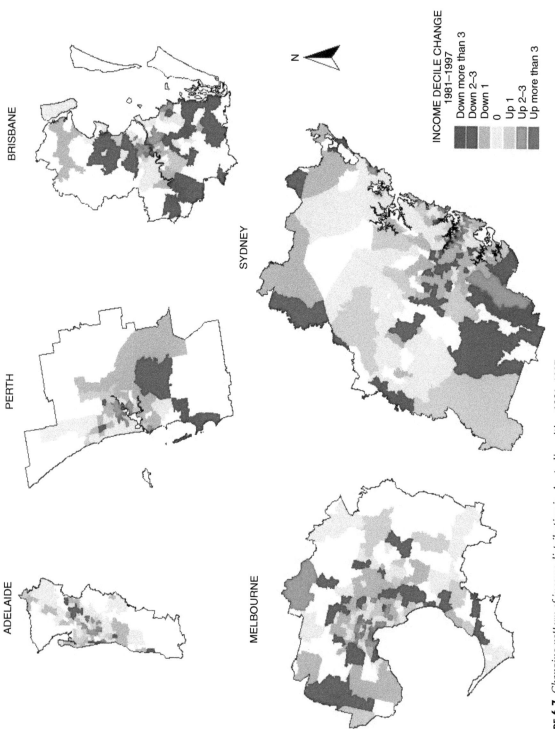

Figure 6.7 *Changing pattern of income distribution in Australian cities, 1981–1997*
Source: author's calculations, based on Australian Taxation Office data

ADELAIDE

PERTH

BRISBANE

SYNDEY

MELBOURNE

INCOME DECILE CHANGE
1981–1997

Down more than 3
Down 2–3
Down 1
0
Up 1
Up 2–3
Up more than 3

N

payments made to beneficiaries generally sit beneath the threshold at which tax is paid. In 1997 personal income under A$21,377 was untaxed. Figure 6.7 compares the shift of average postcode income between decile classes over a period when federal Labor and Liberal–Country Party Coalition governments were in power (1981–97). To make the comparison, 1981 dollars were inflated so they have the same value as 1997 dollars. The statistical analysis on which the maps are based shows the following.

- Average postcode incomes declined in the bottom decile for each city other than Perth, and rose in the top decile. In the case of Sydney, there was a drop from A$26,886–28,670 in 1981 to A$25,309–28,646 in 1997. Roughly, average personal incomes in each city fell by a couple of thousand dollars per worker in the lowest decile. By contrast, average personal incomes in the top decile rose by at least A$5000 in each city. But in Sydney, average personal income in the top decile climbed from A$37,290–51,669 in 1981 to A$50,621–76,027. Consequently, wealthy suburbs are concentrating income.
- Inner-city postcode areas made up the most ground over the 16 years or so. Even though some of these neighbourhoods in Melbourne and Sydney still house poor households in high-rise flats, this is consistent with the impact of gentrification upon working-class communities.
- With a few exceptions on the outskirts of Melbourne and Sydney, many postcode areas in the outer suburbs suffered from falling income during the 1980s and 1990s. Often these suburbs coincide with public housing estates built in the 1960s and 1970s.

When 'polarisation scores' recording the number of decile places each postcode shifted up or down in the income distribution are averaged, it seems that Brisbane, with a score of 1.78, became more polarised between 1981 and 1997, followed by Perth (1.55), Melbourne (1.41), Sydney (1.32) and Adelaide (0.94). Of course, even though their incomes may have risen or fallen further during the period, the most exclusive postcode areas and those with a large public housing presence tended to stay put in the top and bottom income deciles. While not without problems, this simple statistical exercise does tend to bear out what many commentators are saying about the growing socio-spatial polarisation of communities in large urban areas undergoing economic restructuring.

Equally gross disparities exist between the average incomes earned by households in different geographical localities in UK cities. A market research firm that surveys household incomes every three years (the CACI Group's Paycheck Report) established that Temple (EC4Y 7), with an average household income of £51,900, is home to the most highly paid workers in Britain. At the other extreme is Bootle (L 59) with an average household income of only £8200 in 1999 (reported in the *Guardian* 1999). Nine of the 20 poorest postcode areas in the country are concentrated in Liverpool. Well over 80 per cent of all the households in Liverpool's Edge Hill, Birkenhead, Bootle, Middlesbrough, Leicester and central Belfast had to get by on less than £13,000. By contrast, four in every ten households with homes in the central London postcode areas of Barbican, Blackfriars, Belgravia, east and west Temple and Embankment grossed in excess of £50,000.

These disparities affect many aspects of the quality of community life. To take just one example, consider how parents' wealth influences the amount of money schools can raise to supplement the allocation they receive from governments. 'Location, company and trust involvement, and the ability to make effective grant applications all influence the money available to schools' (reported in *The Times*, 12 May 2000, 6). A survey of English schools revealed that 20 per cent of primary schools and 5 per cent of secondary schools raise less than £1000 a year from parents and other private sources. On the other hand, more than 40 per cent of independent schools raised more than £250,000 a year and 21 per cent had long-term development appeals.

A truly divided city: Washington DC

Paradoxically, because it is the capital of the United States, Washington DC represents the extreme version of income polarisation, doubling up as racial segregation, in the one city. 'This is the same city where the President of the United States lives,' said Mike Johnson, a community advocate from Ward 8, the city's poorest section, 'and we're living in a war zone. How is he going to go overseas and tell other people how to live in peace, and he allows this to go on a few miles from the White House?' (reported in the *New York Times*, 26 July 1996, A18). In the early

1970s, the District of Washington was granted a version of home rule that seemed to offer the promise of better things to come whether you lived in Ward 3, the city's richest and whitest, or Wards 6, 7 and 8, the city's poorest and blackest (Table 6.9).

For black Washingtonians, who made up 65 per cent of the city's population, home rule meant the declaration of independence from the white Federal law-makers who had been in charge. The funk-rock group Parliament caught the spirit in their song 'Chocolate City' and celebrated Washington's new home rule as black America's 'piece of the rock' surrounded by 'vanilla suburbs': 'Hey, we didn't get our 40 acres and a mule, but we did get you Chocolate City. You don't need the bullet when you've got the ballot.' But, by the mid-1980s, the dream had soured somewhat, along with a deep recession, a fallen idol in Mayor Barry, and crack cocaine. The divide could not have been deeper by the mid-1990s: west of the Anacostia River there were 30 welfare cases and a handful of homicides; east of the river, 7000 welfare cases and over 400 homicides. In Ward 8, 26 per cent of the households lived at the Federal poverty level, compared with 6 per cent in Ward 3. Ward 8 had only one public library branch and a smaller community library. Ward 3 had four university libraries, three public library branches, plus a larger regional library. Although almost half of the births in the city

take place east of the river, there are only a few paediatricians there willing to accept Medicaid patients. East of the river there are only a couple of restaurants where a family can have a meal. Even McDonald's was closed and boarded up in the mid-1990s. 'Mom-and-pop delis' were also being shut down for selling drugs along with the subs. There was no children's clothing store, no movie theatre and no health club or auto repair shop. But there were plenty of liquor stores.

By contrast, Ward 3, in the north-west of Washington DC, 'remains a sanctuary of middle- and upper-class privilege, of solid public schools and of higher-priced housing enjoyed by some of the most established powers of American government, commerce and culture, and of world diplomacy as well' (reported in the *New York Times*, 26 July 1996, A19). Ward 3 was 88 per cent white in 1990, three out of five adults had college degrees, while median incomes were 60 per cent higher than the city's average (Table 6.9). The north-west was distinguished by upper-income shopping strips, glistening food marts and spacious supermarkets. To sum up:

The splendor of the Northwest makes Washington a graphic tale of two cities. Proud ward residents, endlessly active in civic matters, enjoy the most expensive housing in Washington, with 97 per cent of the single-family dwellings assessed at more than $200,000 and only 1 per cent of

TABLE 6.9 *Ward 8 and Ward 3 population profiles, Washington DC, 1990*

Social indicators	Ward 8	Ward 3	Washington DC
Population 1980	80,937	73,000	
Population 1990	72,221	77,774	
Blacks (%)	91	6	66
Whites (%)	8	88	30
Median household income (US$)	21,312	48,967	30,727
Children under 18 not living with both parents (%)	74	16	66
Births out of wedlock (1989–90) (%)	84	6	67
Violent crimes reported per 1000 residents	40	3	25
Persons over 25 with 4 or more years of college (%)	8	70	33
Single-family homes valued at >US$200,000 (%)	less than 1	97	31

Source: US census statistics held by Office of Planning/State Data Centre; reported in the *New York Times*, 26 July 1996, A18–19

the ward's homes being subsidized. In Ward 8, east of the Anacostia River, 23 per cent of housing is publicly subsidized.

(Reported in the *New York Times*, 26 July 1996, A19)

Locational disadvantage

What are the disadvantaging aspects of place and community? How are people short-changed because of where they live? These are difficult questions to iron out because of the way in which various forms of disadvantage intertwine. When economic disadvantage in the form of relative poverty, and social disadvantage or deprivation, are taken into account, is there any further remaining source of disadvantage? Many urban geographers believe so. There are four qualitatively different attributes that are specific to urban communities that go to make up locational disadvantage. These are as follows.

- Relative accessibility and the attendant costs of overcoming or eliminating the effects of distance (this feeds back into economic disadvantage). Naturally the inconvenience and extra costs of isolation in outer suburbs are compounded for households without access to cars and if local public transport services are infrequent.
- The range and quality of community assets available locally such as educational and medical services, libraries, swimming pools, parks and clubs.
- Environmental amenity/disamenity, including proximity to sources of pollutants and other hazardous agents.
- The reputation of a place. i.e. 'postcode' or 'zip code' discrimination. This can work against local residents in the job market (Chapter 5). It can lead to 'redlining' of suburbs for mortgage finance, which means that banks will not lend in suburbs with a poor reputation. Areas with a higher incidence of crime increasingly attract higher premiums for property insurance. Home-owners in suburbs where there is little demand for housing may find that the capital value of their homes falls in real terms rather than accruing in value over the years.

Ironically, some of the most uninviting, even horrible, places in cities to live are those built by the state in the post-war era. The United States has its 'project' housing, the UK has its 'sink estates', and Australia and New Zealand have their 'dumping grounds' in the outer suburbs. Because of misguided government policies in the 1980s and 1990s, these areas now house more than their fair share of economically marginalised and socially excluded citizens. Although the mix of policies is different in each case, there are recognisable echoes in Nick Davies's (1998) account of the politics of social exclusion in the UK. Tragically, the repercussions of these errant policies all come together in localities ravaged by several decades of industrial decline: 'It's not that the people on those estates are bad, but profoundly that they live in conditions which are bad' (Davies 1998, 32).

As well as the debilitating effects of poverty, Davies singles out the declaration of war on drugs as 'the most serious social policy error since the war'. In his opinion, which is hardly a popular one, the vortex of criminality and violence fostered by the black market on sink estates in UK cities 'has caused far more death and serious illness than the demonised drugs themselves' (Davies 1998, 33). Secondly, he argues that the classic escape route offered to the poor by state education was closed off by the withdrawal of funding and the introduction of 'league-tabling' during the 1990s. The pressure to improve the reputation and performance of a school discourages the admission of troublesome students. Thirdly, according to Davies, the privatisation of council housing has destroyed the communities that once helped to support those who lived there. Fourthly, the 'care in the community' programme has dumped vulnerable and sometimes disruptive patients into the cheap housing on these estates, 'sometimes to sink without trace into a pit of loneliness, sometimes to enrage their neighbours with the symptoms of their illness' (Davies 1998, 34). Finally, there is a cluster of corrosive problems caused by cuts in public services – buses that no longer run, libraries that no longer open, clubs and youth clubs that have folded. A whole layer of social supervision has disintegrated – the park-keeper, the caretaker, the railway porter, the bus conductor – inviting trouble.

CONCLUSION

This chapter has described how housing markets and housing agencies allocate households according to their incomes or needs to different localities within the city. Market income and earnings play an inescapable part in the residential location process; but within the different socio-economic strata and income segments of these submarkets, social and cultural housing preferences that can be traced to demographics, gender relations and ethnic identity lend an added richness to the human landscape of cities. Chapter 7 paints a picture of the social and cultural mosaic of cities, and endeavours to identify the sources of this diversity and difference.

FURTHER READING

Bourne's (1981) text still provides a comprehensive overview of the housing system from a geographical perspective. Recent studies can be sourced from two leading journals, British and American respectively: *Housing Studies* and *Housing Policy Debate*. Similarly, the websites for the two corresponding central government agencies – the Department of Transport, Local Government and the Regions in the UK and the Department of Housing and Urban Development in the United States – provide current information on these national housing systems.

Maclennan and Pryce (1996) provide one of the few analyses that attempts to set out the connections between global economic change, the (UK) labour market and urban housing markets.

Greater detail on the transformation of housing markets arising from the effects of globalisation on world-cities like London and New York can be found in **Fainstein** *et al.* (1992), and **Hamnett** and **Randolph** (1988).

Sharon Zukin's classic monograph on *Loft Living* (1982) was one of the earliest studies of the conjunction of culture and the revival of urban neighbourhoods.

7

SOCIAL AND CULTURAL
MOSAIC OF CITIES

GEOGRAPHIES OF SOCIAL AND CULTURAL DIVERSITY

Urban cultural diversity holds a curious and yet wondrously creative mirror to the paradox of polarisation: while cities become more like other places, they continue to attract the extremes of poor, migrant and footloose populations and the very rich. Their ability to forge 'urban' lifestyles continues to be the city's most important product.
(Zukin 1998, 840)

Sharon Zukin's quote awakens us to one of the defining features of cities caught between the twilight of modernity and the dawn of post-modernism. When writing about the modern industrial city in the twentieth century, researchers gave special emphasis to geographies of class and ethnic segregation. Two of the most widely used textbooks in urban social geography (Knox 1982; Ley 1983), which were both published in the early 1980s, reflect not only the prevailing social character of late-twentieth-century cities in the West but the bias at the time towards positivist social science. Both draw upon studies that use scientific methods to describe residential patterns in cities and behavioural aspects of urban life.

In the intervening years, western cities have undergone a series of quite profound demographic and social changes rivalling the social upheaval that occurred during the course of the Industrial Revolution. In those parts of the New World settled during the eighteenth and nineteenth centuries, indigenous cultures were quickly overwhelmed and pushed to the margins of the colonial city by pervasive European values and customs (Jacobs 1996). In just

a few colonial cities 'outsiders' were occasionally recruited to do the dangerous and dirty work. For example, with the end of the gold rush era in the New World, some Chinese stayed on to lay the foundations for the 'Chinatowns' of New York, San Francisco, Vancouver, Sydney and Melbourne. Also, in the United States after the Civil War, enslaved black Africans were finally free to walk off the plantations in the south and migrate to northern cities. This process quickened during the Second World War when black men and women flocked to factory jobs in America's industrial cities to assist the war effort.

Whereas the cultural legacy of European colonisation was largely that of white domination, in a post-colonial era the internationalisation of migration is increasing the cultural diversity of many western societies. Unlike European migration during the nineteenth century, migrants arriving in cities today are drawn from an extraordinarily diverse array of Asiatic, African, Middle Eastern and Hispanic cultures. Although the use of the term 'multiculturalism' attracts some criticism, it is widely used to describe this phenomenon, and the changes it has brought to the host societies of Europe, North America and Australia. The process started with decolonisation at the end of the Second World War and the resettlement of migrants from former colonies. It has continued with the worldwide movement of foreign workers, displaced persons, political refugees and illegal immigrants (Chapter 1). Regardless of how they enter a country, migrants seek out their own national, linguistic, ethnic, racial and religious groupings in the big cosmopolitan cities. Consequently, the cities of former empires like

London, Paris, Rome and Amsterdam, along with traditional port-of-entry cities like Los Angeles, New York, Honolulu and San Francisco in the United States, Sydney in Australia or Auckland in New Zealand, possess the most multicultural and demographically complex urban populations.

As an illustration of this multicultural diversity, 'Los Angeles is the second largest Mexican, Armenian, Filipino, Salvadoran, and Guatemalan city in the world, the third largest Canadian city, and has the largest Japanese, Iranian, Cambodian, and Gypsy communities in the United States' (Hayden 1995, 84). Auckland, New Zealand, is home to more Polynesians than any other city, many of whom automatically qualify for residence as citizens of former protectorates (i.e. Western Samoans, along with people from the Cook, Tokelau and Nuie Islands). Manukau City in the southern part of the metropolitan region is New Zealand's fastest-growing local authority with a population well over 300,000. A broad breakdown of ethnic composition into European (47 per cent), Pacific Island (20 per cent), NZ Maori (17 per cent) and Asian (9 per cent) does not do full justice to the sprinkling of cultures, language groups and nationalities found in Manukau City at the 1996 census. With their different demographics, the presence of these migrants will begin to slow the ageing of the native-born populations in European, American and Australasian societies.

Expanding opportunities for women outside the home in the new economy are working with broader social forces to dislodge the nuclear family as the dominant household type. In her account of the changing family towards the end of the twentieth century, Judith Stacey (1990, 16) states that men and women 'have drawn on a diverse, often incongruous array of cultural, political, economic, and ideological resources, fashioning these resources into new gender and kinship strategies to cope with post-industrial challenges, burdens and opportunities'. These individual responses, mediated in part by the shift of women into the paid workforce and the challenge this has posed for traditional gender relations within the patriarchal family, are also helping to recast the demographic and social composition of many urban neighbourhoods and suburbs.

The social change that springs from collective organisation and political mobilisation around a shared sense of identity also has the power to transform communities (Castells 1997). The rise of social movements like Civil Rights in the United States, Women's Lib, Gay Liberation and the Greens in Europe, has changed forever the status and role in society of ethnic minorities, women, gay men and lesbians, and environmentalists. In seeking to transform human relationships at their most fundamental level, feminism and environmentalism are proactive social movements. They differ, in turn, from 'a whole array of reactive movements that build trenches of resistance on behalf of God, nation, ethnicity, family, locality' (Castells 1997, 2). Chosen spaces and sites within cities are socially constructed and used by both social movements and cultural minorities to create their own comfort zones and cultural domains (Hayden 1995). Environmentalism can also be reactive when urban communities mobilise to defend their local space against the intrusion of undesirable uses (Chapter 9).

All these disparate strands are coming together in cities like London, Paris, Los Angeles, Vancouver, Miami and Sydney to form the basis of 'new' geographies of cultural diversity and social polarisation. New schools of feminist and post-structuralist theory and cultural interpretation have emerged in the last quarter of the twentieth century to rectify the omissions of earlier research, and to redefine urban societies in terms that attempt to do justice to the nature of the cultural and social transformations taking place (Smith 1999). Early in the 1990s, Sophia Watson and Kathy Gibson (1995) convened a seminar in Sydney to brainstorm the nature of post-modern cities and spaces. The Californian urban geographer Ed Soja (2000) coined the word 'post-metropolis' to capture the essence of the transformation that others describe as post-modern urbanism. His colleagues at USC, Dear and Flusty (1998), go so far as to suggest that this new Los Angeles School of Urban Theory has sufficient cogency to replace the old, classical Chicago School of Urban Theory.

This chapter describes the social and cultural geography of cities caught up in this transition to new forms of urbanism. The main sources of cultural identity and diversity are considered, including race and ethnicity, gender and sexual orientation. They have a powerful bearing, together with class, on the social geography of cities by influencing whom we choose, as social beings, to relate to and interact with – or not – and which social networks and communities we become part of. To begin with, a few select illustrations are provided of the rich cultural and social diversity to be found at a community level in cities undergoing transformation. Then we trace the development of urban social theory and the methods used to account for the residential differentiation of our cities before returning to look more closely at the respective contributions of race and ethnicity, gender and sexual orientation to the make-up of urban communities.

CULTURAL AND SOCIAL SPACE IN CITIES

Cities that have been open to regional and international flows of people harbour a myriad of social and cultural communities that can claim spaces within the city as their own. The cities of former colonial powers now support large immigrant communities. Paris and Marseilles have large Algerian, black African, Moroccan, Tunisian, West Indian and French Indo-Chinese quarters. Important centres that used to link Britain with its colonies (London, Liverpool, Manchester, Birmingham, Leeds, Bradford, Bristol) are now home to black Caribbeans and Africans, Indians, Pakistanis and Bangladeshis. Dutch cities like Amsterdam, Rotterdam, Utrecht and Leiden have strong expatriate communities from the former Dutch West Indies (Indonesia) and Dutch East Indies (Surinam).

International migration is rapidly changing the geography of those globalising cities that are key ports of entry for immigrants from Pacific Rim (Los Angeles, Sydney, Vancouver), Latin American (Miami) and African (Rome) countries. According to the US census, by 1990 the following proportions of the populations of Miami, New York and Los Angeles were foreign born (45.4 per cent, 27.9 per cent, 36.6 per cent) and spoke another language instead of English at home (60.6 per cent, 45.2 per cent, 52.4 per cent). In the space of 30 years, the proportions of Anglos and Latinos in the Los Angeles region were transposed (Table 7.1). In the early 1980s, when Britain negotiated the agreement to hand Hong Kong back to China, Vancouver became the destination of choice for thousands of skilled migrants and entrepreneurs under Canada's business migration programme. Between 1991 and 1996, almost 150,000 migrants from Asian countries settled in Vancouver. Hong Kong was the major source of immigrants with 44,715, followed by China (27,005),

TABLE 7.1 *Changing ethnicity in Los Angeles, 1970–2000*

Ethnicity	1970	1980	1990	2000
Persons (000)				
Anglo	4885	3849	3619	3385
Black	747	929	993	1070
Latino	1024	1918	3230	4805
Asians	234	645	1021	1664
Per cent				
Anglo	70.9	52.4	40.8	31.0
Black	10.8	12.7	11.2	9.8
Latino	14.9	26.1	36.4	44.0
Asians	3.4	8.8	11.5	15.2

Source: US census data prepared by Ong and Blumenburg (1996). Reproduced by permission of University of California

Taiwan (22,315), India (16,185), the Philippines (13,610) and South Korea (6,335). Well-settled, second-generation Turkish communities can be found in European cities like Hamburg, Bremen, Brussels, Stockholm, Paris and Amsterdam – even in the face of sporadic racism and the denial of citizenship (Özüekren and van Kempen 1997).

The scale of Miami's ethnic change between 1960 and 1990 is unprecedented, even for a country dependent on overseas migration for its growth. This is due largely to the inflow of immigrants from Central and South America. It is twice as high as that of New York, and by far the highest of any other city in the United States (Nijman 1996, 284). Whereas immigrants from Latin America made up only 5 per cent of the metropolitan population in 1960, by 2000 they represented more than half of the city's two million people. According to the US census, in 1990 almost 66 per cent of all Latin people were Cubans (428,965), followed by Nicaraguans (67,948), Haitians (45,149), Colombians (43,387), Jamaicans (30,952), Puerto Ricans (18,699), Dominicans (16,394) and Hondurans (15,988). Ethnically, about one-half of Miami's population was Hispanic, about 30 per cent were non-Hispanic whites, and 18.5 per cent were non-Hispanic blacks. Consequently, a city like Miami consists of at least three communities with their own mainstream, hierarchy, social values and, most importantly, their own sense of cultural identity: 'while the Anglos make up Miami's old establishment, newcomers have either not assimilated at all but instead created their own mainstream (Cubans), or they have assimilated to the Cuban mainstream (Nicaraguans), or to the African-American mainstream (Haitians)' (Nijman 1996, 284).

This reflects the extent to which ethnicity and race are also a matter of culture. But from time to time, alliances and cleavages among Miami's Hispanic, Anglo and black communities also take shape around other socio-economic issues. First, both Anglos and blacks have long opposed the massive influx of Cubans and other Hispanics. On the other hand, Anglos and Cubans have shared the fruits of upward mobility, often find themselves in the same class position with respect to jobs and income, and tend to take a dim view of the high unemployment, poverty and crime rates among Miami's blacks (Nijman 1996, 291). There is no doubt that the city's black population has been left behind in the process of economic restructuring and internationalisation. Finally, blacks and recent Hispanic immigrants often bear the brunt of the old Anglo establishment's hostility or discrimination.

For quite different reasons, the recognition achieved by social movements like feminism, gay liberation and so-called grey power, and people with disabilities, has given rise to other communities in cities where the unmet needs of women and children, gays and lesbians, the aged and those with disabilities are properly catered for. There are other, subterranean, spaces in cities that are a by-product of the ill-treatment of people at the margins of society. They provide sanctuaries for indigenous minorities like the Aboriginal community in Sydney's Redfern, or the homeless men and women who gravitate to New York's 'skid row' on the Lower East Side (the Bowery), or the 'street urchins' of cities like Bucharest, Bogota, Caracas, Ankara or Manila. These people, who 'sleep rough' and are often moved on by city law enforcement officers, have a very tenuous hold on such space in the city.

Yet other spaces in the city attract people who are drawn by the opportunity to live in hybrid communities with a vibrant cultural and ethnic mix. For over a century in New York City, 'downtown' has been synonymous with certain neighbourhoods below 14th Street on Manhattan Island and any other part of town that is cool, bohemian and a little dangerous:

There is the downtown of the fin-de-siècle Bowery with its dance shows and beer parlours; the 40s West Village with

its writers; the Cedar Tavern of the 50s replete with drunken abstract expressionists; the 60s, when Warhol brought his Factory scene to SoHo, and the 70s return to the Bowery with punk and CBGB's.

But the decade that seems to define downtown for most of us today is the go-go 80s when everything heated up. Wall Street money poured in. Loft prices soared. Starving artists suddenly achieved rock-star status. Drugs were plentiful and strong. Clubs like Area and Palladium became the places to watch bohemia, money and the underworld mix and combust.

(Lewine 1997, 13)

In the late 1990s, with Manhattan's property market heating up, the trend-setters were driven across the East River to edgier parts of Brooklyn. Old Williamsburg on the Brooklyn foreshore, with its abandoned docks, Schaeffer brewery and sugar refineries, became home to 9000 artists – the highest per capita ratio of bohemians in New York (according to *Time Out*). In London, Notting Hill began to take off in the 1960s and Islington in the late 1970s. In the mid-1990s, 'Cool Britannia' was a byword for Camden and Islington:

To be 'cool', all these villages had to strike a delicate balance between hold[ing] on to enough of the danger, the seediness and the ethnic and cultural mix that made them chic in the first place, while surrendering enough of it to make it safe for incomers to enjoy. The moment it becomes too comfortable, trendification leads inexorably to gentrification, accompanied by rapid house-price inflation. At which point the party moves on …

(Reported in *The Economist*, 15 April 2000, 95)

These few opening illustrations serve to suggest, therefore, that the social and cultural space of cities can be highly variegated, particularly when minorities are pressed by mainstream society to segregate as a way of accommodating difference.

SOCIAL INTERACTION, FORMATION OF IDENTITY, URBAN COMMUNITIES

Social interaction and social networks in cities

Despite where it lies on a spectrum from homogeneity to heterogeneity, the demographic, social and cultural distinctiveness of any urban neighbourhood or suburb depends on how people who differ according to class, race, ethnicity, gender and sexual orientation relate to one another and interact socially. In complex urban societies, the possibilities for social interaction are vastly increased beyond the primary relationships that typify village life. Ferdinand Tönnies (1887) was the first theorist to realise that two general forms of social interaction take place in western societies. He proposed that the first of these, *Gemeinschaft*, was characteristic of agrarian communities in which the basic social unit is the family or kin-group. Social relationships tend to be close-knit, highly interdependent, and long standing – maybe over several generations in the same farming community – because of the low levels of mobility. The second, *Gesellschaft*, arose with industrial urbanism and the proliferation of more specialised social roles, where adherence to rational and efficient decision-making and contractual obligations takes precedence over solidarity with kin or loyalty to 'local folk'. Australian students who grow up on a farm in 'the Bush', or leave a small rural town to go to a university in a capital city, have a good sense of these differences in the nature of social interaction. No doubt many Canadian students from the Prairies, or American students from the mid-west, also have the same experience. Because of this, sociologists observing the changes that flowed from the movement into cities in the late nineteenth century began to question whether the close-knit ties and social cohesion we associate with community could be sustained in cities.

Yet primary relationships are the building blocks of our core social networks irrespective of where we might live. These are the social ties based on blood and duty that exist between relatives, or upon the mutual interests and loyalties that bind friends together. Secondary relationships emerge out of the roles we assume in the process of voluntarily joining sporting or social clubs, or service organisations like Kiwanis or Meals on Wheels. They are also characteristic of the more formal social involvement that accompanies membership of a business or trade association, a professional institute or a trade union, or a resident action or environmental lobby group.

Although primary and secondary forms of social interaction usually give a degree of cohesion to the community life of neighbourhoods and suburbs, the social networks of individuals in large urban regions are equally likely to extend well across town. In 1964, an urban planner called Melvin Webber, teaching on the Berkeley campus of the University of California, coined the term 'non-place urban realm' to describe social contacts and networks that are no longer confined to an urban locality. Since then, with the advent of high-volume, inter-continental flight and the Internet as a medium for social interaction, social networks now extend well beyond the confines of physical contact within cities. With the electronic linking of work or leisure roles, 'virtual communities' are being created without the need ever to meet, although sometimes e-mail communication can lead to the social interaction that cyber-enthusiasts call 'face-time'. In the network society, these communities of interest 'may have more in common with one another than with their next-door neighbours' (Cairncross 1997, 242).

'Community'

Primary and secondary relationships form the basis of community where it still exists in contemporary urban societies. While it is sufficient here to note that the term 'community' enjoys everyday use, the construct has become quite problematic for sociologists. The hallmarks of the 'ideal' community include:

- the sociability and 'face-to-face' interaction that characterises neighbouring
- a sense of solidarity and a sense of belonging
- an attachment to place that comes from a shared sense of local identity
- the presence of groups with a strong civic or community service focus.

Notwithstanding the scepticism about the existence of 'community' in contemporary cities, when sociologists have gone looking they have been able to establish that there are neighbourhoods and suburbs where there is still a strong sense of community. For example, in *The Death and Life of Great American Cities*, Jane Jacobs (1961) insists that the city is an inherently human place, where sociability and friendliness can be observed alive and well on the streets of her New York City neighbourhood, Greenwich Village. According to Suttles (1972), a city tenement block, such as those found on the Lower East Side, where neighbourly interaction takes place round the clock, as adults come and go and children play in the streets, is a microcosm of community. Other studies have confirmed that when immigrants from rural societies resettle in a big city they create the same cohesive social networks and sense of collective identity typical of life in the villages they leave behind. Social interaction among the Italians who settled the north-west corner of inner Boston so resembled village life that Gans (1962) called his study *The Urban Villagers*.

Long-established working-class neighbourhoods with a very settled population have also yielded classic studies of community. One of the first, and still the best known, of these is Young and Wilmott's study of Bethnal Green before the East Londoners who lived there were relocated to the New Town of Dagenham further down the Thames. For them, the key to the sense of community was 'a feeling of solidarity between people who occupy the same territory' (Young and Wilmott 1957, 89). They observed at first hand the maintenance of strong and frequent contact between relatives living in the vicinity, supported by social interaction among neighbours on the shop floor, in the high street, down at the pub and at the local football club. By contrast, this strong sense of community, which used to be synonymous with traditional working-class suburbs, has all but collapsed on many of the council housing estates at the centre and outer edges of British cities (Davies 1998). Community development and 'capacity-building' on these 'sink estates' has become a priority, not only for the Blair government but for many states in the USA and Australia with a similar legacy of neglect.

There is now a planning movement called 'New Urbanism', which aims to recreate the idyllic community for 'frightened' middle-class Americans fleeing old industrial cities. While the violence, deteriorating conditions and rising taxes of the cities are the catalysts, New Urbanism has its basis in nostalgia for the Arcadian vision of eighteenth-century village life and in the communitarian agenda promoted by the social philosopher Amitai Etzioni (1995). It finds its realisation in the spread of 'common-interest developments' and gated communities in the United States. By the early 1990s about one in eight Americans lived in the 150,000 master-planned communities, condominia, gated villages and other versions of the common-interest development (McKenzie 1994). By the late 1990s, there were 30,000 gated communities serving as home to over eight million Americans (Blakely and Snyder 1997). Essentially walled compounds with security guards and surveillance systems, gated communities are

proliferating in those states that are the main points of arrival and concentration of Latinos (southern Florida; Orange County; California; Texas; Arizona) or outside 'black' Washington DC (Montgomery County, Maryland).

In the United States, at least, the new breed of community builders of the twenty-first century use restrictive covenants to carry out and enforce their vision of community. What sets these common-interest developments apart is the degree to which they are privately owned and operated by community associations that enforce the by-laws in order to defend property values and preserve the pre-ordained 'character' of the community. They make their own community policing arrangements, and traditional community services like schools, parks, cultural centres and libraries, and even street cleaning and maintenance, are privatised. In this way, these citizens avoid paying taxes to maintain governmental and city services outside their community. While Celebration, Disney's model community in Florida, is the most talked-about post-suburban community of its kind (Ross 1999), the 17,000-resident common-interest development built by Arizona's Delway Corporation in the late 1990s is a more representative version of what McKenzie (1994) calls 'privatopia'. The residents buy in on the basis of owning the education system from 'K thru' 8th Grade', the parks and gardens, and their own neighbourhood vehicle, which is not 'street legal' outside the complex, at a cost of US$15,000. What concerns many thinking Americans is that, in the process of building their own version of community, these citizens are self-consciously eroding the foundations of civil society.

Social categorisation and the formation of identity

The processes of social categorisation and the formation of identity also influence the residential composition and dynamics of urban communities. People are grouped into social categories and can have their identities ascribed in keeping with societal norms and prejudices (male/female, young/old, able-bodied/disabled, 'black'/'white', heterosexual/ homosexual). These norms are constructed to secure and perpetuate the dominant institutional and organisational arrangements of society. 'They are a product of how power and resources – which may be real (money, cars, homes) or symbolic (a question of how people think, and what they take for granted) – are struggled over and manipulated' (Smith 1999, 12). In her Ph.D., Kay Anderson, now a Professor of Human Geography at Durham University in the north of England, argues that Vancouver's Chinatown exists as much as a racial category in the minds of other Canadians as a socially constructed place. According to Anderson (1998), the idea of Chinatown as a racial category reflects the power of place and institutional practice over time in Vancouver. Other aspects of identity might be taken on by choice in the process of defining 'self', rounding out personal experience and giving meaning to life (language, ethnicity, nationality, religion).

Most identities are not set in stone but are fluid, many-faceted and interchangeable. A Londoner may feel English in Britain, British in Europe and European in Asia. Even though we may project a singular identity, it is much more likely to be multifaceted, depending on whom we are interacting with in time and place. Consider the perspective of someone of mixed-race parentage:

As far as mainstream British society is concerned I am black, but sometimes it helps to make the distinction as to what kind of black that is. I don't think that to choose one identity is to reject the other part of your past. It doesn't mean I love my mother any less because I say I'm black and she's white … I've just had a different racial experience from her.

(Darryl Slater, reported in *Guardian Weekly*, 1 June 1997, 23)

This process, whereby people are assigned to, or choose to identify with, one facet of identity rather than another, 'can have a profound impact on what we get out of life' (Smith 1999, 12). Moreover, the possible fragmenting of cultural, sexual or national identity as we respond to 'labelling', or in trying to reconcile this with the multiple roles of partner, parent (mother/father), neighbour, employee, party member or citizen, can be 'a source of stress and contradiction in both self-representation and social action' (Castells 1997, 6).

This is no more evident than in the disparate groups that rallied around Pauline Hanson's 'One Nation' Party in late-1990s Australia. They included supporters variously in favour of the preservation of family values, cutbacks to Aboriginal welfare and the migrant intake, and support for the small business sector. This was also cross-cut by opposition to free trade and the loss of jobs, unchecked foreign investment, political correctness and 'multiculturalism', the movement for an Australian republic and the banning of weapons in Australia. While One Nation was a powerful lightning conductor to identity-based issues dear to conservative rural voters, the movement was started by a Brisbane shopkeeper and drew as much popular support from post-war suburbs where 'Aussie battlers' have lost jobs to factory closures. The overarching causes that unified these voters in 'the Bush' and the cities were fears about the fragmenting of the traditional Australian identity, and a shared sense of unfairly bearing the brunt of economic restructuring in the 1980s and 1990s. The restoration of a collective identity around nostalgia for a lost past and 'pride of place' put these forgotten-about rural communities and peripheral suburbs at the forefront of the politicians' minds.

People consciously use space, and the material and symbolic differences between places to construct their own identities and to identify with others. In *Soft City*, a Londoner describes how he rarely trespasses outside the limits of an area defined by the Thames to the south, by Hampstead Heath and Highgate Village to the north, by Brompton cemetery on the west and Liverpool Street Station on the east, and

when I do I feel I'm in foreign territory, a landscape of hazard and rumour. Kilburn, on the far side of my northern and western boundaries I imagine to be inhabited by vicious drunken Irishmen; Hackney and Dalston by crooked car dealers with pencil moustaches and goldfilled teeth.

(Raban 1974, 161–2)

The distance between places in urban areas can be used as a means of living apart from those who are relegated to the category of 'Others'. The kinds of distancing and exclusions that can arise apply to the 'western suburbs' of Sydney, Melbourne and Adelaide. Once constructed as 'the *other* side of town', the rich mosaic of ethnic communities is all but airbrushed away in media representations of a stereotypical 'Westies' subculture (Powell 1993).

Alternatively, communities are formed around the efforts of people to establish a sense of collective identity in cities. Toronto and Montreal, for example, have very cohesive Jewish communities that 'seem more strongly related to an internal desire to maintain values and identities' than to being treated as 'outsiders' (Peach 1996, 395). And in the case of Barnet, north London, and Midwood, Brooklyn (NY), by congregating within the boundaries of an *eruv*, Hasidic Jews can still observe the strict codes and customs of the sabbath without being housebound. While they are not allowed to drive on the sabbath, within the *eruv* it is possible to walk to the synagogue, carry money, do the shopping, push prams and the like. Accordingly, in 1999, a 15 km religious boundary known as an *eruv* was created around the prosperous neighbourhoods north-west of Hampstead Heath in north London: 'Most of the eruv will be invisible, marked by existing boundaries such as major roads. To complete the circle according to religious law, however, strands of nylon fishing line are to be strung between 88 telegraph pole-sized posts dotted along the eruv border' (reported in *Guardian Weekly*, 15–21 July 1999, 10).

Most people have no need to demarcate community space in order to anchor their identity in such a definite way. Yet social spaces in the city serve almost by default as a loose kind of 'container' for social and cultural differences (Smith 1999). At the same time, the neighbourhoods and suburbs that people gravitate to can also have a direct bearing on their lives. As we saw with the cases of 'spatial mismatch' and 'spatial entrapment' in the urban labour market (Chapter 5), the internal geography of cities can assist or hinder people to varying degrees by mediating for differences in class, race or gender. So, while there is nothing *intrinsically* oppressive or emancipatory about space, there are communities in parts of every city where, depending on our self-identity, we will be recognised and accepted, or rejected and unwelcome.

America's ethnic gangs provide perhaps the most graphic illustration of both how cultural identity is reified in the act of inscribing urban space – that is, 'turf warfare' – and how that in turn impinges upon people's lives. 'Blacks fighting blacks, turf encroachments by Hispanic rivals and vicious forays by gangs with ethnic roots from Asia to South America, are tearing non-white America apart at the seams' (Woodley 1998, 11). Los Angeles has about 1000 street gangs with national and international connections to a membership of over 200,000. The largest is the 18th Street gang with 6000 Hispanic 'homies' – homeboys or gang brothers – spread from Mexico to

Canada. Their main rivals are the 2000-strong Mara Salvatrucha, which started in LA in 1984–5 when El Salvador was a war zone. While gangs are 85–90 per cent male, female gang membership is on the rise, reflecting the feminist allusions that crept into gangsta rap in the late 1990s. More than 30 all-female gangs have been identified, including the South Side Sissors, Wally Girls and the Lady Rascal Gangsters.

In the 1940s, the wartime economy and mechanisation in the south drove blacks north and west for work. In 10 years the African-American population of LA climbed from 10,000 to 100,000. *De facto* and *de jure* segregation was in full force, but the black ghetto eventually expanded into white and Italian areas, taking over the 30 blocks at the heart of south-central Los Angeles (Fig. 7.1). With the police failing to protect black residents from the white backlash, blacks began to form gangs to defend their neighbourhoods. They then fought among themselves, with the FBI and the Los Angeles Police Department (LAPD) supplying the arms until south-central was politicised by the Black Panthers (Woodley 1998, 13). Baby Panthers recruited in suburbs like Compton, Watts and Inglewood provided the cradle for the modern south-central gangs, the Crips and the Bloods. The Spike Lee movie *Boyz N the Hood* portrays south-central LA as a crack cocaine 'heaven and hell', and the improbable source of an export culture – manifest in dress, music and attitude – that has swept the world.

Appearances in the 'hood' can be deceptive. Young mothers and children perched on veranda steps a few blocks from neon-lit main-street liquor stores and taco bars on a balmy suburban evening deflect attention from the boarded-up windows blown out by drive-by assassins. This culture of violence kills over 800 youths in Los Angeles each year. Almost 80 per cent of documented gang members come from homes where the father is dead, in jail or missing. Life expectancy for an African-American male growing up in south-central LA is 21 years. Those black gang members that don't die young have a one-in-three chance of spending at least one year in prison, and a Hispanic a one-in-six chance, whereas, over his lifetime, a white male in the United States has a 1 in 23 chance of being imprisoned (Wacquant 1997, 8). Although some LA police admit to being 'the biggest gang of all' and to having reservations about 'going to war on children' (Woodley 1998, 14–15), the Californian prison system had grown to

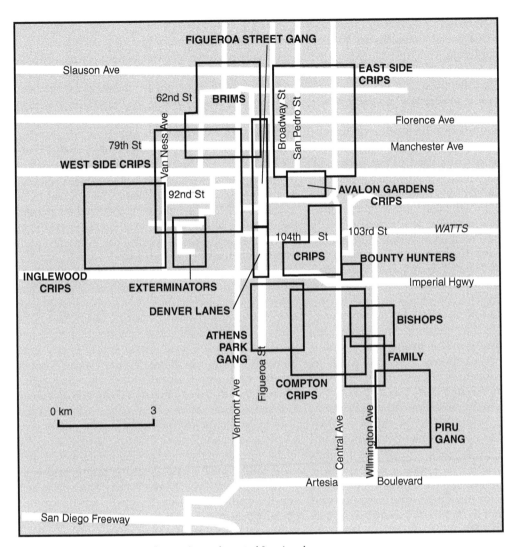

FIGURE 7.1 *Turf boundaries of gangs in south-central Los Angeles*
Source: Davis (1990). Reproduced by permission of Verso

over 200,000 prisoners by 1998. A total of 10 per cent of those imprisoned were held in Los Angeles county jails. The only way for Spike Lee's young hero to escape this fate was to work towards a college education as a way out of the ghetto. Some places have the power to stunt our lives.

Therefore, although geographers must always take care to guard against spatial determinism, as individuals we consciously use the differences between social spaces in cities as an expression of our identities – remembering that our identity is part given (male/female, black/white, parent/child, able-bodied/disabled, widow/widower), part chosen (couple/single, self-employed/employee, believer/agnostic), or partly a mix of both (qualified/unqualified, mono-lingual/bi-lingual). But it goes further than that. These social spaces take shape around the commitment of individuals to a collective identity, and it is this that helps to differentiate the community space of cities. While some of us seek out socially mixed and culturally vibrant corners of the city with a reputation for being on 'the edge', others want to be part of an orderly, mainstream, suburban community with the social networks to support family life or senior living.

Quite crucially, how we seek to use space in (re)constructing our identities also plays a role in the ongoing reshaping of society. 'The organisation of residential space is an expression of deep-seated socio-economic and political inequalities in racially divided societies' (Smith 1999, 13). This may be quite overt when space is used to cut people off from one another and to foster separate social development. The building of the Berlin Wall entrenched two social systems – capitalism and communism – side by side in the same city for four decades. In the late 1990s, Israel's Likud government proposed annexing more land to the west of Jerusalem on learning that the Palestinian community could grow to a projected 45 per cent of the city's population by 2020. By redrawing the boundaries so as to extend administrative powers over nearby Jewish towns, including some in the occupied West Bank, the government of Binyamin Netanyahu aimed to shift the population ratio between Jews and Arabs in the expanded Jerusalem by 2020 to 70:30 (reported in *Guardian Weekly*, 28 June 1998, 12).

A policy of separate development, or apartheid, was imposed in South Africa by creating tribal 'homelands' and issuing a residential permit – the notorious pass – to all black and coloured South Africans (Lemon 1991). In a speech to parliament in 1952, the then Minister of Native Affairs, Dr Verwoerd, set out the regulations that would control urban development in South Africa until apartheid was overturned in 1994:

- every town or city, especially industrial cities, must have a single corresponding black township
- townships must be located an adequate distance from white areas and separated by buffer zones to be given over to industry or open space
- settlement planning should minimise travel through the group area of another race
- existing wrongly situated areas should be moved.

(Williams 2000, 167)

Under these regulations, labourers from the homelands working in Johannesburg, Durban, Elizabeth and Cape Town were banned from residing in the suburbs and made to travel long distances to wretched townships like Soweto, 20 km to the south-west of Johannesburg.

Social divisions between classes, races and religions are likely to persist even after other prohibitions have been removed, because of the way housing markets operate in capitalist societies. Because of this, in 1954 the US Supreme Court ordered that, where there was a gross racial imbalance in the school population, white and black students be bussed out of their areas to integrated schools. The desegregation order applied to more than 500 school districts (Hillson 1977). It was overturned in 1992 by a more conservative bench arguing that 'racial balance is not to be achieved for its own sake' (eight votes to nil). The Supreme Court ruled that racially segregated schools could be tolerated if this were the result of demographic shifts or voluntary population movements, rather than administrative intent. With America's ethnic groups settled into new patterns of largely black inner cities and new black middle-class suburbs, this will leave predominantly white suburban and edge city communities free to run almost entirely white schools. At the time, the judgement left the civil rights movement in two minds. While the National Education Association voiced concern that it could lead to resegregation across the country, other civil rights groups embraced the radical view that all-black schools with their own curricula would ultimately do more for black learning and self-esteem.

The relationship between society and space, and the interaction between the two, has fascinated social scientists, including urban geographers, for well over a century. The next section outlines some of the key developments in urban social theory during the twentieth century. This body of urban social theory and supporting case studies is the main source of our accounts of the social (and cultural) geography of cities.

URBAN SOCIAL THEORY

Urbanism as a way of life

The migration of the European peasantry in the second half of the nineteenth century into industrialising cities in Europe and the New World sparked enormous curiosity among social philosophers like Durkheim, Weber, Simmel and Tönnies (Sennett

1969). They set about postulating how living arrangements and social organisation would be changed by the movement of so many people into the burgeoning cities of Europe and America. Simmel, for example, stressed the importance of the differences in the built environment of cities in a paper with a highly suggestive title, 'The metropolis and mental life' (see Sennett 1969, 47–60). The implication was that city life and the physical setting in which it takes place can have a profound bearing on social and psychological well-being. Simmel's ideas were elaborated by one of the founding members of the famous Department of Sociology at the University of Chicago in a now classic paper, 'Urbanism as a way of life' (Wirth 1938).

Wirth postulated that under the most intensely urban conditions, such as those found in central Chicago or New York in the 1920s and 1930s, the incidence of social and psychological disorders would increase among people overwhelmed by city life and prone to 'psychic overload'. According to Wirth, the pressures of urban living posed a challenge to social norms – *anomie* – from people less able to conform, or from people intent on anti-social behaviour. And, even though urban industrial societies have devised forms of social control such as a criminal code, policing and welfare agencies to deter deviance, they could not match the 'moral order' of close-knit communities. But the other possibility is that, rather than causing mental illness or social deviance, the inner city with its cheap lodgings – doss houses, single room occupancy (SROs) hotels, overnight shelters – has traditionally attracted people who want to be left alone to live anonymously. These very complex links between urban environment, population density and human behaviour have given rise to an important field of social research (Bell and Tyrwhitt 1972; Walmsley 1988).

Chicago's urban ecologists and 'natural areas'

In the 1920s, the Department of Sociology at the University of Chicago pioneered the study of subcultures, and social life and organisation within industrial cities. From the perspective of the Chicago School, the interdependence between urban society and the geography of the city held the key to observing and learning about the social problems that surfaced with increasing urbanisation. In a paper called

'The urban community as a spatial pattern and a moral order', one of the leaders of the Chicago School, Robert Park (1925), asserts 'the importance of location, position and mobility as indexes for measuring, describing and eventually explaining social phenomena'. While Park was not about to reduce society to space by studying how different groups make use of the built environment in cities, the Chicago School aimed to improve its understanding of the social processes and problems of the day. A city bursting at the seams and teeming with immigrants, like Chicago in the 1920s, provided an ideal laboratory.

With their students, Chicago's urban ecologists undertook extensive fieldwork in hand-picked districts that served as supportive niches for quite different forms of social activity and community life. Among the best known of these case studies are Wirth's (1928) study of *The Ghetto* and Zorbaugh's (1929) study of *The Gold Coast and the Slum*. Anderson's (1923) study of *The Hobo: The Sociology of the Homeless Man*, Thrasher's (1927) investigation of *The Gang: A Study of 1313 Gangs in Chicago* and Shaw's (1929) work on *Delinquency Areas* investigate the symbiosis between forms of deviant behaviour and spatial context.

The Chicago School looked to natural ecosystems and Darwin's theory of evolution, or competition between the species, to provide an analogue of the processes producing natural areas in cities. Park (1925) noted that, while language, culture, religion and race provided the motivation for residential segregation, geographical barriers and physical distance, along with improved mobility, provided the means to practise it. According to Park, competition for suitable housing in a city like Chicago could lead to the 'invasion' and 'succession' of a neighbourhood just as competition between species might in plant communities. Thus ecological dominance was a necessary precursor to the evolution of natural areas. It was the social and cultural homogeneity of areas like the Black Ghetto, the 'Gold Coast', Little Italy, Towertown or Chinatown that made them distinguishable from other residential areas in Chicago. Figure 7.2 illustrates the application of indicators like the presence of notable people listed in Chicago's Social Register and recipients of welfare to differentiate the Gold Coast and the slum.

However, with the pressure brought to bear upon local housing markets by successive waves of

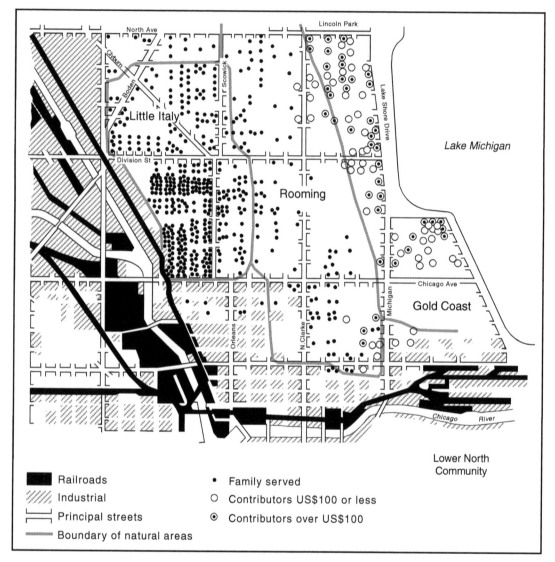

FIGURE 7.2 *Chicago's Gold Coast and the slum in the 1920s*
Source: Zorbaugh (1929). Reproduced by permission of University of Chicago

immigrants arriving in a city like New York, the social ecology of neighbourhoods can change many times over. In a meticulous study of ethnic succession, the urban geographer Barny Warf (1990) shows how Brooklyn neighbourhoods have periodically been transformed by immigrant arrivals seeking work and shelter. As each ethnic group established a toehold, and then tightened its grip on a piece of Brooklyn real estate, its respective worlds of work, social life and culture often collided in social space.

Tightly bound communities could negotiate these inter-group tensions from a position of strength, and keep everyone informed about accommodation and jobs coming on to the market. By 1870, class-differentiated English, German and Irish communities were established on the Brooklyn foreshore (Fig. 7.3a). In the 1890s, Brooklyn was invaded by several hundred thousand Italians fleeing from the poverty of the Mezzogiorno. Then, with the Brooklyn waterfront growing in stature as a ship-building centre,

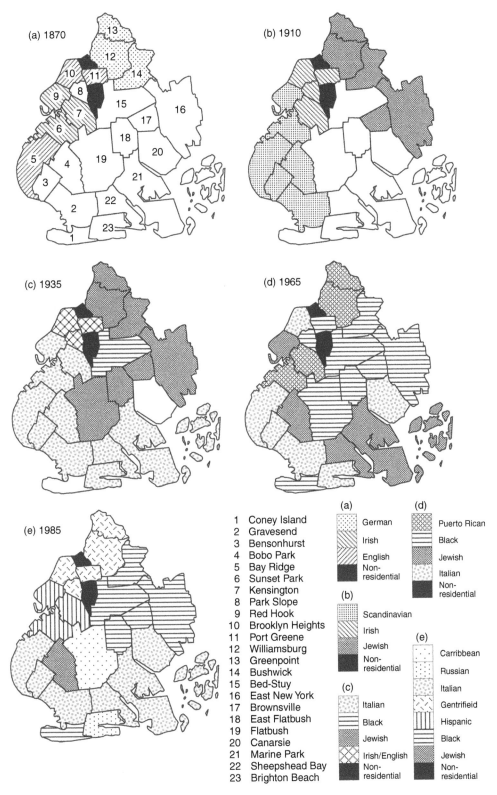

(a) 1870

(b) 1910

(c) 1935

(d) 1965

(e) 1985

1	Coney Island
2	Gravesend
3	Bensonhurst
4	Bobo Park
5	Bay Ridge
6	Sunset Park
7	Kensington
8	Park Slope
9	Red Hook
10	Brooklyn Heights
11	Port Greene
12	Williamsburg
13	Greenpoint
14	Bushwick
15	Bed-Stuy
16	East New York
17	Brownsville
18	East Flatbush
19	Flatbush
20	Canarsie
21	Marine Park
22	Sheepshead Bay
23	Brighton Beach

(a)
German
Irish
English
Non-residential

(b)
Scandinavian
Irish
Jewish
Non-residential

(c)
Italian
Black
Jewish
Irish/English
Non-residential

(d)
Puerto Rican
Black
Jewish
Italian
Non-residential

(e)
Carribbean
Russian
Italian
Gentrifieid
Hispanic
Black
Jewish
Non-residential

Figure 7.3a–e *Ethnic succession in Brooklyn, New York, 1870–1980s*
Source: Warf (1990). Reproduced by permission of Pion Ltd

Scandinavian immigrants made the journey across the Atlantic (Fig. 7.3b). By 1925, the Brownsville and East New York areas were home to 285,000 Jews, who were mainly second-generation émigrés moving up and out from the garment-making lofts of the Lower East Side. The emergence of the black community in the Bedford-Stuyvesant dates from the availability of blue-collar jobs and the opening of the 'A' subway line linking Brooklyn with Harlem in the mid-1930s (Fig. 7.3c).

The greatest shift in Brooklyn's ethnic composition occurred between 1940 and 1980 when, along with de-industrialisation, it lost 750,000 whites (especially Irish, Italians and Jews) and gained 650,000 blacks (Warf 1990, 110). Near the tail end of this period of industrial closure and exodus (around 1965), Brooklyn became home to the world's largest community of Orthodox Jews and received its first airborne immigrants from Puerto Rico (Fig. 7.3d). By 1970 the Puerto Rican population had grown to 270,000. Lastly, in the 1980s thousands of young careerists working in Manhattan's financial district were driven across the East River by New York's real estate prices to launch a 'brownstone revival' in north-western localities like Brooklyn Heights and Williamsburg. Other newcomers to seek refuge from the exorbitant rents of Manhattan include a 600,000-strong Caribbean community and about 50,000 Russians, who have recreated their own 'Little Odessa' at Brighton Beach (Fig. 7.3e).

While the term 'natural area', with its connotations of social Darwinism, is no longer used, Park's (1925) original pronouncement on the relationship between social distance and the physical separation of groups has since given rise to wide-ranging research on urban segregation. Dissimilarity indices are used to measure ethnic, religious and social segregation, the distribution of minority groups, residential propinquity and distances between marriage partners (Peach 1975).

Urban residential patterns: the Burgess and Hoyt models

The fieldwork by the Chicago School on natural areas was brought together with the account of invasion and succession by E.W. Burgess (1925) to explain the growth and residential organisation of industrial cities. Burgess noticed that Chicago's outward growth was fuelled by the pressure that

successive waves of immigrants placed on the innermost housing space. As the real incomes of migrants with jobs improved, they were able to afford the working man's fare to leave behind the overcrowded neighbourhoods. According to Burgess, these pressures in the housing market set up a ripple effect similar to a stone being dropped into a pond. This gives his model a definite zonation corresponding to the occupational mobility and ability to save of immigrant households at different stages of adjustment. Because of the tendency for ecological processes to sort similar households, he felt justified in generalising about the residential composition of any zone at a given distance from the Loop, or elevated railway, that rings the centre of Chicago. Figure 7.4 displays Burgess's now-classic model of concentric zones with residential densities falling away from the tenements of the ghetto, through the apartment and two-flat areas, to the bungalows occupied by city commuters and their families. Embedded within the zonal pattern, where language, race and culture come together, are natural areas like Little Sicily, Chinatown and, further out, Deutschland.

This zonal representation of urban residential patterns, which Burgess generalised to other cities undergoing industrialisation, was called into question by a land economist in the 1930s. In the process of providing advice to the Federal Housing Administration on mortgage insurance, Homer Hoyt became convinced that residential growth in American cities was more axial than concentric:

rent areas in American cities tend to conform to a pattern of sectors rather than of concentric circles. The highest rent areas of a city tend to be located in one or more sectors of the city. There is a gradation of rentals downward from these high rental areas in all directions.

(Hoyt 1939, 76)

These conclusions are based on his maps of the shifts in the location of high-rent areas in six US cities between 1930 and 1936 (Fig. 7.5). These maps suggested to Hoyt why these fashionable areas tend to follow the main commuter lines, whilst placing as much physical distance between the residential elite and industrial zones as possible. He concluded that the answer lay in the dynamics of the housing market, which transfers obsolete housing from higher-income groups, who vacate in order to occupy stylistically fashionable dwellings, to middle-income

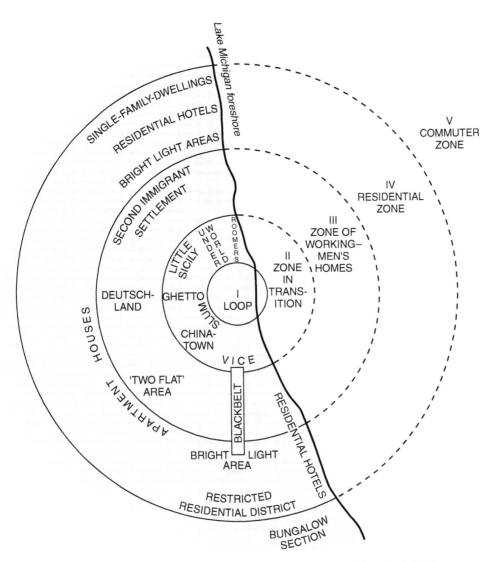

FIGURE 7.4 *The concentric zone model of social ecological areas demarcating Chicago in the 1920s*
Source: Burgess (1925)

groups. The vacancies they create are, in turn, filled by lower-income households. With consecutive building cycles adding new housing to the edge of an expanding city, this leaves the housing vacated by the wealthier households to filter down to the less affluent families. Unlike Burgess's invasion and succession model, the impetus for moves through the housing stock comes from the creation of new dwellings on the city's outskirts. With further refinement, the mechanics of Hoyt's hypothesis have

come to form the basis of 'filtering theory' in housing economics.

Social area analysis

For a long time it was thought that these two models of urban residential patterns were diametrically opposed. But, with the development of social area analysis, the relationship between the concentric zone model and the sectoral hypothesis

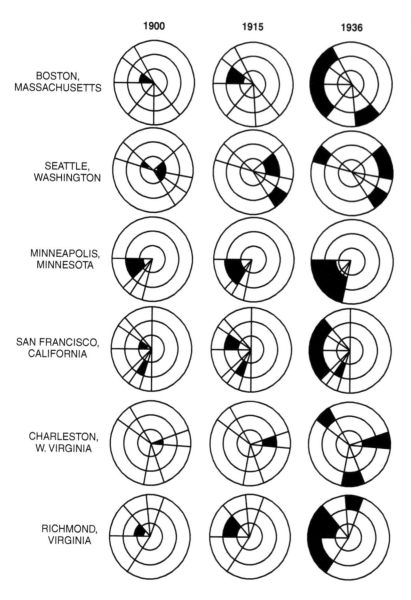

	1900	1915	1936
BOSTON, MASSACHUSETTS			
SEATTLE, WASHINGTON			
MINNEAPOLIS, MINNESOTA			
SAN FRANCISCO, CALIFORNIA			
CHARLESTON, W. VIRGINIA			
RICHMOND, VIRGINIA			

FIGURE **7.5** *Sectoral shifts in the location of high-rent residential areas in six US cities between 1930 and 1936 Source*: Hoyt (1939)

has been clarified. Shevky and Bell (1955) are the two sociologists most closely associated with the development of social area analysis. They wanted a technique that would help them differentiate between 'community areas' in Los Angeles and San Francisco in an objective way. A start was made by assuming that it should be possible to detect the effects of industrial urbanism in American society, as initially postulated by Louis Wirth (1938), in the residential profiles of 'community areas' or urban social areas.

Wirth attributed the changes taking place in urban society in the first decades of the twentieth century to:

- growth in the sheer size of the largest American cities
- the much greater crowding and population densities present in these cities
- the added demographic and social heterogeneity of urban populations.

Shevky and Bell then selected corresponding indicators of the three major trends that Wirth suggests should accompany 'increasing societal scale': (i) change in the range and intensity of social relations; (ii) greater functional specialisation of employment; (iii) greater social diversity and complexity. These rather abstract processes are evident respectively in the shift taking place in the division of labour (from manufacturing to service occupations), in the composition of urban households (as more women take on paid work and as 'alternative' living arrangements are adopted), and in the widening disparity between suburbs (with greater in-migration and residential mobility producing more social and ethnic segregation).

Shevky and Bell (1955) hand-picked census variables to represent each of these social trends in American society, which they characterised as 'social rank', 'urbanisation' and 'segregation' – although Bell preferred the labels 'economic status', 'family status' and 'ethnic status'. By standardising these variables, they were able to sum the unweighted averages and obtain a composite score with respect to social rank and urbanisation for each census tract in San Francisco. When these scores are mapped in a social space diagram, a social geography of 'community areas' in 1950s San Francisco is obtained (Fig. 7.6). Its main features include: an extensive belt of working-class families (low social rank, low urbanisation) living adjacent to the industries that used to face the Bay; corresponding tracts of middle-class, family-oriented suburbs (higher social rank, low urbanisation) to the south-west of San Francisco; a characteristic zone in transition sloping down to the northern waterfront with high numbers of floating 'loners' (low social rank, low urbanisation); and the north-eastern quadrant with its strong, family-centred, immigrant communities (low social rank, low urbanisation, high segregation scores).

What social area analysis and factorial ecology, which followed it, left in no doubt is that the zonal and sectoral models of residential patterns are not mutually exclusive. On the one hand, the Burgess model focuses on the differentiation of communities that arises from family-dominated ('suburbanism' and outer suburbs) as opposed to 'alternative' ('urbanism' and inner-city) living arrangements. And, on the other hand, Hoyt's model differentiates communities according to social class and status,

and in so doing reveals the extent to which a city is spatially polarised.

FACTORIAL ECOLOGY

Research on social areas in cities was broadened in the 1960s and 1970s by the urban geographer Brian Berry, working with a group of graduate students at the University of Chicago (Berry and Rees 1969). By combining the newly developed power of high-speed computers and a multivariate statistical model called factor analysis, they were able to analyse and classify small area data from the US census for cities as never before (Berry 1971; Rees 1971). This technique for mapping the social geography of cities is known as 'factorial ecology'. When the findings from countless analyses of North American cities are compared (Berry and Kasarda 1977), the overwhelming tendency is for the distribution of socio-economic status or class to be sectoral, while the mapping of the urbanisation construct – 'family or household status' – is zonal. It is this confidence in the generality of the Chicago model depicted in Figure 7.7 that led Berry and Kasarda to claim that

An orderly social ecology results through like individuals making like choices, through regularities in the operation of the land and housing markets, and through the collaboration of similar individuals in excluding those of dissimilar characteristics from their neighborhoods or in restricting certain minority groups to particular areas.

(Berry and Kasarda 1977, 130)

On the contrary, the application of factorial ecology methods to cities in Europe and Australasia (Timms 1971) confirmed just how different urban residential patterns can be when the state is a major social housing provider. In the British context, for example, Robson (1969, 132) concluded that, 'The appearance of such large areas of local authority housing has made nonsense of the rings or sectors of the classical ecological theory. Indeed, the game of hunt-the-Chicago-model seems to be exhausted as far as the analysis of model developments in British urban areas is concerned', whereas in Australia and New Zealand, the scale of post-war suburbanisation has been such that the prime residential areas are often outflanked by state housing estates built on the cheapest land at the edge of Sydney, Melbourne, Adelaide, Auckland and Wellington.

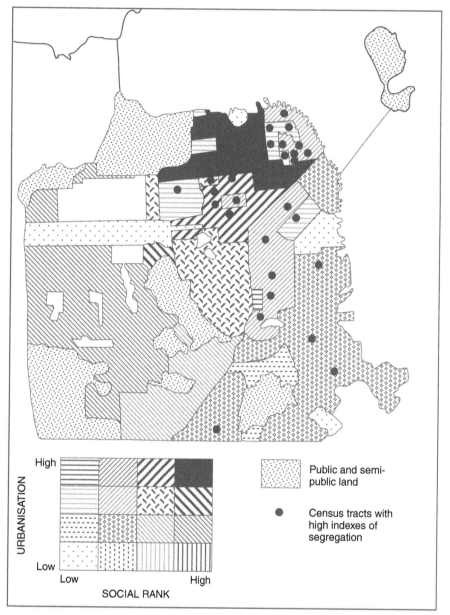

FIGURE 7.6 *Social geography of 'community areas' in San Francisco in the 1950s derived by social area analysis*
Source: Shevky and Bell (1955). © Board of Trustees of the Leland Jr University, renewed 1983

Consequently, the distribution of socio-economic status and income in European and Australasian cities does not unerringly comply with the Burgess–Hoyt prediction of an upward gradient from the city centre. These differences in class and income patterns left such a strong impression on the urban

sociologist Ivan Szelenyi when he was in Australia that he sought to represent them schematically (Fig. 7.8). But, since then, the middle-class takeover of the inner suburbs has so altered the income curves that it is not unreasonable to suggest that a 'cone of wealth' is forming within the inner zones of

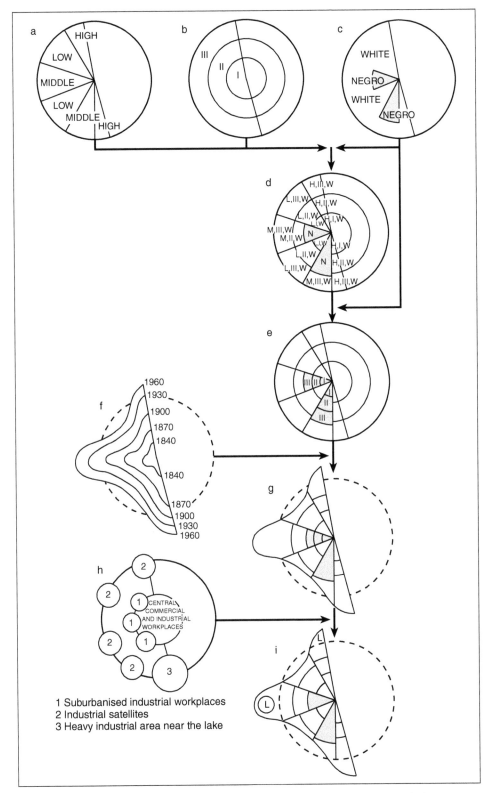

FIGURE 7.7 *An integrated social space model of Chicago obtained by factorial ecology techniques*
Source: Berry and Kasarda (1977, 130). Reproduced by permission of Macmillan

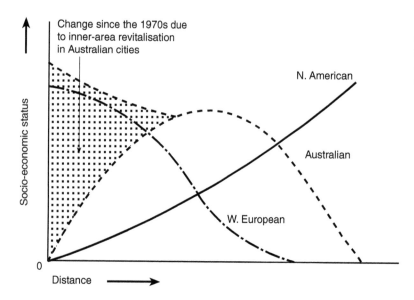

Change since the 1970s due to inner-area revitalisation in Australian cities

N. American

Australian

W. European

Socio-economic status

0

Distance

FIGURE 7.8 *Comparison of socio-economic gradients typically found in West European, N. American and Australian cities*
Source: after Szelenyi (1978)

Sydney, Melbourne, Adelaide, Brisbane and Perth (Badcock 2000).

The other criticisms levelled at factorial ecology relate to the thinness of its theoretical insights and to the failure of its users to consider vital aspects of social relations that are not amenable to measurement. On the first count, the factorial ecology studies leave us none the wiser about the social 'mechanisms' that translate social differences (categories and identities) into spatial patterns in cities (Smith 1999). By this, Smith means the racist, gendered and sexual prejudices that often operate as barriers to access in the labour and housing markets of cities: 'direct and indirect, personal and institutional discrimination open up spaces in which to live, work, learn and be creative to some people, and close them down to others' (Smith 1999, 14). And on the second count, because it treats the household as the principle unit of analysis, factorial ecology disregards the unequal social relations that exist between men and women – not only within society, but within the household as well.

RACIAL AND ETHNIC DIVERSITY IN CITIES

The presence of racial and ethnic minority communities in many western societies lends a rich patina to the residential fabric of host cities. Even in cases

where they constitute a relatively small minority nationally, as they do in Britain (Table 7.2), such communities have a considerable impact upon social, political and economic life in the areas where they choose to live. For example, at the 1991 census, 28 per cent of the populations of Leicester and Slough were from minority ethnic communities (mainly Indian). The corresponding figures were: 22 per cent for Birmingham (mainly Pakistani); 20 per cent for Luton (mainly Pakistani); 20 per cent for Greater London (all communities). In some of the London boroughs, the degree of concentration exceeds 30–40 per cent: 34 per cent for Hackney (mainly black Caribbean); 36 per cent for Tower Hamlets (mainly Bangladeshi); 42 per cent for Newham (mainly Indian); 45 per cent for Brent.

But these levels of concentration do not approach the prevailing levels of segregation in US cities like Detroit (Deskins 1996), Chicago (Abu-Lughod 1997) or Los Angeles, for that matter (Ong and Blumenberg 1996). Comparative research on ethnic segregation in cities makes use of the index of dissimilarity to measure the degree of intermixing and dispersal of minorities (Fig. 7.9). It lies within a range between 0 (no segregation) and 100 (complete segregation). Studies undertaken in the United States, Canada, Australia, the Netherlands, Belgium, Germany and Israel reveal very similar patterns of resettlement in these contrasting urban settings. North-west Europeans (British, Irish,

TABLE 7.2 *Population of Great Britain by ethnic group and age, 1996–1997*

Ethnic group	Percentages				All ages (= 100%) (m.)
	Under 16	16–34	35–64	65 and over	
White	20	27	37	16	52.9
Black					
Black Caribbean	23	33	36	8	0.5
Black African	30	41	27	2	0.3
Other black groups	46	39	13	–	0.1
All black groups	29	37	29	5	0.9
Indian	27	32	36	5	0.9
Pakistan/Bangladeshi					
Pakistani	38	35	24	2	0.6
Bangladeshi	42	35	20	3	0.2
All Pakistani/Bangladeshi	39	35	23	3	0.8
Other groups					
Chinese	17	41	38	4	0.1
None of the above[†]	44	30	23	2	0.6
All other groups[†]	40	32	26	2	0.8
All ethnic groups[*]	21	27	37	15	56.3

[†]Includes those of mixed origin.
[*]Includes ethnic group not stated.
Source: Office for National Statistics (1998)

Germans, Scandinavians, Dutch) have the lowest segregation scores, followed by southern and eastern Europeans (Italians, Greeks, Poles), then Hispanics or Latino immigrants and finally African-Americans with segregation scores typically in the 70s and 80s.

Therefore, while many minorities in British and European cities have inferior access to good-quality housing, only London's Bangladeshi community in Tower Hamlets approaches the levels of hyper-segregation reported for African-Americans or Hispanics in many of the older industrial cities in the United States (Fig. 7.9). Apart from that, the only place in Europe where a minority, in this case Turkish, exceeds more than 50 per cent of the population is the three blocks of Kreuzberg in Berlin. One important difference is that ethnic minorities in Europe and Australia are working migrants on the whole, and are not subjected to anywhere near the

same barriers that black Americans encounter (Marcuse 1996; O'Loughlin and Friedrichs 1996).

At the same time, an underlying ethnic transition with broad cultural and political ramifications is taking place in the United States. The 2000 census revealed that the number of Americans describing themselves as Hispanic grew by 60 per cent in the inter-censal decade and totaled 35.3 million people (reported in *Guardian Weekly* 15–21 2001, 36). Hispanics are those people who trace their ancestry to a Spanish-speaking country. They now match the African-American population in strength and are set to become the country's largest ethnic minority grouping. The census also revealed a higher-than-expected number of blacks – 1.76 million people – who identified themselves as of mixed parentage. With those counted who checked black and another race on census night, the black population totalled 36.4 million in 2000.

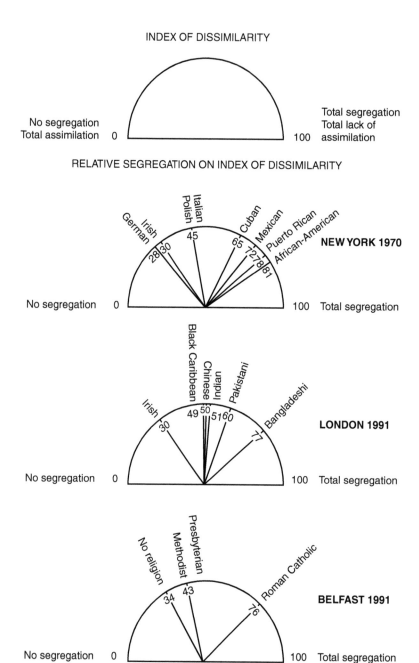

FIGURE 7.9 *Differences in the index of dissimilarity: New York, London and Belfast*
Source: Peach (1996). Reproduced by permission of Taylor & Francis Ltd

With the combined minority populations – Latinos, Asians and blacks – growing at seven times the rate of the non-Latino white population during the 1980s, the patterns of residential segregation are destined to evolve within US cities. Figure 7.10 classifies 318 metropolitan areas in terms of their ethnic mix. Frey and Farley (1996) have calculated the change in the segregation scores between 1980 and 1990 for all those 'multi-ethnic' metropolitan areas with more than one of these minorities (i.e. >3 per cent of the

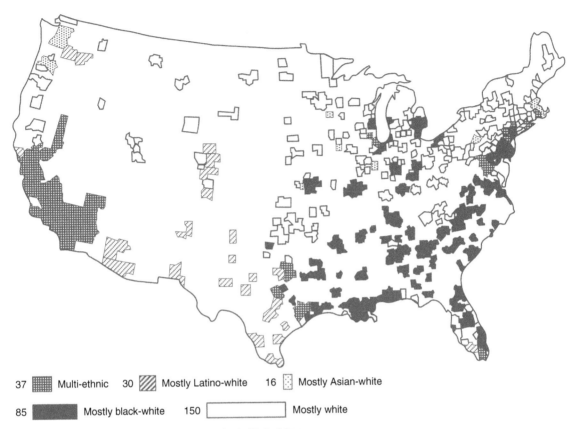

| 37 | Multi-ethnic | 30 | Mostly Latino-white | 16 | Mostly Asian-white |

| 85 | Mostly black-white | 150 | Mostly white |

FIGURE 7.10 *Multi-ethnic metropolitan areas in the United States*
Source: Frey and Farley (1996). Reproduced by permission of Demography

metro population in 1990). They found that, while Latinos and Asians are significantly less segregated than blacks, with average scores in 1990 of 43, 43 and 64 respectively, only African-Americans registered a slight fall in their segregation score – from 68.8 in 1980 to 64.3 in 1990. But in a number of multi-ethnic 'new southern' cities (like Houston and Dallas) and western cities (like Los Angeles, Las Vegas and Riverside) the fall in the 'black/non-black' score exceeded −10 in the 1980s (Frey and Farley 1996, 42–3). The indication is that multi-ethnic metropolitan areas with growing Latino and Asian populations are more open to mixed-race and mixed-ethnicity co-residence than most northern and old southern cities with their rigid housing markets and lingering racial intolerance.

Two urban geographers, James Allen and Eugene Turner (1989), set out to identify the most ethnically diverse urban places in the United States (Table 7.3). They established that in 1980 more than half of

the top 100 most diverse urban areas were located in California, with 20 in Los Angeles alone. The LA cities of Gardena and Carson head this list, followed by Los Angeles, West Carson, Monterey Park, Pomona and Cerritos. Cities near the top of the list in Table 7.3 come close to having populations with roughly equal proportions of whites, blacks, Asians and Latinos. For example, the Gardena area has supported a growing Japanese community since the 1920s, including a cluster of Hawaiians of Japanese origin. The Korean and Chinese populations almost match the Japanese enclave in size and Filipinos are also settling in the area. And, while the growth of Gardena's Latino population has been inversely related to the out-migration of Anglos, both the white and black populations seemed to have stabilised at the 2000 Census at just under one-quarter of the total (Soja 2000, 295–7). But whether residential propinquity translates into deeper cultural mixing in these

TABLE 7.3 *The 20 most ethnically diverse cities in the United States*

Gardena, CA	2.00	Langley Park, MD	1.74
Carson, CA	1.99	Pittsburg, CA	1.72
Daly City, CA	1.85	San Francisco, CA	1.71
Union City, CA	1.80	Pomona, CA	1.70
National City, CA	1.77	Oakland, CA	1.70
Marina, CA	1.76	Gallup, NM	1.68
Seaside, CA	1.75	Cerritos, CA	1.67
Los Angeles, CA	1.75	Passaic, NJ	1.67
West Carson, CA	1.75	Jersey City, NJ	1.66
Monterey Park, CA	1.74	Richmond, CA	1.64

Note: An index of 1 indicates an ethnic mix corresponding to the US population. Values greater than 1 indicate increasing levels of diversity.
Source: Allen and Turner (1989). Reproduced with permission

ethnically diverse cities depends on willingness to inter-marry or co-habit. Data collected by Allen and Turner (1989) indicate that rates of inter-marriage in Los Angeles County vary from 58 per cent and 34 per cent for Puerto Ricans and Cubans to a range between 4 and 7 per cent for Cambodians, Vietnamese, non-Hispanic whites, Koreans and blacks.

In the UK, cultural hybridisation in cities is taking on a different complexion as white and second-generation black Britons from differing ethnic backgrounds choose 'other race' partners (Policy Studies Institute 1997). By the late 1990s, half of British-born Caribbean men, a third of Caribbean women, and a fifth of Indian and African-Asian men had a white partner. As a consequence, the number of multiracial children in Britain grew by about 40 per cent in the 1980s, and even faster during the 1990s. About 80 per cent of 'Caribbean' children now have a white parent. They are predominantly urban, since this is where most black Britons live, but are more likely to have moved beyond the inner city to better housing than that in which their non-white parents grew up. What concerns some of those Commonwealth immigrants who arrived in the 1950s and 1960s is that future generations will know very little about the cultural identity of their forebears.

This intermingling of people with different cultural backgrounds is beginning to leave its mark on community life and the residential fabric of multicultural cities, especially in the changing neighbourhoods favoured by immigrants. In a city like Los Angeles,

The mix of cultures projects itself on to the urban landscape quickly … overlapping groups produce an energetic, chaotic street scene, full of surprises. Jewish restaurants offer kosher burritos; a restaurant in Little Tokyo displays a plastic pepperoni pizza in the front window as if it were sushi … An Anglo supermarket becomes an American Christian church, with a billboard advertising grace instead of cantaloupes. An African American home for young working women becomes a shelter for homeless Central American refugees. An Anglo American apartment complex in the San Fernando Valley is taken over by one thousand Cambodian residents, who reorganize it as if it were a traditional village. Laundry is shared and residents tend herb and vegetable gardens in the main courtyard.

(Hayden 1995, 84–5)

Similar forms of ethnic mixing and cultural adaptation can be observed throughout large Australian cities, where the settlement process and local housing markets have acted to congregate newcomers. For example, on high-rise public housing estates in inner Sydney and Melbourne, migrants from every continent are scattered through each tower block and local schools, get involved in tenant associations and share community gardens.

THE THREE-GENERATION MODEL OF URBAN RESETTLEMENT

A three-generation model has been developed to summarise the process whereby first-generation

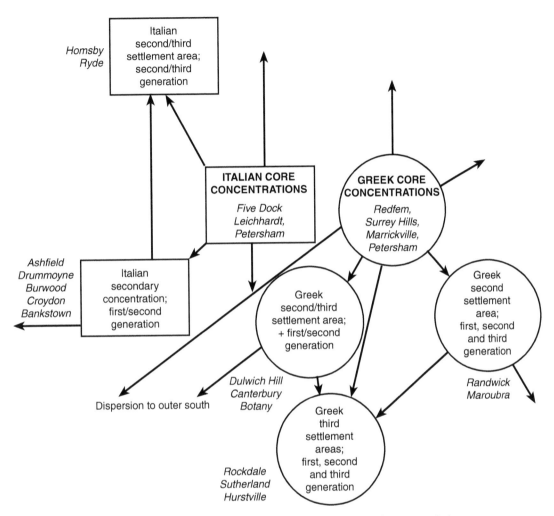

FIGURE 7.11 *The settlement stages of successive generations of overseas migrants in post-war Sydney*
Source: Burnley (1996). Reproduced by permission of P.W. Newton and M. Bell

immigrants establish a highly segregated beachhead in the cheapest rental housing that was traditionally available in the zone in transition around the city core. With better employment prospects, the second generation begins to buy their own homes and create a less segregated ethnic village apart from the original ghetto. Third-generation movers disperse further afield to the suburbs, implying that some degree of social assimilation is taking place at the same time. The sequence of moves from the original beachheads established by the Greek and Italian communities in Sydney, to these second- and third-generation settlements in the suburbs, is depicted in Figure 7.11. Both communities, and others like

them, now contribute to the increasing cultural diversity of Sydney suburbs as the degree of residential segregation between different national groups gradually weakens (Table 7.4). However, as noted above, the residential dispersal of southern Europeans did not proceed as far as for northern Europeans over the 30-year period 1961–91.

Leichhardt and Marrickville are two inner-Sydney municipalities with strong residual Italian and Greek communities, which are now being gentrified. Other middle-distance suburbs, like Campsie and Lakemba, where second-generation Italians have settled, also have Lebanese, Vietnamese, Korean and Chinese enclaves. Bankstown and Cabramatta in the mid-west

TABLE 7.4 *Indexes of dissimilarity between European-born and Australian-born population in metropolitan Sydney, 1961–1991*

Birthplace	1961	1966	1971	1976	1981	1986	1991
Austria	50.5	27.5	29.9	23.1	21.4	20.0	19.2
Czechoslovakia	47.7	34.7	39.6	29.3	27.1	27.2	23.2
Germany	27.7	25.1	24.2	19.1	15.1	12.9	12.7
Greece	56.7	54.7	55.1	52.6	49.4	50.3	49.7
Hungary	38.3	38.9	36.9	35.3	34.4	34.0	33.4
Italy	35.9	32.5	33.4	34.3	34.3	35.0	35.6
Malta	53.0	49.4	47.7	41.4	39.5	40.6	39.8
Netherlands	28.9	28.7	24.4	20.5	16.2	16.7	15.8
Poland	33.6	33.7	34.4	34.6	35.3	33.3	31.0
Spain	n.a.	54.1	56.2	n.a.	41.6	39.8	36.9
UK and Ireland	24.0	18.4	19.2	10.1	9.7	10.6	11.7
USSR	23.3	32.1	32.8	34.7	37.4	34.8	35.3
Yugoslavia	36.1	32.9	40.6	37.8	35.9	36.9	37.3

Source: Compiled from Australian Bureau of Statistics census files by Burnley (1996). Reproduced by permission of P. Newton and M. Bell

are close to post-war migrant hostels. Because war refugees arriving from Russia, Latvia, Poland and the Ukraine in the 1950s, Vietnam in the 1960s and 1970s, and Lebanon in the 1980s and 1990s, spent their first six months in these hostels, they have become part of the surrounding multicultural patchwork. Bankstown and nearby Auburn also have large German-Austrian and Turkish communities. The City of Fairfield, centred on Cabramatta, was once a market gardening area, which explains the considerable Italian and Croatian presence. In addition, Fairfield is home to 35,000 residents born in Southeast Asian countries such as Vietnam, Laos and Kampuchea (Burnley 1989; Wilson 1990; Dunn 1993).

Hence, any account of this complex cultural mix in Sydney's mid-west has to look beyond the operation of a three-stage resettlement process (for Greeks and Italians), to the location of the Commonwealth migrant hostels (eastern Europeans, Vietnamese, Lebanese), and to market gardens in the area dating from the 1920s (Turks and Croatians). While they co-exist residentially, each ethnic community in Sydney remains quite tight-knit in order to preserve its cultural identity. It is only on rare occasions that the rivalry between football clubs founded by migrants from different countries spills over into an outbreak of hostility between 'ethnic' gangs on the streets of Sydney or Melbourne.

Congregation and segregation

The persistence of ethnic segregation in cities is not solely due to the antipathy held towards subordinate minorities by a dominant white population. By congregating together as an identifiable community, immigrants can keep alive and pass on their language, cultural values and traditional customs. Ethnic clusters have stronger social networks and informal arrangements for self-help, and are able to support a much greater range of own-language services (shopkeepers, doctors, clinics, teachers) and institutions (churches, banks, clubs, libraries). As well as providing a cultural and religious focal point for community life, Sikh temples and Muslim mosques in British cities are also a source of food, shelter, schooling and recreation.

It may be the case, then, that, while a highly segregated minority like the Bangladeshis in Britain undoubtedly encounter racism in their daily lives, there are additional factors at work. Even though the Commission for Racial Equality uncovered clear evidence of a racially biased housing allocation policy in the London Borough of Tower Hamlets, this does not explain why half of London's Bangladeshi population has chosen to live in a single borough where, incidentally, they also minimise contact with fellow Muslims from Pakistan – with whom they once shared a common nationality (Peach 1996, 391). On the one hand, Bangladeshis are not compelled to live in Tower Hamlets; and, on the other, the legacy of political division outweighs the unifying force of a common religion.

In other urban settings, the segregation of minorities that share the same national identity can be directly attributed to religious conflict. For example, Robinson (1996) obtained an index of dissimilarity of 85 for India's Gujerati Muslims and Gujerati Hindus living in Britain in the 1980s. Some of the highest levels of segregation in the UK are to be found in Northern Ireland, where sectarian violence has confined Roman Catholics to traditional Catholic areas like the Shankill Road. With the escalation of violence in the 1990s, public housing estates became the objects of paramilitary struggles, ethnic cleansing and scorched earth policies to prevent the occupation of abandoned space (Peach 1996). Hence they tend to be even more segregated than the population as a whole.

The ghettoisation of black Americans

'The ghetto is often thought of as an inevitable first stage of accommodation of ethnic minorities migrating to a foreign country' (Peach 1996, 383). However, the experience of African-Americans definitely does not fit the expected transition from ghetto through ethnic village to suburban dispersal. Although black home-ownership is on the rise and middle-class blacks are beginning to move to the suburbs in growing numbers, in many US cities the more recently arrived Latino and Asian populations are less segregated than the African-American population. This leads Ceri Peach to conclude that there is a fundamental difference in the processes responsible for ethnic settlement patterns in US cities. American futurologists were quite wrong in the 1950s to regard the black ghetto as a beachhead leading next to the urban village, and then onwards and outwards to suburbanisation and the crabgrass frontier: 'The European model is essentially voluntaristic and protective; the African American model of segregation is negative and destructive' (Peach 1996, 385).

The black ghetto occupies a place in the American psyche ranging from 'pathological underclass' (Jencks and Peterson 1991) to 'institutionalised racism' (Wacquant 1997). Economic restructuring has kept Chicago's black population more centralised than in any other comparable US city (Abu-Lughod 1997). At first those who migrated from the Deep South were confined within the 'black belt' on the southern side of the city (Fig. 7.4). But, with the movement of white households to the suburbs, blacks now constitute over 85 per cent of the city of Chicago's population. Chicago was particularly hard hit by job losses in the 1980s, with city neighbourhoods suffering much more than suburban Cook County. And black residents account for less than 2 per cent of the population in those affluent northern and north-western municipalities that enjoyed net job gains in the late 1990s. This leaves African-Americans in Chicago's poor neighbourhoods isolated from employment and business opportunities, and trapped in communities branded as too risky for business investment on the one hand, and housing submarkets characterised by deteriorating or abandoned housing, bank aversion to home loans and home foreclosure, on the other. These problems, coupled with the increasingly segregated, underfunded and deteriorating educational system, 'portend continued disaster for the Chicago region's African-American and Hispanic communities' (Abu-Lughod 1997, 361).

A Marxist urban geographer like David Harvey would go even further than this. Writing in the 1970s (Harvey 1973), he suggested that the 'black ghetto' is not so much a problem for American society as a solution; this is to say that

If one views ethnic minorities as forming the reserve army of labour, representing a threat to the employment of established workers if they were to unionise and force a redistribution of surplus value, then the ghetto reinforces the visual threat of this group. Further, it allows easier policing of the perimeter by leaving the ghetto to look after itself. If it riots, it burns down only its own facilities. Race and the ghetto, in these terms, becomes the solution rather than the problem.

(Peach 1996, 385)

Indeed, this was the fate to befall Watts in south-central Los Angeles in April 1992 (Fig. 7.1). Five days of rioting followed the court verdict that four LAPD officers were not guilty of beating Rodney King: 'Scores of people were dead;

thousands had been injured; and property damage was estimated in the billions of dollars' (Dear and Wolch 1992, 917).

Aboriginal Redfern

The movement of indigenous people off their tribal lands in the 1950s and 1960s led to the formation of precarious Aboriginal communities in the bigger Australian cities. Although the neighbourhood of Redfern is home to only about 400–500 people of Aboriginal identity, and contains no more than 5 per cent of South Sydney's population, it has achieved notoriety among white Australians as the closest thing Australia has to a 'black ghetto'. 'Redfern is indeed a racialised place ... it is widely held by white and some black Australians that blight, crime, poverty, substance abuse, truancy, vandalism, youth disaffection, and despair have found their natural habitat on that district's streets' (Anderson 1998, 213).

Redfern is an inner-city block of about 70 terraced houses that was acquired by the Commonwealth on behalf of the Aboriginal Housing Company (AHC) in 1973. As well as addressing the housing needs of a growing number of destitute Aborigines in central Sydney, the grant-in-aid symbolised the Whitlam government's commitment to self-determination for indigenous people. Since 1974, the AHC as landlord has endeavoured to renovate the housing stock and improve the local residential environment of 'the Block'. Despite its efforts, housing conditions range from properties in need of minor repairs to out-and-out squalor. According to Anderson (1998, 217), about two-thirds of the tenants are women heading households comprising four to five children or young adults. When relatives arrive from 'the country', they invoke traditional rights to stay over in Redfern and this contributes to overcrowding. A total of 50 per cent of the tenants were unemployed and 'on welfare' in the mid-1990s; a tiny minority of long-standing residents were permanently employed; the remainder receive allowances from local Aboriginal enterprises.

Sydney's Redfern may be a microcosm of the ghetto experience, but it simply does not feature in that city's social geography to the same extent as the black ghetto does in northern cities of the United States. In fact, with gentrification closing in on all sides in inner Sydney and the approach of the 2000 Olympic Games, the NSW government stepped up

the police presence and tidied up 'the Block' to avoid offending visitors to Sydney.

CHANGING GENDER ROLES AND DIVERSIFYING HOUSEHOLDS

Increasing numbers of neighbourhoods and suburbs in post-industrial cities are also acquiring greater social and cultural diversity as a consequence of the changing economic role of women together with the shrinking of the traditional family. As Castells (1997, 222) points out, 'What is at issue is not the disappearance of the family but its profound diversification, and the change in its power system.'

These social solvents spring from an awakening of female consciousness and the felicitous 'fit between women's working flexibility, in schedules, time, and entry and exit to and from the labour market, and the needs of the new economy' (Castells 1997, 173). Although the argument is tinged with functionalism, the growth in part-time and casual jobs that followed in the wake of labour market deregulation (Chapter 5) gave women more room to negotiate their multiple roles in the patriarchal household – the 'daily, quadruple shift', as Castells would have it, of paid work, home-making, child-rearing and 'night duties' for the husband. Women awakened to subversive ideas outside the home began to question traditional family values and unsettle gender relations. And, because of the decisiveness of women's earnings for the household budget and access to housing (Chapter 6), female bargaining power has gathered strength. This has led to a steady rise in the divorce rate in most post-industrial societies – although to nothing like the same extent in Catholic Italy (Table 7.5).

Households are now much more varied in their composition. With increasing numbers of adults seeking different living arrangements, the family unit is declining in importance as a household type within most post-industrial societies, with the notable exceptions of Japan and Spain. The United States leads the trends in marriage break-up and the formation of single-parent households (Table 7.6). Figure 7.12 traces trends in the changing composition of US households over a 35-year period, 1960–95. The archetypal nuclear family, or 'married couples with children', dropped from 44.2 per cent of households in 1960 to 25.5 per cent in 1995.

TABLE 7.5 *Trends in divorce rates per 100 marriages in developed countries*

Country	1970	1980	1990
Canada	18.6	32.8	38.3
Czechoslovakia	21.8	26.6	32.0
Denmark	25.1	39.3	44.0
England and Wales	16.2	39.3	41.7
France	12.0	22.2	31.5
Greece	5.0	10.0	12.0
Hungary	25.0	29.4	31.0
Italy	5.0	3.2	8.0
Netherlands	11.0	25.7	28.1
Sweden	23.4	42.2	44.1
United States	42.3	58.9	54.8
(former) West Germany	12.2	22.7	29.2

Source: Adapted by Castells (1997) from the original by A. Monnier and C. de Guibert-Lantoine. Reproduced by permission of Blackwell Publishers

TABLE 7.6 *Trends in single-parent households (%)*

Country	Early 1970s	Mid-1980s
Australia	9.2	14.9
France	9.5	10.2
Japan	3.6	4.1
Sweden	15.0	17.0
United Kingdom	8.0	14.3
United States	13.0	23.9
(former) USSR	10.0	20.0
(former) West Germany	8.0	11.4

Source: Adapted by Castells (1997) from the original by A. Burns. Reproduced by permission of Blackwell Publishers

More and more families are headed by single parents. Those headed by a female increased by 90 per cent in the 1970s and 21 per cent in the 1980s, so that by the mid-1990s, one in four children in the United States was growing up in a fatherless household (Castells 1997, 225). Among the 'other families with children' in Figure 7.12 are 'blended' households formed by divorcees with children from previous relationships. The number of 'blended' families in the USA, including children from a previous marriage, grew from four million to five million between 1980 and 1990.

Non-family households in the United States doubled as a proportion of all households between 1960 and 1995 – from 14.7 per cent to 30 per cent (Fig. 7.12). These include people living alone, or joining with other males and females to 'house share', perhaps in a gay or lesbian relationship. In breaking with the family unit, non-family households fabricate a pattern of living that provides networks of support, increasing female-centredness, and a succession of partners and options throughout the life cycle (Castells 1997, 227–8). Thus one-third of middle-class divorced couples in the suburbs of San Francisco maintain ties with former spouses and their relatives in order jointly to shoulder the responsibility of raising their children (Stacey 1990, 254).

The survey of Worcester households (Hanson and Pratt 1995) begins to reveal the greatly expanded range of domestic strategies adopted by working couples trying to manage the demands of paid work and parenting. Over three-quarters of mothers had taken, or were currently taking, a break from paid work to care for children. However, in the absence of maternity leave, women in lower-status jobs were forced to quit their jobs at the time of childbirth. When they returned to work: roughly half of the 492 women in the survey took a job that was closer to home; one in five suffered a fall in occupational status in the process; and 61.5 per cent had to settle for part-time employment (Hanson and Pratt 1995, 130–9). While some working-class couples organise their shifts sequentially to be at home with children, the partner of a well-paid manager or professional has a range of options including full-time housewife/husband. On the other hand, in countries with a supporting parent's benefit, single parents with young children may find that the tax system acts as a 'poverty trap' and discourages part-time employment.

With the upswing in female participation rates (Chapter 6), the suburbs of the 1960s and 1970s, with their patriarchal families, low densities, detached housing and reliance on cars, are poorly suited to the needs of 'careerists' working downtown and to female-headed families dependent on welfare.

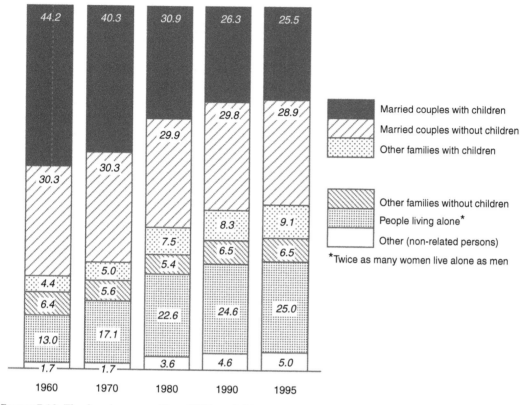

FIGURE 7.12 *The changing composition of US households, 1960–1995*
Source: adapted from Figs 4.12a and 4.12b in Castells (1997)

Women's support networks are crucial to single mothers, as well as to mothers with full- or part-time work outside the home. In North America and Europe, low-income families headed by women are more likely to live in central cities, to rent rather than own, and to reside in multi-unit apartments. Here they have access to women's networks for child care, and a wider range of the social services that 'they are dependent [upon] for their very survival' (Wekerle 1984, 11). Likewise, the rejuvenation of the central city housing market is partly due to the attraction it holds for affluent city workers, especially those households containing professional women ('double-income no-kids'; 'singles'). It is now widely accepted that 'the search for neighbourhoods and locations supportive of dual-career families has been a major impetus for gentrification and the revitalisation of inner city neighbourhoods' (Wekerle 1984).

Spaces supportive of gay and lesbian communities

The assertion of sexual identity by growing numbers of gays and lesbians in the second half of the twentieth century has given rise in many cities to self-contained communities that are supportive of their lifestyle and offer a protective residential enclave. The most celebrated communities in North America include San Francisco's gay Castro and lesbian Mission districts, Montreal's lesbian district, Seattle's Capital Hill, Houston's Montrose section, Cincinnati's Liberty Hill and Washington's Dupont Circle. Originally, these were inner-city neighbourhoods with gay and lesbian bars and clubs, which slowly developed a reputation as a 'liberated zone'. In each case, 'a community in transition became the destination for gay and lesbian migrants, whose influx increased property values and who became increasingly open in their lifestyles as their numbers increased. One gay rights advocate summed it up by saying, 'We're here. We need a realtor' (Abrahamson 1996, 111).

San Francisco's Castro and Mission districts are two of the most self-contained communities of their kind in the United States. As a major naval base and destination for military personnel on leave during the Second World War, San Francisco became a magnet for homosexuals discharged from the armed services. Bit by bit, during the 1950s and 1960s, the gay 'scene' moved southward from the docks to the Tenderloin district, then to south of Market, and finally to the Castro district. In the late 1960s it was a somewhat blighted and run-down, Irish working-class neighbourhood in decline as local factories closed. By the mid-1970s an estimated 30,000 homosexuals from elsewhere in San Francisco and around the country had moved into the district. These included gay physicians and psychologists, specialising in the treatment of gay patients; gay lawyers handling discrimination and child custody cases; a gay-owned savings and loan association, which did not discriminate against gay applicants; and travel agencies and insurance brokers that were sensitive to gays' needs.

Today, Castro is an affluent middle-class neighbourhood of renovated Victorian terraced apartments, which are noteworthy for their pastel-coloured paintwork. Residential streets are lined with all the boutique shopping, professional services and institutions serving a gay clientele. About 2 km east of Castro is the Mission district, which features retailing and services catering for the local lesbian community. Valencia Street is the commercial and institutional hub of the most fully developed lesbian enclave in the United States. However, while both these enclaves are home to large numbers of gays and lesbians living alone, in pairs or in groups, they also are very racially and economically diverse communities where no single lifestyle or household type predominates. According to Abrahamson (1996, 115), only a relatively small segment of Mission's population is lesbian owing to the fact that the whole city's openness to alternative lifestyles tends to blunt their identification with a specific neighbourhood.

Sydney, which is now known internationally for its annual Gay and Lesbian Mardi Gras,

has at least one well-defined, visible gay male commercial and residential district in and around Oxford Street and the neighbouring suburbs of Darlinghurst, Paddington,

and Surrey Hills, as well as two or three somewhat less visible and more 'mixed' ones in the modestly gentrified suburbs of Glebe, Annandale, and Newtown.

(Knopp 1998, 164)

Some of these same suburbs in Sydney's inner west – Newtown, Stanmore, Erskineville, Redfern, Leichhardt, Annadale and Balmain – are places chosen by many women living alone or with other women. Leichhardt, Newtown and Erskineville are sometimes referred to as the 'lavender triangle'. 'Women alone or living with a partner or friends share impulses, needs and desires with gay men: proximity to work, a community with which to identify, a neighbourhood which is friendly to non-traditional households and affordability' (Murphy and Watson 1997, 137).

CONCLUSION

This chapter has moved from describing the social regularities observed in twentieth-century cities to highlighting the present-day cultural diversity of urban communities. As Murphy and Watson (1997, 140–1) point out for Sydney, just as a city can be mapped across cultures so also it can be mapped across genders and bodies. This is to say that women and men may inhabit different spaces in the city in pursuit of different lives. Urban space can also be differently lived by women and men,

depending on whether they are rich or poor, old or young, black or white, with children or without, established or recent immigrants, gay or straight. Sex/gender forms part of the warp and weft of the city. There is no static sexed/gendered urban subject in relation to urban space. The patternings, interconnections, disjunctures of women's lives are formed within, and themselves form, the geography and shape of Sydney. Sexed bodies are produced in the very skin and bone of its built environment, and the patterns they make can at once be fleeting and temporary, contingent or seemingly locked in disadvantage and marginality.

(Murphy and Watson 1997, 141)

What Murphy and Watson are highlighting is the incessant social and cultural transformation that goes on as men, women and children habitually renegotiate their use of space in cities. These are the human processes that ultimately give shape to the social and cultural mosaic of cities. Maybe, if we can apply Susan Smith's words to the urban context, 'the quest for belonging, the search for similarity, the hunt for commonality and community are what contain the next critical geography of society and space' (Smith 1999, 22).

FURTHER READING

The new edition of *Urban Social Geography: An Introduction* (**Knox and Pinch** 2000) is the best reference to the social geography of cities.

Susan Smith's (1999) essay on 'society-space' provides a very accessible discussion of recent debates on the nature of social space and culture.

Aspects of social and cultural identity in a wide variety of urban communities are considered by authors like **Castells** (1997), **Sandercock** (1998), **Abrahamson** (1996), **Fincher** and **Jacobs** (1998), and **Hayden** (1995).

Compilations and review articles on ethnic segregation and social polarisation in cities are available in a special issue of *Urban Studies* (**van Kempen** and **Özüekren,** 1998), **Peach** (1996), **O'Loughlin** and **Friedrichs** (1996), and **Marcuse** and **van Kempen** (2000).

8

CITY ENVIRONMENTS AND LIVING CONDITIONS

INTRODUCTION

From the mid-nineteenth century onwards, polemic tracts against the city raised awareness of mounting environmental and social costs as people flocked into Victorian cities (Mumford 1961). Before the development of the suburb there was no escaping their slums, the poverty, the crime and violence, the cholera, the smog-ridden skies:

The industrial cities of the nineteenth century were hell: they suffered extremes of overcrowding, poverty and ill-health. Stinking open sewers spread cholera and typhoid; toxic industries stood side by side with overflowing tenements. As a result, life expectancy in many of the industrial cities of Victorian England was less than twenty-five years. It was precisely these hazards and basic inequities that led planners like Ebenezer Howard in 1898, and Patrick Abercrombie in 1944, to propose decanting populations into less dense and greener surroundings: Garden Cities and New Towns.

Today, by contrast, dirty industry is disappearing from cities of the developed world. In theory at least, with the availability of 'green' manufacturing, virtually clean power generation and public transport systems, and advanced sewerage and waste systems, the dense city model need not be seen as a health hazard.

(Rogers 1997, 2.32–3)

The above extract is from a great little book, *Cities for a Small Planet*, and the author, British architect Richard Rogers, chaired the Urban Task Force appointed by the incoming Blair government in 1997 to devise a strategy for reviving Britain's ailing cities. Higher urban densities in place of the spreading blanket of suburbs in the south-east English

countryside are part of the blueprint for 'urban renaissance' in Britain (Urban Task Force 1999).

The environment of nineteenth-century cities was especially hazardous because workers had to live within walking distance of their jobs in the factories and coalmines, and industrial pollutants and household waste would not bear the cost of transportation into the countryside (Chapter 4). The one exception was where the city was sited on a river into which wastes could be discharged. Improvements in waste treatment and transport technologies steadily changed that for urban populations during the twentieth century. This is why the eminent urban planner and one-time urban geographer Peter Hall (1999, 66) regards suburban development and the mass transit systems that made it possible as 'the twentieth century success story'. Similarly, the application of technology to waste disposal and pollution dispersion provided industrialists and city managers with the means to shift the environmental burden of cities out into the countryside, where it becomes someone else's problem.

Environmental management in First World cities is currently geared to waste reduction and recycling, and to cleaning up 'brownfield' sites in abandoned industrial districts. This goes hand in hand with the mitigation of the most damaging effects of urban growth and consumption upon the countryside. Because affluent societies are demanding that cities as living environments meet their much higher expectations with respect to quality of life (Eckersley 1998), urban managers now have to satisfy citizens that the planning and design of future cities is ecologically sustainable. The challenge is to devise policies for sustainable urban development while designing 'energy-efficient, ecologically friendly' cities. But here it pays to note that the 'Green and clean' agenda of prosperous cities in Europe and North America does not contain the same environmental priorities occupying rapidly developing cities in Asia, Africa and Latin America. 'Brown and mean' is a much more accurate summation of the environmental problems besetting not only the mega-cities, but also many small and medium-sized cities in the South. In this respect, living conditions for the urban poor in developing countries are often no better than those that awaited immigrants crowding into the tenements of nineteenth-century London, Manchester, Glasgow, Paris or New York.

This brief overview introduces some of the environmental concepts and themes to be explored further in Chapter 8. Having defined the challenge before urban managers and planners, we will return in the latter stages of Chapter 9 to look at some of the strategies that are being developed to make for more sustainable cities in the future.

WORKING CONCEPTS

Environmental limits

Cities are undoubtedly expressive of the highest artistic and scientific endeavour. In seeking progressively to improve the structure and form of cities as a habitat for the human species, architects and engineers have not only modified the natural environment; at times they have sought to control the environment. Strengthening the city's defences against natural disasters is just another way in which humans like to think they have established mastery over Nature. Anyone who lives at the heart of giant cities like Tokyo, New York or Berlin will be aware that this can create a false sense of the city as a self-sustaining system.

Ecologists are insistent that it is high time that people living in cities fully appreciated the extent to which urban existence is sustained by Nature. This depends on the continuing ability of the environment to provide energy and food to urban populations, absorb their waste and undertake basic 'life-support' functions such as temperature maintenance and protection against solar radiation.

Each city is an ecosystem that in turn depends upon the larger biosphere of which it is only a tiny

part. The combined effect of human activities in cities around the globe is placing an intolerable strain on the biosphere. 'There is increasing evidence to suggest that we are breaking, or risking breaking, some important global carrying capacity thresholds' (Expert Group on the Urban Environment 1996, 40). The contribution to global warming of greenhouse gas emissions originating in cities is the most alarming demonstration of the interconnectedness of human activity and natural ecosystems.

Determining where the critical 'environmental limits' lie – the Earth's carrying capacity – requires exacting scientific research. For example, it has taken some time for the link between greenhouse gases and global warming to be firmly established. Ecologists advise that, where there is lingering doubt about the impact of human activity upon the environment, the 'precautionary principle' should apply. Significantly, the Maastricht Treaty, which provides the legal foundation for the European Union, stipulates that 'very substantial weight must be given in decision taking' to avoiding 'potentially critical risks to the physical ecosystem' (Expert Group on the Urban Environment 1996, 40). This is to give nature the benefit of the doubt.

Gigantic cities like New York, Shanghai or Mexico City combine the most extreme forms of human modification of the natural environment. Investors and property developers have provided the motive, and engineers and architects the means, to dominate nature. But this is only a surface reflection of a deeper rupture between current economic practices under capitalism and the dynamics of natural ecosystems. 'Green' economists like Daly and Cobb (1990) expect that because they are no longer sustainable many of these practices will eventually prove uneconomic. 'Nowhere is this more evident than in cities, where the prevailing patterns of production, distribution and consumption are associated with the rapacious demands for non-renewable energy resources, and where major environmental stresses arise because of the prodigious generation of waste products' (Stilwell 2000, 207–8). For example, because of the concentration of economic activities in cities and their significantly higher population densities, built-up areas occupying less than 10 per cent of the land area account for 75–80 per cent of the energy consumed within the European Union (Expert Group on the Urban Environment 1996, 113).

Cities are ecosystems

Too often in the process of building cities nature has been overtaken by the built environment. 'Urban development usually reduces biomass and biodiversity by building over land and displacing animal and plant populations' (Expert Group on the Urban Environment 1996, 45). Cities need to be viewed as ecosystems in their own right. 'Since they build themselves by destroying nature', in the past cities were never thought of as ecosystems. But the opportunities exist for city-builders of the future to redeem themselves by seeing to it that the built environment is more completely in harmony with nature. In an effort to 'Green' cities, officials working with local communities are putting projects in place to 'revegetate' industrial wastelands, protect remnants of native bush, and create wildlife sanctuaries and natural wetlands.

Urban metabolism

In order to explain this view of the city as an ecosystem, Herbert Giradet (1992) likens the city to a 'metabolism' into which energy flows, undergoes various forms of conversion and consumption, and is then discharged to waste or emitted as pollutants (Fig. 8.1). Cities with a linear metabolism consume and pollute at a high rate. Cities with a circular metabolism minimise the new inputs and maximise recycling. 'Since the large majority of production and consumption takes place in cities, current linear processes that create pollution from production must be replaced by those that aim at a circular system of use and re-use. These processes increase a city's overall efficiency and reduce its impact upon the environment' (Rogers 1997, 2.32). In urban management terms, this means legislating and pricing to encourage local enterprises and consumers to recycle materials, reduce waste,

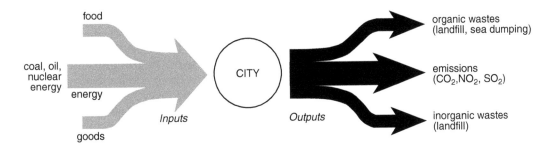

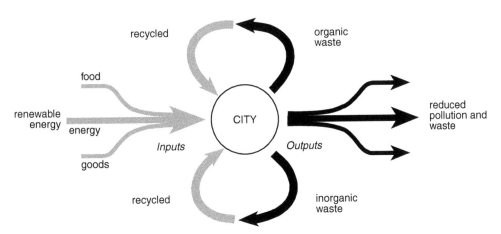

FIGURE 8.1 *The flow of energy and wastes in cities with linear and circular metabolisms.*
Source: Giradet 1992. (Redrawn by Rogers 1997.) Reproduced by permission of Faber & Faber

conserve exhaustible energies and tap into renewable ones. Other pro-business commentators believe that market competition between the big energy users and makers of environmentally harmful products will eventually force them to introduce Green technologies of their own accord.

Ecological footprint

William Rees (1992), a community planner teaching at the University of British Columbia in Vancouver, developed the idea of an 'ecological footprint'. This refers to the resources a city's population consumes in excess of the carrying capacity of the site the city occupies. This is expressed as an 'ecological deficit' per person of the land area needed to support the population. The ecological footprint also draws attention to the 'interaction field' beyond the city from which it takes those resources, and over which it distributes its waste and pollution. Chapter 2 noted how dependent Chicago's nineteenth-century industries were upon its hinterland for raw materials and how, in turn, this demand impacted environmentally upon the soils, forests and fisheries. Nowadays, a city of 10 million may typically need to import at least 6000 tonnes of food every day to feed its population. This dependence upon the countryside has been reduced somewhat in a few nonwestern societies. In China, up to 20 per cent of the food needs of cities have traditionally been met by urban farming. Havana provides almost 5 per cent of Cuba's food. Therefore cities impact upon ecosystems and natural landscapes, not only within the built-up area but also within a surrounding region.

Historically, if the local ecosystem supporting a city became degraded, the prosperity of the city suffered. In extreme cases, the very viability of a city could be threatened. This was the eventual fate of Rome, even though at the height of the Empire the Romans were able to draw timber, grain, ivory, stone and marble from North Africa (Chapter 2). Alternatively, the Chinese developed quite sophisticated solutions for disposing of city waste on the flood-prone plains to avoid polluting their crops and fisheries. In the past, a city's size and economic base were limited by transport costs to the resource endowments of its hinterland.

The railways, transoceanic shipping, trucking and airfreight broke the reliance of cities upon the natural productivity of their surrounding areas.

High-energy transmission lines enable power generated in remote rural areas to feed the needs of urban industries and households. With the capacity of city dumps close to exhaustion, solid waste that cannot be recycled or incinerated has to be transported further out into the countryside. As accessibility improves, more and more city dwellers opt for a 'lifestyle block' or a 'weekender' in the countryside, or venture into remote areas in search of recreation – and take their habits as consumers with them. Cities concentrating wealthy, cosmopolitan shoppers now import exotic consumer goods from distant places. To remain competitive, farming has to meet the demands of urban consumers for highly processed, permanently available, packaged food. Consistency and standardisation take precedence over nutrition, animal welfare, resource efficiency, diversity, regional differentiation and even taste (Expert Group on the Urban Environment 1996, 44). All these constitute forms of urban spillover that can be likened to an ecological footprint.

In calculating the ecological footprint of an urban population, Rees (1992) develops the ecological equivalent of 'carrying capacity'. Ecologists define this as the population of a given species that can be supported indefinitely in a given habitat without permanently damaging the habitat it depends on. Estimates are derived of the energy consumed in the process of feeding, housing, transporting and providing an urban population with goods and services. These estimates are summed and converted into a measure of the land area required per person to sustain this level of resource utilisation by an urban resident.

For example, Wackernagel and Rees (1996, 86) calculate that the Lower Frazer Valley, which extends eastward 144 km from Vancouver BC, over an area of about 4000 square kilometres and is home to 1.8 million people, has an actual ecological footprint of 77,000 square kilometres. In other words,

the Lower Frazer Valley requires an area 19 times larger than its home territory to support its present consumer lifestyles, including 23,000 square kilometres for food production, 11,000 square kilometres for forestry products, and 42,000 square kilometres to accommodate energy use. This figure represents the region's 'ecological deficit' with the rest of the world.

(Wackernagel and Rees 1996, 87–8)

In another case, Folke *et al.* (1996) investigate the appropriation of renewable resources like wood, paper, fibres and food (including seafood) by 29 cities on the shores of the Baltic Sea. Their results suggest that, to satisfy the consumption requirements of these cities, resources are appropriated from an ecological footprint that exceeds the combined urban areas by 200 times. Significantly, because of the role marine resources have traditionally played in the economies of the Baltic States, the appropriated ecosystem area (AEA) within the Baltic Sea greatly exceeds (by 133 times) the equivalents for forests (by 50 times) and agricultural land (by 17 times).

Wackernagel and Rees (1996, 85–97) report that the ecological deficits of urban-industrial countries vary from a low of 2–3 ha per person for Japan and South Korea, to a high of 4.3 and 5.1 ha per person for Canada and the United States. Urban residents of most European countries lie somewhere within this range, with 3–4 ha footprints. Globalisation processes like the lowering of barriers to free trade now enable affluent consumers in North American or European cities, for example, to source their energy supplies and consumer goods from every corner of the globe. There are growing signs that the unequal terms of trade between the North and South are actually accelerating the depletion of essential natural capital, thereby undermining global carrying capacity (United Nations 1996, 150). In its annual Human Development Report for 1998, the United Nations noted that 20 per cent of the world's population, and overwhelmingly concentrated in the industrial cities of the North, accounted for 86 per cent of global consumption (reported in *Guardian Weekly*, 20 September 1998, 11). And this is notwithstanding the growth of an increasingly affluent middle class in the cities of the developing world.

ECOLOGICALLY SUSTAINABLE URBAN DEVELOPMENT

There is a growing acceptance of the responsibility that the present generation has towards future generations for custody of the environment. This principle is integral to the generally accepted definition of sustainable development as first enunciated in the Brundtland Report: 'Sustainable development is development that meets the needs of the present without compromising the ability of future generations to meet their own needs' (World Commission on Environment and Development 1987, 43).

The overarching goal of sustainable development in a rapidly urbanising world will require a fundamental break with the way cities have been developed and managed in the past. However, as Eckersley (1998, 4) notes, 'sustainable development offers an alternative to conventional growth as a path to progress, but exactly what it means and how it can be achieved remains unclear'. For some, sustainability is a vague and contested concept that defies definition. To others that matters less than the debate and the new modes of enquiry and learning it has engendered. Sustainability is a key concept in environmental ethics and Green politics. Deciding what is sustainable raises 'leading' questions about ecology, social equity and efficiency. For future urban development to be sustainable, resource use, economic policies, technological development, population growth and institutional structures will have to be harmonised.

The difficulty lies not only in spelling out how ecologically sustainable urban development might be approached, but even more in building the political support to achieve it (Chapter 9). For example, having frankly admitted that since taking office the environment had taken a back seat to Labour's priority issues of health, education and crime, the British prime minister went on to explain that,

There are points of real conflict between the interest of consumption and the longer term interests of the environment, and between politicians' need to woo the electorate as well as lead it ... We should build a business case for the environment, working to harness clean technologies, seeing business as part of the answer rather than the problem.

(Tony Blair, reported in *Guardian Weekly*, 2–8 November 2000, 9)

But the marriage of business interests and the pursuit of environmental sustainability has led to an uneasy suspicion of the motives of some of those corporations that are quick to capitalise on the opportunity to promote a clean, Green image for their products in the marketplace. According to many environmentalists, this new form of whitewash is tantamount to 'Greenwash'.

There is a sense in which the concept of 'ecologically sustainable urban development' is an oxymoron – large dense urban settlements like today's mega-cities and the future prospect of them being sustainable is so much at odds. Therefore, in the context of ecologically sustainable development, sustainable cities are something to be strived for 'because we all know that cities which import their energy, raw materials and food, and export their wastes, cannot, by definition, be sustainable'. Rather, 'We must strive to achieve cities which are less unsustainable' (Troy 2000, 1).

Ecological sustainability imposes three stringent conditions. Firstly, there should be no further depletion of biodiversity. This serves as a limiting factor to uncontrolled urban growth. Secondly, the ecological integrity of the environment must not be endangered. In the case of cities, this means operating within the coping capacity of the natural biological mechanisms that disperse – winds and tides – or break down organic materials. Thirdly, there is an obligation to future generations to leave the natural environment in the same, if not better, condition in which we found it. In urban environments, this principle of 'intergenerational equity' has led to the cleaning-up and restoration of brownfield sites, and the declaration of park systems to protect remaining natural habitat.

Achieving greater self-sufficiency in food production, energy generation, water harvesting and sewage processing represents one of the most promising avenues for making the transition to ecologically sustainable urban development. Troy (2000, 2–4) offers three practical steps that can be, and are being, taken by urban authorities and citizens to lessen the presently unsustainable impact of cities upon the natural environment:

- reduce the average household consumption of water
- reduce per capita energy consumption
- reduce per capita generation of waste.

However, this is hardly a prescription for those poor households in Third World cities that lack piped water, or electricity and gas, and rely for their livelihood on recycling the waste dumped on the municipal tips. Even though the technologies now exist to make the leap to cleaner production in developing cities and some advanced waste processing systems have been installed, their budgets do not extend to coping with the environmental overload that will be generated along with any improvement in the living standards of the urban poor. Nevertheless, there will be some environmental gains with improved living standards in developing economies. For example, in Chinese cities the switch from coal to natural gas and other cleaner fuels for domestic cooking and heating is beginning to take place.

Lastly, there is broad agreement among ecologists and urban planners that the complex array of environmental problems caused by urban development has to be approached holistically if there is to be any appreciable progress towards sustainability (Stilwell 2000). 'A growing interest in the sustainability of cities requires consideration of socio-economic as well as environmental factors, and an exploration of new planning models and approaches designed to enhance sustainability' (Hutton 1992). Gains from waste reduction and pollution control in cities will only accrue from the adoption of economically sustainable practices. In an era of economic restructuring, the combined effects of urbanisation, consumerism and social polarisation create tensions that work against social sustainability in cities.

ENVIRONMENTAL CONDITIONS IN CITIES

The adverse effects of the overconcentration of polluting and waste-creating activities in urban areas are not restricted to spillover effects on the environment beyond cities. Environmental conditions in densely populated cities have deteriorated to the point where most residents suffer from background exposure to traffic noise and air pollution. High concentrations of nitrogen dioxide or contaminated water circulating within the urban environment may also place an entire population at risk.

But some environmental hazards originate from a single source – industrial pollutants or waste discharge – and can be avoided if a household has the income to locate in a 'safe' part of the city. This leaves poorer communities exposed to petrochemical and metallurgical processes that are extremely hazardous from an environmental health point of view. For example, when a number of the chemical storage tanks in Melbourne's Coade Island tank farm caught fire in the mid-1990s, several inner western suburbs had to be evacuated to escape the poisonous fumes. The landscapes of cities dependent on smelting copper, lead and zinc are dotted with slag heaps laced with lead, arsenic and cadmium. Whether it's Port Pirie in South Australia or Karabash in the Urals, for a generation or more children have been poisoned by the heavy metal particulates that rain down on the soil from the smelters' chimney stacks. They exhibit higher rates of congenital defects, disorders of the central nervous system and diseases of the blood, glands and the immune and metabolic systems than children from control groups. And in one of the most horrifying industrial accidents, the accidental release of methyl iso-cyanate in the Indian city of Bhopal caused the death of over 3000 people and permanent injuries to 100,000, while 200,000 residents were evacuated.

The environmental hazards that are the most pervasive in terms of their impact on the health of city dwellers arise in connection with the supply and removal of water and sewage, the production and disposal of solid waste, ambient air pollution and the generation of harmful noise. Some of the environmental hazards that urban populations face are the result of natural events, but the level of risk and the numbers of residents exposed to the risk are often magnified by uncontrolled urban development. Poor environmental health may be traceable to a single cause depending upon the susceptibility of particular individuals in the urban population (respiratory complaints among babies and the elderly), or it may be more complex in its aetiology. For example, from the case studies below it is apparent that the incidence of some communicable diseases is associated in cities with poverty, substandard housing, overcrowding and unsanitary living conditions.

Supply and removal of water and sewage

With the growth of European cities during the nineteenth century, hydraulic engineering and public health systems were developed to combat the outbreak of water-borne infectious diseases among urban populations. Reservoirs were built in catchment areas to collect potable water, which was then piped to urban consumers. The domestic water supply was then used to transport waste out of the city and back into the environment (Fig. 8.2). From an 'urban metabolism' perspective, this is a linear flow system. As cities have grown in size and spread within the surrounding bioregion, the impact upon the environment of excessive water extraction combined with the discharge of increasing volumes of effluents and stormwater has worsened. In the European city, for example, liquid waste can

include final washes from metal works, refineries, food-processing plants, tanneries and textile plants, along with the waste from homes, hospitals and other institutional users. The challenge for urban managers and city engineers is to convert linear water services, where they exist, to a circular water flow model by better re-use, recycling and onsite management.

Much of the rapidly expanding informal settlement encircling cities in the South lacks closed drains and garbage-collection services. Consequently, 'Most rivers in cities in the South are literally large open sewers' (United Nations 1996, 146). In Brazil, for example, 40 per cent of São Paulo's 17 million residents and 56 per cent of Rio de Janeiro's households have no proper sewerage connections. Where treatment plants are inadequate, or where the discharge is untreated or only partially treated sewage, river and coastal ecosystems are being destroyed. In the summer of 2000, the long-term effects of environmental abuse were visible on the beaches of Rio de Janeiro:

Soaring sewage levels have prompted bathing bans on its most famous beaches, a large oil spill caused environmental havoc on its northern shores, dismembered corpses have been swept in by the tide at Ipanema, and more than 100 tonnes of dead fish have been removed from the city's uptown lagoon.

(Reported in *Guardian Weekly*, 30 March–5 April 2000, 3)

But an environmental crisis of these proportions is not unknown in cities of the North. Between July and September 1998, Sydney's water was undrinkable on a number of occasions owing to a series of cryptosporidium outbreaks. Although initially attributed to dead livestock, a wide-ranging enquiry identified many possible sources of contamination and ruled that Sydney Water lacked sufficient regulatory control of the catchments to guarantee safe drinking water. The Sydney Catchment Authority was subsequently established in July 1999 to manage the water supply and protect a catchment area of some 370,000 ha, containing the main reservoirs.

Production and disposal of solid waste

Major health and environmental problems can also occur when insufficient care is given to the handling and disposal of the solid wastes produced by urban households and industries. Mexico City, with an

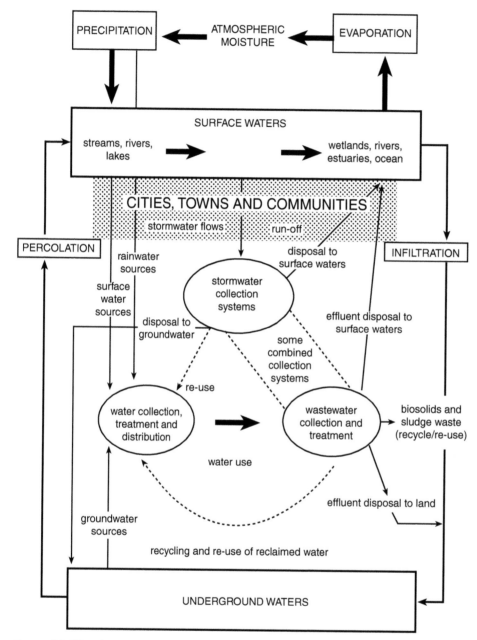

FIGURE 8.2 *The urban water cycle*
Source: Parliamentary Commissioner for the Environment (2000, 3)

estimated population of 22 million, generates 12,000 tonnes of garbage a day, which helps to sustain about seven rats per person. Solid wastes comprise organic materials that are biodegradable because they eventually break down and decompose in landfills, sewage treatment plants or aquatic receptors. The cities of southern Europe produce a greater proportion of their solid waste in this form, as do Australian households. For example, garden waste accounts for about 16–18 per cent of waste disposed to landfill in Melbourne.

For the most part, industrial waste and construction materials from building and demolition sites in cities are non-biodegradable. In cities undergoing major development, or even redevelopment, building waste can account for as much as 40–50 per cent of the landfill by volume. Non-biodegradable products, mainly in the form of plastics but especially PVC, cause major environmental problems owing to their continuous accumulation and the dioxins that are released if PVC is mixed with household refuse and incinerated at a low temperature. Because large urban incinerators also contribute to the greenhouse effect, scientists now question the wisdom of burning solid waste. But the use of landfill sites for hazardous or toxic materials exposes people living close by to the risk, and endangers the ecosystem if they leach into the water table.

In the early 1990s, the European Commission set a solid waste disposal target that will stabilise the quantity of waste generated by the average urban household at the 1985 EC level of 300 kg per capita. The respective estimates obtained in 1989 for Australian consumers, city by city, are very high by international standards: 422 kg per capita in Adelaide; 649 kg per capita in Melbourne; 837 kg in Sydney; and 1066 kg in Brisbane (AURDR 1995, 72). Paper, aluminium, tin and glass containers, and packaging make up 30–40 per cent of the volume of the solid waste deposited in urban landfill sites. It moves quickly along the production and distribution chain to urban households, but then takes a long time to decompose.

Ambient air pollution

Although few large cities have acceptable air quality, the severity of air pollution and the mix of the pollutants vary enormously from city to city. The mix and concentration of air pollutants in many cities in the North and the South are already high enough to cause added distress to people with asthma or chronic breathing problems. Environmental health studies undertaken in several cities in Europe and the United States have found that an increase in high pollution days is accompanied by a rise in respiratory diseases, and even in the death rate (United Nations 1996, 143–4). In the United States deaths due to cardiac and respiratory disease were 37 per cent higher in the most polluted city compared with the least. The death rate in London

increased by more than 10 per cent in a four-day smog episode in December 1991.

Most of the ambient air pollution in urban areas comes from the combustion of fossil fuels to power motor vehicles, to meet industrial needs and to generate heating and electricity. Coal is the traditional source of energy, and in Chinese cities where it is still extensively burned in domestic stoves is the major source of air pollution. World Bank estimates suggest that sulphur dioxide from China's high consumption of 'dirty' coal causes 50,000 premature deaths and 400,000 new cases of chronic bronchitis every year in 11 of the largest cities. But in cities with high vehicle usage, including those in the South that are quickly catching up, the main pollutants – photochemical pollutants, lead and carbon monoxide – come primarily from motor vehicles. Hence cities like Tokyo, New York and London have relatively low levels of sulphur dioxide and suspended particulates but all have problems with motor vehicle emissions – as do many larger cities in the South (Table 8.1).

Many cities in Russia's heavily industrialised Urals region are being forced to make a choice between keeping dirty factories in production to feed their population, and jeopardising the health of their children in the process. Factory closures have improved air quality in some cities, but 65 million people, or 44 per cent of the population, still live in cities that exceed the Russian government's strict limits on air pollution (reported in *Guardian Weekly*, 8–14 July 1999, 33). Environmental research conducted in 1995 revealed that more than 13 per cent of the soil tested from kitchen gardens was contaminated with heavy metals, oil, pesticides or other harmful substances. Residents of heavily polluted urban areas suffer from higher rates of diseases of the blood, lungs and glands, and from higher rates of nervous system disorders and congenital defects. In an equally polluted industrial region – Poland's Upper Silesian Industrial Region – urban male life expectancy fell during the last 15 to 20 years of the twentieth century. In the city of Katowice, the 1989 infant mortality rate was 25.5 per 1000 live births, which compares unfavourably with the Polish average of 16.1 (United Nations 1996, 144–6). In several central and eastern European cities, where exposure to lead-emitting industries remains high, it is estimated that the mental development of at least 400,000 children may have been impaired.

TABLE 8.1 *Air quality in 20 of the world's largest cities*

City	Sulphur dioxide	Suspended particulates	Airborne lead	Carbon monoxide	Nitrogen dioxide	Ozone
Bangkok	Low	Serious	Above guideline	Low	Low	Low
Beijing (Peking)	Serious	Serious	Low	(no data)	Low	Above guideline
Bombay	Low	Serious	Low	Low	Low	(no data)
Buenos Aires	(no data)	Above guideline	Low	(no data)	(no data)	(no data)
Cairo	(no data)	Serious	Serious	Above guideline	(no data)	(no data)
Calcutta	Low	Serious	Low	(no data)	Low	(no data)
Delhi	Low	Serious	Low	Low	Low	(no data)
Jakarta	Low	Serious	Above guideline	Above guideline	Low	Above guideline
Karachi	Low	Serious	Serious	(no data)	(no data)	(no data)
Manila	Low	Serious	Above guideline	(no data)	(no data)	(no data)
Mexico City	Serious	Serious	Above guideline	Serious	Above guideline	Serious
Rio de Janeiro	Above guideline	Above guideline	Low	Low	(no data)	(no data)
São Paulo	Low	Above guideline	Low	Above guideline	Above guideline	Serious
Seoul	Serious	Serious	Low	Low	Low	Low
Shanghai	Above guideline	Serious	(no data)	(no data)	(no data)	(no data)
Moscow	(no data)	Above guideline	Low	Above guideline	Above guideline	(no data)
London	Low	Low	Low	Above guideline	Low	Low
Los Angeles	Low	Above guideline	Low	Above guideline	Above guideline	Serious
New York	Low	Low	Low	Above guideline	Low	Above guideline
Tokyo	Low	Low	(no data)	Low	Low	Serious

Note: These figures are based on a subjective assessment of monitoring data and emissions inventories.
Source: United Nations (1996); extracted from UNEP/WHO report on Urban Air Pollution in Megacities. Reproduced by permission of Oxford University Press

On the other hand, air quality problems in some of the world's largest and wealthiest cities are due to the build-up of carbon dioxide and ozone in the atmosphere. Ozone is caused by reactions between nitrogen dioxide, hydrocarbons and sunlight. It is present in photochemical smog along with other hazardous chemicals. Although the data displayed in Figure 8.3 are very general, they highlight the nature of the relationship between carbon dioxide emissions per capita and private versus public

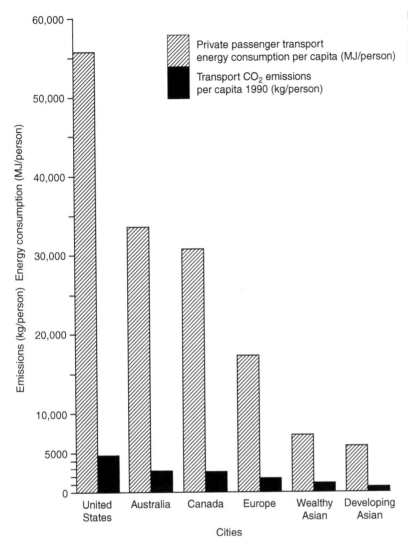

FIGURE 8.3 *Air quality in cities in relation to private vehicle dependency* Source: based on Newman (1999, Figs 3 and 4)

transport usage in different groupings of cities. As well as identifying those societies with the most car-dependent urban transport systems – the United States, Australia and Canada – Figure 8.3 points to the critical environmental problem that lies ahead should more commuters in Asia's developing mega-cities turn to private means of transport. At present, car usage in Chinese cities is typically about one-twelfth to one-fifteenth of the levels found in US or Australian cities. Whereas Singapore, Tokyo and Hong Kong greatly increased the capacity of their mass transit systems during the 1990s, cities like Bangkok, Jakarta and Kuala Lumpur have favoured road-building.

A study conducted by 11 member cities of the International Council for Local Environmental Initiatives (ICLEI) compared total carbon dioxide emissions for five European and six North American cities (ICLEI 1993). Per capita emissions from the European cities averaged only 56 per cent of the North American cities. The difference was attributable to the higher motor vehicle use in North America. Transport emissions ranged from 0.93 tonnes of carbon dioxide per capita for Helsinki to 7.28 tonnes per capita in Minneapolis–St Paul, an almost 800 per cent variation (AURDR 1995, 50).

The cities with the highest ozone concentrations tend to be those with high per capita vehicle

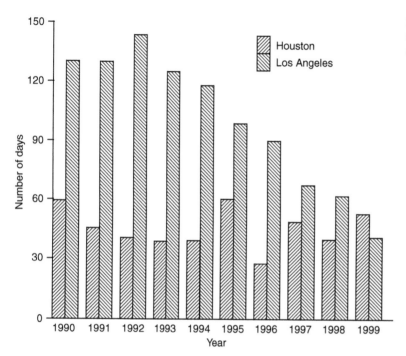

ownership *and* high sunshine hours. Although Los Angeles immediately comes to mind, cities like Mexico City, São Paulo and Tokyo also have serious ozone problems partly because they are situated in saucer-shaped basins (Table 8.1). In recent years ozone levels have begun to climb in US cities like Houston, owing to the number of 'grandfather' petrochemical and industrial plants present (Fig. 8.4). Grandfather plants were built before the introduction of the Clean Air Act in 1971 and are bigger polluters than newer plants. In 2000, Houston's grandfather plants accounted for 36 per cent of the chemicals Texas released into the atmosphere annually. Significantly, 'Since Mr Bush took office [as State Governor], the number of days when Texan cities have exceeded federal ozone standards has doubled, and Houston and Dallas currently face federal deadlines to make sharp cuts in air pollution or risk losing federal money for their roads' (reported in *The Economist*, 22 July 2000, 31).

Recent research on variations in urban air quality has drawn attention to the contrast between the sprawling form of car-dependent North American and Australasian cities, and compact European cities with their well-patronised public transport systems (AURDR 1995, 50–6). Comparative analysis

of the relationship between urban densities and the quality of air in cities suggests that three factors are important discriminators: car ownership levels; fuel consumption per capita; and the share of the journey to work taken by public transport, walking and cycling (Newman and Kenworthy 1999). More extensive research is tending to confirm the view that, while people in densely settled cities might be expected to make less use of cars, and walk or ride to work, fuel consumption per capita is also sensitive to other factors like fuel and land prices.

Harmful noise

In more of the world's cities it is growing increasingly difficult to escape the nuisance of intrusive, and sometimes harmful, noise. High noise levels can lead to hearing loss, but other less traumatic effects include disturbed sleep, reduced productivity in the workplace and increased anxiety. If a noise threshold is chosen that is well above an acceptable level – say 70 dB(A) – then 47 per cent of Sofia's population and 25 per cent of Budapest's population are exposed to excessive noise (United Nations 1996, 147). Elsewhere in eastern Europe, the figures fall to 20 per cent and 19 per cent for St Petersburg and Crakow respectively, but these levels are still well above those recorded in cities like Copenhagen (4 per cent) and Amsterdam (2 per cent). An Australian study indicates that in 2000 an estimated 500,000 Sydney residents, or about 12 per cent of the population, were exposed to traffic noise levels greater than 65 dB(A), and 1.5 million in the range 55–65 dB(A) (AURDR 1995, 66).

There are four main sources of noise pollution within the urban environment: industrial equipment, construction work, road traffic and aircraft. Many international airports are situated in the midst of densely populated areas, and, although some are governed by night curfews, people in cities like New York, Hong Kong, Sydney, Mexico City, Bombay, Los Angeles and Bangkok live with the incessant noise of jet aircraft taking off and landing. Noise from major highways and some sections of the elevated rapid transit system is also a problem in those cities where transport corridors have been pushed through heavily built-up areas. In fast-growing Asian cities like Shanghai and Bangkok the noise levels from trucks, buses, motor cycles and even high-speed water craft can reach 80–90 decibels in some central locations at rush hour. Not surprisingly, it tends to be the urban poor living in illegal or informal settlements squeezed between major highways or around the perimeter fence of airports that have the highest exposure to traffic noise.

Uncontrolled urban development and expansion

The problems associated with uncontrolled urban development have been exaggerated in some cities by the relief, fragility and instability of the natural environment they occupy (Fig. 8.5). For example, extreme climatic events during the winter of 2000 served to emphasise the consequences of permitting spillover growth on flood-prone areas in the south-east of England. Mexico City lies in the Valley of Mexico, perching at 2250 metres above sea level. It straddles an unstable fault line and is surrounded by volcanoes and a mountain chain that for centuries blocked adequate drainage and turned the Valley of Mexico into a huge swamp. The surrounding peaks and temperature inversions prevent emissions from escaping, transforming the valley into

a giant bowl of yellow-grey gunk. More than 3.6 million vehicles clog the roads and, combined with 32,000 industrial plants, spew more than 12,000 tons of pollutants into the air every day. Air quality is unsatisfactory by international standards 324 days of the year.
(reported in *Guardian Weekly*, 21 December 1997, 17)

The land supporting Mexico City is primarily reclaimed marsh and mushy soil, which magnifies the vibrations and destruction when earthquakes strike. A massive tremor in 1985, which measured 8.1 on the Richter scale, and the associated aftershocks, killed at least 10,000 people, injured 50,000 and made 250,000 homeless. Because of the earthquakes, rather than building up, the city spread, contributing to the sprawl and impoverished *barrios* that circle the city – the so-called Rings of Misery.

A wave of urbanisation during the 1990s in Turkey increased the population of cities in the north-west and created a demand for cheap housing. Industrial investment poured into places with a high earthquake risk and apartment construction often proceeded without enforcing proper engineering standards. An earthquake registering 7.4 on the Richter scale in August 1999 levelled countless buildings and killed over 10,000 people. According to Turkish seismologists, if the earthquake's epicentre had been in Istanbul, up to a million of the city's 10 million residents could have perished.

San Francisco had to be rebuilt after the earthquake and Great Fire of 1906. Los Angeles is also located in a very unstable and fragile environment (Davis 1999a). Except for their vulnerability to earthquakes, Southern Californians have much in common with other urban communities that have settled 'Mediterranean' climatic zones. Along with the classical Mediterranean littoral, central Chile, and the coastal zones of South Africa's Cape Province, and West and South Australia, these environments experience hot, dry summers, and are prone to fire and floods.

With the continuing pressure of population growth, urban development in the LA region has pushed further out into bushfire and mudslide country. A Los Angeles postcard depicts California's four seasons as 'Earthquakes, Mudslides, Fires (and Riots)'. In less than three years, Los Angeles had three of the ten most costly national disasters since the US Civil War. The damages bill after the earthquake in 1994 exceeded US$42 billion. The bushfires in 1993 cost US$1 billion. This was also the estimated repair bill following the riots that broke out in south-central

FIGURE 8.5 *Flood waters spilling through a barrio neighbourhood in Caracas*
Source: unknown

Los Angeles in 1992. The city's reliance on subterranean water drawn from the great artesian basins beyond the coastal mountain ranges represents a further source of vulnerability for Angelenos.

UNHEALTHY AND UNSAFE URBAN ENVIRONMENTS

In the field of environmental health, the conditions in which people live are recognised as one of the underlying causes of disease. Epidemiologists try to isolate the background factors involved in diseases that are transmitted within populations. These can range from genetic effects to inadequate income, poor diet, lack of education, access to good-quality health care, overcrowding, substandard housing, poor hygiene and cultural practices that unwittingly increase the risk of contracting disease. Therefore, although the causation is often very complex, the crowding of people into densely populated cities and the unhealthy conditions that exist in some of the worst neighbourhoods, are known to contribute to the spread of communicable diseases and a higher incidence of serious illness.

The cholera epidemics that swept through Europe during the nineteenth century were undoubtedly accentuated by the massing of people in cities. Although other diseases caused by contaminated water (typhoid, typhus and diarrhoea) or by people

TABLE 8.2 *Access to safe drinking water and sanitation services in urban centres, selected Third World countries, 1980–1990*

Country	Urban access to safe drinking water[†] (percentage of population)		Urban access to sanitation services[*] (percentage of population)	
	1980	1990	1980	1990
Africa				
Angola	85	71	40	25
Burundi	90	99	40	71
Congo	36	92	17	–
Egypt	88	95	45	80
Ethiopia	69	91	96	76
Nigeria	60	81	30	40
Asia				
Bangladesh	26	82	21	63
China	79	87	85	100
India	77	87	27	53
Indonesia	35	68	29	64
Philippines	65	85	81	78
Thailand	65	89	64	80
Vietnam	70	39	23	34
Latin America				
Bolivia	69	77	37	40
Brazil	80	95	32	84
Cuba	99	100	–	100
Ecuador	82	63	39	56
Mexico	64	81	51	70
Peru	68	77	57	77
Venezuela	91	89	90	97

[†]Access to an adequate amount of safe drinking water located within convenient distance from the user's dwelling.
[*]Access to a sanitary facility for human waste disposal in the dwelling or within convenient walking distance.
Source: United Nations (1996); compiled from World Health Organization and UNICEF survey data. Reproduced by permission of Oxford University Press

living in close proximity (tuberculosis and acute respiratory infections) had a more serious health impact, cholera was so contagious that it hastened investment in the first municipal water supply and waste removal services, and the establishment of great metropolitan hospitals. Hence the number of infant deaths attributed to communicable diseases has fallen to 0.4 per 1000 live births in western Europe, where a highly urbanised population enjoys

decent housing and access to good-quality health care (United Nations 1996, 134). But in urban settlements in Africa, Asia and Latin America, where housing and health care are very basic, the rates are typically several hundred times higher.

Table 8.2 compares the progress made between 1980 and 1990 in extending access to safe drinking water and sanitation services in the urban areas of a selection of African, Asian and Latin American countries. On the basis of the information supplied by member states to the United Nations, it is clear that the rate of progress is very uneven between countries, and that, in the case of Kenya's cities, the rate of urbanisation has exceeded the capacity to catch up with the backlog from 1980. 'Where water supplies and provision for sanitation are inadequate for high proportions of the entire population, diarrhoeal diseases can remain one of the most serious health problems within city-wide averages' (United Nations 1996, 135–6). For example, in Ghana's capital, Accra, a study of living conditions linked the high incidence of childhood diarrhoea to insufficient toilets, fly infest-ation, interruptions to water supplies, risky techniques for storing water and the tolerance shown to children defecating in the yard or street (McGranahan and Songsore 1994). In São Paulo, which, along with Jakarta, was part of the same study, those households without easy access to piped water were 4.8 times more likely to die from diarrhoea than those with water piped to their house. Many other studies of poor urban districts confirm that diarrhoeal diseases are a major cause of premature death among infants and serious illness. According to the World Health Organization, 'reductions of between 40 and 50 per cent in morbidity from diarrhoeal diseases is possible through improved water and sanitation' (United Nations 1996, 136).

As piped drinking water and sanitation services are gradually extended to more urban households in the South, other causes of death and injury, like the risk of being involved in a traffic accident on a crowded city street, will assume greater importance. Death and disability caused by traffic accidents in cities in the South are set to soar in line with an estimated doubling of world traffic volumes between 1990 and 2020. By contrast, in European and North American cities, deaths on the road have been in decline for 30 years. For example, between 1965 and 1994, the number of vehicles on Britain's mainly urban roads doubled, but the accident rate more than halved.

The International Federation of Red Cross and Red Crescent Societies predicts that, by 2020, traffic accidents will displace respiratory diseases, tuberculosis, war and HIV as the leading cause of death and disability – though not ahead of heart disease or cancer (reported in *Guardian Weekly*, 5 July 1998, 23). Worldwide, road accidents are now the single biggest cause of death of men aged 15 to 44 years.

By 1990, accidents between pedestrians, cyclists, rickshaws, and buses and cars, in the crowded streets of Indian and Bangladeshi cities were claiming 40 and 77 deaths respectively each year per 10,000 vehicles. By comparison, road deaths in Japanese and Australian cities averaged about two or three persons per 10,000 registered vehicles.

The association of poor health and bad environmental conditions at the neighbourhood level in cities is borne out by social health atlases that map patterns of deprivation and disease (Glover *et al.* 1999). In wealthy societies there tends to be a higher incidence of the so-called diseases of affluence like heart disease and cancer, which are typically associated with the lifestyle and hazards of urban and industrial development. Nevertheless, the effects of living in deprived neighbourhoods are still apparent in the health gaps that can be observed in England between the inner areas of London, Liverpool, Manchester and Newcastle upon Tyne, and the Home Counties that attract prosperous commuters from London. Table 8.3 indicates that the local authorities in which these neighbourhoods are to be found are among the most deprived in England. Towards the end of the 1990s, infant mortality rates were at least twice as high in these inner-city neighbourhoods; and during the decade, premature death rates for all children, and adults under 65 years, had climbed at 2.6 times the rates recorded for places like south Suffolk, Buckingham and Wokingham, Berkshire. The worst health areas had 4.2 times as many children living in poverty as the best, 3.6 times as much unemployment and 1.5 times as many GCSE exam failures (reported in *Guardian Weekly*, 9–15 December 1999, 9).

Diseases that have long since been brought under control in countries with high living standards can sometime resurface as isolated outbreaks if conditions permit. This happened in the late 1990s in a number of New Zealand's most deprived suburbs, when a sharp rise was reported in the so-called diseases of poverty – such as rheumatic fever (the main cause of chronic heart disease in children), tuberculosis and meningitis. The worst-affected suburbs of Otara, Mangere and Glen Innes make up part of the southern sector of the Auckland metropolitan area. South Auckland has a very young, ethnically diverse population of over 350,000 and is growing at a rate of 2 per cent annually. As communities, Otara – where the film *Once Were Warriors* was shot – and Mangere are noted for their entrenched welfare

TABLE 8.3 *The 20 most deprived local authority areas in England*

	Area	Score
I	Liverpool	40.07
2	Newham	38.55
3	Manchester	36.33
4	Hackney	35.21
5	Birmingham	34.67
6	Tower Hamlets	34.30
7	Sandwell	33.78
8	Southwark	33.74
9	Knowsley	33.69
10	Islington	32.21
11	Greenwich	31.58
12	Lambeth	31.57
13	Haringey	31.53
14	Lewisham	29.44
15	Barking and Dagenham	28.69
16	Nottingham	28.44
17	Camden	28.23
18	Hammersmith and Fulham	28.19
19	Newcastle upon Tyne	27.95
20	Brent	26.95

Source: Power and Mumford (1999); extracted from Department of Environment, Transport and Regions Index of Deprivation

dependency. Many households are recent migrants – Maori and Pacific Islanders, and refugees from Southeast Asia – who live in overcrowded, damp and poorly heated state houses.

'Green' and 'Brown' Agendas

Cities in the North concentrate most of the world's middle- and upper-income households with high-consumption lifestyles. Although cities like Manila, Rio de Janeiro and Jakarta are not without high-living elites who match the extravagance of the inhabitants of Santa Barbara in southern California, or the Riviera in southern France, there is a crucial difference. The middle class is expanding in newly industrialising countries, but per capita levels of consumption are still much lower overall.

Thus there is a glaring contrast between the environmental priorities of cities in OECD countries with

their 'Green and clean' agendas, and Third World cities struggling to keep pace with a 'Brown Agenda' (Leitmann 1994). A group of analysts working for the World Bank first drew attention in the early 1990s to the need for a different environmental agenda for cities in the South (Bartone *et al.* 1994). They called for a shift in funding to begin to deal with the unsafe drinking water, the untreated sewerage, the polluted air, and the makeshift shacks, unlit alleyways and open stormwater drains of the shanty towns. 'There is still a tendency for Northern and middle-class perceptions as to what is "an environmental problem" to bias environmental priorities' (United Nations 1996, 156). Ambient air pollution in cities, and the pollution of rivers, lakes and estuaries, often receive more attention than water shortages, and poor sanitation and drainage in local communities – even though these environmental problems take a far greater toll on the health of the urban poor in the South.

ENVIRONMENTAL PRIORITIES OF AFFLUENT SOCIETIES: EUROPEAN CITIES

Notwithstanding the history of environmental abuse that accompanied urban development during the nineteenth and twentieth centuries, huge strides were made in the area of public health in cities. For instance, life expectancy in French cities increased from 32 years to 45 years from 1850 to 1900, principally because of the improvements made to water supply and sewage disposal. A concerted effort in Europe since the mid-1970s to reduce the amount of sulphur dioxide released from dirty factories into the atmosphere has brought the problem of 'acid rain' under control. According to the Global Environment Outlook 2000, the level of discharge had fallen to 25 per cent of what it was in 1980 (reported in *Guardian Weekly*, 23–29 September 1999, 7).

Current environmental priorities for affluent European cities revolve around closing the energy and nutrient cycles to the point where urban ecosystems are self-sufficient enough to be sustainable (Expert Group on the Urban Environment 1996, 100–31). In the case of water management, this involves: controlling sources of water pollution; harvesting and returning stormwater to natural aquifers; recycling 'grey water'; and installing biological treatment plants for urban waste. Savings of non-renewable energy are being achieved in European cities by applying Green technology in the areas of co-generation of electricity and heat, by using the excess heat generated from industrial processes, and by developing wind farms. Energy is also being conserved through improvements to the efficiency of commercial buildings and housing. Approaches to the management of solid waste generation in European cities include: charging for excessive packaging; incentives for biodegradable, reusable or recyclable packaging; maximum separation of biodegradable waste, paper, glass, metals and toxic substances at source; local composting of household and garden waste; regulations on the use, re-use and recycling of building materials; and penalising the dumping of solid waste in landfill sites.

ENVIRONMENTAL PRIORITIES OF THE POOR

The Brown Agenda covers the most immediate and critical environmental problems facing cities in developing countries: a lack of safe drinking water, sanitation and drainage; inadequate solid and hazardous waste management; uncontrolled emissions from motor vehicles, factories and low-grade domestic cookers; accidents linked to congestion and overcrowding; and land degradation. For example, in 1995 one-fifth of the urban poor in the Third World did not have access to safe drinking water. According to a study by the Stockholm Environment Institute, 90 per cent of urban sewage in poor countries is released untreated, usually into an ocean, river or lake (reported in *Guardian Weekly*, 29 June 1997, 12).

As a consequence of biases held by western analysts, throughout the 1970s and 1980s most aid agencies and development banks gave a low priority to water and sanitation in both rural and urban areas in the South. Yet, once the emphasis began to shift towards investing in Brown Agenda projects, a new set of problems emerged, owing to the underfunding of running and maintenance costs. 'Many internationally funded water-supply and sanitation projects in the late 1980s and early 1990s were to rehabilitate or repair water and sanitation systems installed by these agencies not too many years earlier' (United Nations 1996, 156). The practical reality is that for the foreseeable future the massive operational effort required simply to provide basic

services to the South's urban poor will leave little time and few resources with which to address the issues that currently disturb the environmental sensibilities of the North.

Bangkok: a case study

A range of rather unique water-related environmental problems, which gives an added urgency to the Brown Agenda, are to be found in estuarine cities like Bangkok, Dhaka, Calcutta and Shanghai. With a population of nine million in 1995, the Bangkok Metropolitan Region is expected to grow to 14 million by 2006. The economic growth that occurred during Thailand's boom in the 1990s was overconcentrated in the metropolis and placed the regional ecosystem of the Chao Phraya River and delta under enormous environmental stress. Prior to westernisation in the mid-nineteenth century, the local economy and culture were uniquely oriented around water and the annual flooding of the river system. An extensive canal system was dug to supplement the natural waterways (*klongs*), and to assist with irrigation and drainage. But, as western merchants and diplomats became more numerous, they petitioned for roads so that they could use horse-drawn carriages for commerce and leisure. As a result, much of the traditional canal and drainage system was filled in to create avenues (Fig. 8.6). With Bangkok's growing dominance of the Thai economy, environmental degradation had accelerated to the point where the ecosystem that the city depends upon can no longer cope (Setchell 1995).

LOSS OF AGRICULTURAL LAND

The built-up area of the Bangkok Metropolitan Region (BMR) has trebled in size since 1974, resulting in the conversion of highly productive rice fields. Between 1974 and 1984, 45 per cent of the land conversion took place within a 11–20 km belt around the city centre. The zone of vigorous urban development pushed outwards to beyond 30 km between 1984 and 1994, when a number of really colossal, fully integrated high-rise satellite towns mushroomed on the edge of Bangkok. But much of the new development is haphazard and extravagant, and adds to the infrastructure costs and energy consumption as commuting increases.

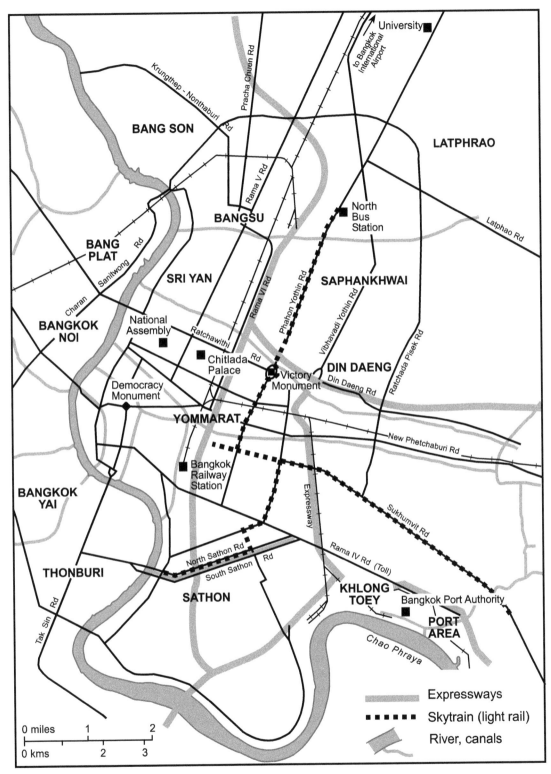

FIGURE 8.6 *Transport corridors and waterways in the Bangkok Metropolitan Area*
Source: compiled from Bangkok Metropolitan Area Map

Water quality and ground subsidence

The Chao Phraya River and the 900 or so canals form a vital part of the Bangkok Basin's ecosystem. Those canals that have not been filled in now receive untreated industrial, restaurant and domestic waste. Estimates indicate that about 1.5 million cubic metres of waste water are discharged into the city's waterways each day. Less than 2 per cent of Bangkok's dwellings are connected up to a mains sewage system. Run-off from the BMR's three open dumps enters the surface and groundwater systems. No one really knows what toxic or hazardous wastes are mobilised and infiltrate the groundwater system in this way. Thus the Chao Phraya River typically has a level of dissolved oxygen approaching zero. Because so many households are living alongside the canals, and farmers are still drawing their irrigation water from the system, this is having a serious impact on the health and livelihood of communities that depend on 'floating markets'.

The environmental problems that can be linked to Bangkok's watery ecosystem are exacerbated by the pumping of water for industrial and domestic use. A study by the Asian Institute of Technology estimates that Bangkok has sunk by about 1.6 metres since 1960. This subsidence, combined with rising sea levels, increases the vulnerability of Bangkok to flooding. Groundwater pumping – industries get about 90–95 per cent of their daily needs this way – has also contributed to the leakage of salt water from the Gulf of Thailand into the local aquifers. This has affected agriculture within the Bangkok region and driven some market gardeners to sell up.

TRANSPORT-RELATED ENVIRONMENT PROBLEMS

By the mid-1980s, according to an Asian Development Bank survey, transportation accounted for 56 per cent of the energy consumed in Thailand. The traffic snarl-ups in Bangkok are notorious. Notwithstanding the conversion of canals, only about 8 per cent of the land area of Bangkok is dedicated to roads compared with about one-third for a city like London. Between 1978 and 1991, car ownership rose from 505,000 to 2.6 million, and by 2000 was an estimated 4 million. Although an aboveground mass transit system bringing commuters in

from the northern and eastern suburbs – with a shorter cross-town line linking up the tourist hotels – opened early in 2000, Bangkok's traffic engineers have been wedded to expressways as the most promising solution for too long (Fig. 8.6). This encourages more development further out in the BMR and raises the expect-ations of affluent commuters who look to the air-conditioned, hermetically sealed car as the only escape from the heat and noise.

Unhealthy living conditions

Motor vehicle emissions currently account for 60–70 per cent of all air pollution in the Bangkok region. In the late 1980s, average values of 277 were measured on an air quality index where 100 is an acceptable standard. More than 10 per cent of Bangkok residents suffer from some form of respiratory illness. Children are now reported to have the highest blood lead levels in the world, even exceeding levels found among children living in Mexico City.

Migrants from the countryside, often displaced from the land, look for a toehold when they first arrive in a big city. Until they get a job, or establish a niche in the informal sector, they want to avoid paying rent. This finds newcomers in a city like Bangkok squatting, perhaps illegally, near to factories, canal or road easements, or rubbish dumps where the threat of eviction is reduced but the risk of exposure to environmental hazards is heightened. Where the public health risk is so high, poor households are caught in a bind. They cannot afford the most basic water supply or sewage, so are forced to buy drinking and rinsing water if they are to avoid using the contaminated canal water. Then they are left with no alternative but to flush their own domestic waste back into the canal at their doorstep.

The Brown Agenda, centred upon sanitation, has not had such a high priority in the minds of Thailand's political elite, or Bangkok's environmental management authorities, because it doesn't impact so directly as a public health issue on the growing middle class. For them, solutions to traffic congestion have a higher priority. Even though the next-generation mass transit system in Bangkok will run underground, the planned construction of expressways, towering high-rise housing and extensive factory zones will put the regional ecosystem under further stress.

Vulnerability to rising sea levels

Significantly, there are a number of other mega-cities that, like Bangkok, are located at the confluence of great rivers or on deltas. Such cities were pivotal in the early establishment of water-based trading patterns (Chapter 2). They include Dhaka (Ganges, Bramaputra), Calcutta (Ganges), Cairo (Nile), Shanghai (Whangpoo and Yangtze), Quangzhou (Pearl) and Karachi (Indus). This historic, even ancient, legacy not only makes for a much more complex ecological footprint, but it makes these vast urban populations especially vulnerable to any future rise in sea level.

Cities and global warming

By increasing the atmosphere's ability to absorb infra-red energy, greenhouse gas emissions are disturbing the way the climate maintains its balance between incoming and outgoing energy. This trapping of excess energy in the lower layers of the atmosphere contributes to global warming. Greenhouse gases include the carbon dioxide released by the burning of fossil fuels in cities and the removal of forests, and the methane and nitrous oxide produced by paddy rice and livestock. The weight of scientific evidence indicates that, without taking steps to slow greenhouse gas emissions, global temperatures could increase by between 1 and 3.5°C over the next 100 years (Climatic Change Secretariat 1999). At its Shanghai meeting in January 2001, the UN-affiliated Intergovernmental Panel on Climatic Change warned that under 'worst case' conditions this figure might rise to 5.8°C over the next century (reported in *Guardian Weekly*, 25–31 January 2001, 1). In turn, this could reduce the rate at which the earth sheds energy into space by about 2 per cent, and, because energy cannot simply accumulate in the lower atmosphere, the climatic system has to adjust.

Because of their insatiable appetite for energy, urban industries and centres of population produce large quantities of greenhouse gas emissions. Cities are a major source of life-threatening greenhouse gases, but the contribution each makes to global warming varies greatly depending upon the sophistication of the technology employed in industrial and waste management, and the consumption patterns of their population. Table 8.4 relates the levels of urbanisation, per capita energy consumption,

car dependency and fuel economy to greenhouse gas emissions produced by the most industrialised economies. When the Kyoto Protocol to the Convention on Climatic Change was adopted in December 1997, the respective contributions to global carbon dioxide emissions in 1990 were as follows: the United States, 36.1 per cent; the Russian Federation, 17.4 per cent; Japan, 8.5 per cent; Germany, 7.4 per cent; the United Kingdom, 4.3 per cent; Canada, 3.3 per cent; Italy, 3.1 per cent; Poland, 3.0 per cent; France, 2.7 per cent; and Australia, 2.1 per cent (Climatic Change Secretariat 1999, 33–4). Table 8.4 reveals that the greatest overall differences are in the average amounts of energy (GJ) consumed, in the carbon dioxide emissions produced by industry and power generation, and in the car-based urban travel undertaken by North Americans and Australians compared with Europeans and the Japanese.

Greenhouse gases are dispersed globally by atmospheric circulation so that particulate matter from the coal-fired power stations supplying China's big coastal cities, for example, is transported to Hawaii. But, more generally, 'Northern urbanites, wherever they are, are now dependent on the carbon sink, global heat transfer, and climate stabilisation functions of tropical forests' (United Nations 1996, 150). Conversely, the clearing and burning of tropical rainforests in Indonesia not only release stored carbon into the atmosphere; they also aggravate the smog levels for the residents of already heavily polluted cities in the immediate region – especially cities like Singapore, Jakarta and Kuala Lumpur. Atmospheric circulation means that everyone living in cities ultimately has a vested interest in reducing greenhouse gas emissions.

CONCLUSION

The history of disease control and environmental health in cities demonstrates that, when faced with life-threatening events, humans are capable of organising and devising the technical means to ensure their survival. However, in the case of sanitation, solid waste disposal, air pollution and uncontrolled urban development, this has typically involved imposing environmental externalities created in cities upon the surrounding countryside and nearby rural communities. As more of the world's expanding population moves into cities, urban communities are being

TABLE 8.4 *International comparisons of energy use and greenhouse gas emissions*

Indicator	Year	Aust.	USA	Canada	Europe	Japan
Urban popn as % of total	1993	85.2	76.2	78.1	75.0	77.9
Energy (GJ) consumed per capita	1991	215	320	325	134	140
Energy (MJ) consumed per US$ of GNP	1991	17	17	21	7–19 (varies by country)	6
(Percentage change since 1971)		(−8)	(−27)	(−18)	(+8 to −41)	(−25)
Energy and industry-related CO_2 emissions per capita (t)	1991	51.1	19.53	15.21	8.2	8.79
Overall vehicle km/capita	1990	9160	13,270	8490	(OECD) 7630	
Urban car use (pass. km/capita)	1980	10,729	12,586	(Toronto) 9850	5600	(Asia) 1799
Average new passenger vehicle fuel consumption (litres/100 km) (tests may vary)	1988	9.1	8.2	8.1	7.9 (W.Ger.) 6.8 (Italy) 7.4 (Spain) 8.2 (Sweden) 7.4 (UK)	8.6
Av. pass. vehicle fleet consumption (litres/100 km)	1988	11.8	10.8	11.84 (1986)	10.7 (W.Ger.) 7.6 (Italy) 8.5 (Spain)	10.7 (1986)

Notes: GJ = Giga-joule; MJ = Mega-joule; t = tonne
Source: AURDR (1995), based on data provided by OECD and World Resources Institute. Reproduced by permission

urged not only to adopt less unsustainable forms of urban development but to reduce the coverage of their ecological footprints. The discussion of urban politics and the management of cities in the next chapter broadens out to include recent shifts in environmental policy and the actions undertaken by growing numbers of local communities in support of ecologically sustainable cities.

FURTHER READING

The **United Nations** (1996) volume on urbanisation contains tables of environmental indicators for major international cities and an accompanying commentary on the management of urban environments *Cities for a Small Planet* is a beautifully crafted and richly illustrated essay written by a man who cares about cities (**Rogers** 1997).

Official publications such as those authored by the **Expert Group on the Urban Environment** (1996) and the Australian Urban and Regional Development Review (**AURDR** 1995) provide state-of-the-environment reports for European and Australian cities.

The best summary of writing on the ecological footprint of cities is contained within **Wackernagel** and **Rees** (1996), both of whom are associated with the development of the construct.

Haughton and **Hunter's** (1994) book on sustainable cities presents the key principles and strategies for sustainable urban development.

9

URBAN POLITICS AND THE MANAGEMENT OF CITIES

INTRODUCTION

This chapter covers the business end of managing cities and the role that urban politics plays in that. From an urban geographer's perspective, knowledge of the political system of governance and management that operates at the local level in cities is vital for two main reasons. First, it holds the key to understanding how the built environment is created and modified over time. In large metropolitan regions like Tokyo, New York, Los Angeles, Mexico City and Manila, all the major strategic decisions about funding and assigning priority to land release for housing and jobs are made in the political arena. The same goes for the development of the metropolitan transport system, and necessary public utilities like electricity, gas, water and sewage. Urban development on this scale not only requires massive public investment, but has the potential greatly to enhance the wealth of property developers and commercial firms who stand to gain from increased land values and/or improved business prospects. Because of the economic growth, jobs and higher revenue from property rates that new investment in cities promises to bring, 'business elites' are able to cultivate close links with City Hall.

Secondly, the institutions of the state and those who govern have jurisdiction over the creation of wealth and its redistribution in cities. They also determine the priorities for service provision and where services are located. The political process is also the means by which citizens' groups and urban communities repel unwanted development, or contest access to local services and environmental conditions that meet with their expectations. In the case of the former, this could be the threat of a toxic waste dump because councillors are prepared to waive environmental safeguards in a poorer part of town to get the business, while, in the latter case, during a period of fiscal austerity during the 1980s and 1990s, countless communities had to battle government agencies to resist the 'rationalisation' of services. Throughout North America, Europe and Australasia there are inner-city neighbourhoods or suburbs that have experienced school closures, the shutting-down of hospital wards, the loss of police and fire depots, and the withdrawal of public transport routes. As we will see later in this chapter, this assertion of citizens' rights and political mobilisation around an urban issue may partially draw upon the strengths of existing social movements like women's rights, the Greens, gay liberation or grey power. Alternatively, it might be a local cause that temporarily galvanises people living together in the same community even though they have quite different class and cultural backgrounds. In this case, Castells (1997, 60–4) suggests that 'territory' provides an added source of identity.

For these reasons, governance and politics have a dual role, along with markets, in producing an urban geography of winners and losers. These are the twin conceptual pillars of 'agency' and 'structure' that are always present in theoretical analysis – if not stated, implicitly at least. This is the essence of *urban political economy* as practised by good thinkers like Michael Smith (1988) and Frank Stilwell (1992).

CITY, STATE AND MARKET

Local authorities in cities are just one of the governing institutions that have responsibility for managing urban development, providing services, redistributing revenue and framing social and environmental regulations on behalf of business interests and the citizens of cities. But the relative powers and resources that any urban government commands depends upon the structure of governance adopted by the nation-state. Each nation-state has a central political locus with overriding powers – the central state. Unitary systems of government like those of the UK or New Zealand have only two principle tiers – central and local – with some limited functions devolved to specific regional bodies like the new governing body of metropolitan London, or New Zealand's regional councils for environmental management. Federal systems like those of the United States, Germany and Australia have three tiers of government – central, state and local – which makes for a significant difference in the command over power and resources exercised by local politicians and managers in cities. The creation of the European Union has introduced a supra-national government with further power- and resource-sharing implications for cities in the member states. For example, special programmes now enable those European cities that qualify to obtain significant additional resources for community development and urban regeneration schemes.

While local authorities in cities constitute one of the lower tiers of governance exercised within the nation-state, this is the level of government where politicians, managers and planning staff normally intervene in property and labour markets on behalf of competing interest groups (Chapter 6). But the ability and readiness of local authorities to intervene in these markets vary from city to city depending, first, on the power and resources that the central state is prepared to concede to the local state. The most celebrated case of centralist intervention occurred in 1986 when the UK's Thatcher government abolished the Greater London Council (GLC) along with the other upper-tier councils in the English conurbations:

When he led the Greater London Council in the early 1980s, Ken Livingstone [who became London's first *elected* mayor on 4 May 2000] defined himself by opposition to Margaret Thatcher, who was so enraged that she ultimately abolished the GLC. Mr Livingstone's instinctively oppositional stance will be reinforced by the fact that his new job has very few real powers, making it hard for him to initiate policy even if he wanted to. On the main issue of the election campaign – the future of London's underground system – it is central government which sets the policy.

(Reported in *The Economist*, 6 May 2000, 33)

In Australia, states have the powers under their respective local government acts to install a city administrator if council politics have deteriorated to the point of stalemate, or if corruption is endemic. Both the Sydney and Melbourne City Councils have had to suffer the indignity of a city administrator running council business in the post-war period. Significantly, in their independent capacity, though no doubt acting under instructions from the governments of New South Wales and Victoria, both gave approval to the fast-tracking of many of the building applications for office blocks in the central city that had been held up by warring factions on the council.

Secondly, the extent to which governments at all levels regulate for market imperfections or distortions depends on the prevailing climate for politically driven urban change. During the course of the 1980s, the political mood in most welfare-state societies switched from middle-class support for progressive social and urban policies to a conservative form of neo-liberalism. The relaxation of market regulations ranged from the extremes experienced in New Zealand to much more benign measures under social democratic governments in Europe. Through the 1980s and 1990s, for example, central governments, especially in the United States and New Zealand, but also in Australia, progressively withdrew support from the urban and social housing sectors in terms of policy, programmes and funding levels. However, the picture is more complicated in the UK, where the Conservative government severely reduced the urban regeneration and social housing budgets during the 1980s at the same time as it dismantled parts of the local planning system and privatised council housing.

'Politics do matter'

Although hegemonic tendencies within the global economy towards the end of the twentieth century pushed analysis in the recent past strongly towards structural accounts of urban development, *politics do matter* in making sense of cities. This re-assertion of political agency in theoretical accounts of urban change has been part of a wide-ranging critical reaction within social and philosophical writing to structural analysis. Feminists and cultural theorists led the way during the 1980s and 1990s in developing the post-structuralist challenge to what was dominantly Marxist political economy.

For a time in the 1970s and 1980s when de-industrialisation was at a high point, there was a strong feeling that cities and local communities in highly developed economies were under siege. The local politicians elected to govern and manage cities appeared incapable of halting the corporate flight, and loss of investment and jobs that ensued following the takeover of local business by global conglomerates. Neo-Marxists writing about cities and housing then were convinced that in urban development ... *capital rules* (Harloe 1977; Peterson 1981; Harvey 1985). At the same time, the post-war record of welfare states helped to re-inforce the impression that the problems of unemployment, poverty, substandard housing and declining urban services in many older European and North American cities are largely structural in origin and unlikely to be fixed with 'band aid' urban programmes. Hence, policy analysts like David Donnison (1998) propose that 'urban problems' are of society's making and need to be tackled at their root cause. This is the alternative to holding back and treating the symptoms once they have emerged at a local level in cities.

The grounds on which, and the interest groups on whose behalf, urban politicians decide to underwrite private investment for, or grant approval to, a development proposal, or allocate urban services, or regulate to control for environmental effects, form a key theoretical question in urban studies (Judge *et al.* 1995). For example, is there a systemic or structural bias in political decisions favouring a power elite like a pro-growth business lobby in cities? Or do politicians respond to competing interest groups on an issue-by-issue basis? How has the emergence of social movements based on differences in gender, sexuality, ethnicity and commitment to the environment cut across normative precepts like the 'public interest' and the 'common good' in urban politics and planning (Sandercock 1998)? Answering these kinds of questions requires at least a rudimentary grounding in the different ways that

political theorists conceive of the dominant power structures in society, and the relationships and struggles between the key political actors involved in shaping urban development.

The distinguishing features of six separate bodies of political theory are outlined in the first part of this chapter. Pluralism, structural explanations, business elites, mode of social regulation, urban regimes and 'identity politics' represent six alternative strands of theory that have been utilised to illuminate recent developments in the way cities are governed and managed. The second part of the chapter examines how politics, governance and community action influence the strategic response to some of the critical social and environmental issues caused by the pressures of urban decline and growth in highly industrialised and developing economies. Increasingly in affluent societies, the management of cities is directed to ensuring that urban development is ecologically sustainable – or less unsustainable in Troy's (2000) terms – and that urban environments are liveable.

POLITICAL THEORIES USED IN URBAN ANALYSIS

In recent years, concerns about the way globalisation is impacting upon cities and reshaping community life in most welfare state societies have forced a rethinking of some of the political theory formerly applied to urban governance and management. As the leaders and citizens of cities adjust to the realities of global pressures generated well beyond the local state, there is a need to take account of a much broader set of underlying processes. Increasingly, urban politics involves working to manage the complex tensions and harness the opportunities arising from the:

- changing nature of the relationship between global capitalism and the nation-state
- shrinking involvement in urban affairs of the central government in many welfare-state societies
- growing competition and rivalry between cities and regions for 'roving' global investment (Chapter 3)
- more active role of citizens and the intrusion of 'identity politics' into urban affairs.

Pluralism

The classic pluralist interpretation of urban politics, and the reigning wisdom for 30 years, was developed by American political scientists like Dahl (1961) and Banfield (1961) in the 1950s. They set out to decide who governs a city. Dahl's conclusions were based on a study of New Haven, Connecticut, and Banfield's on Chicago. They contend that the city politics proceed on an issue-by-issue basis, with elected politicians weighing up the respective merits of the claims presented by various groups with an interest in the urban development process. These could be investors, employers, workers, voluntary agencies, consumers, environmentalists, community activists or residents rallying round a local cause. But once they are represented by formal organisations like the Chamber of Commerce or a trade union, or have acquired the status of a non-governmental organisation (NGO) or community association, pluralists argue that these stakeholders have a more or less even say as lobbyists in the politics of urban development.

More recent revisions to pluralism – neo-pluralism – stress that the terrain of urban politics has become so fragmented and unpredictable owing to the increasingly pluralist nature of society that the effective management of cities suffers as a result (Judge *et al.* 1995). For example, the enactment of 'no-growth' provisions by communities around the edge of US cities has blocked the provision of affordable housing closer to suburban jobs, and undermined the regional coordination of infrastructure and services. In the process, these NIMBY-inspired campaigns have taken on the very powerful development lobby. NIMBY stands for 'not in my backyard' and is typically associated with the defence of property by middle-class home-owners in the suburbs. On rarer occasions, very poor and culturally mixed inner-city communities have successfully campaigned for equal treatment by the urban planning system – often at the expense of the grandiose schemes of city 'boosters'. According to neo-pluralists, these unlikely successes stem from the fact that in local government elected politicians and the officials they appoint are held directly accountable by citizens for their decisions.

Structural explanations

According to theorists who are wedded to a Marxist model of class structure, a pluralist interpretation underestimates the capacity of the business sector consistently to advance its class interests. Neo-pluralists accept that the political process can be used to concentrate resources and power but deny any consistent class bias on the part of governments in their dealings with lobbyists and other pressure groups. Structural theorists, on the other hand, insist that capitalist societies are structured, of necessity, in such a way as to further capital accumulation (Clark and Dear 1981). They argue that organised business has a dominating position at the apex of the power structure in capitalist societies. Miliband (1969), one of the most prominent of the structuralists, goes so far as to propose that the state as an autonomous set of institutions is effectively 'captured' by the ruling class and used as an 'instrument' to serve capital's ends. Urban texts with titles like *Captive Cities* (Harloe 1977) and *City Limits* (Peterson 1981) reflect the appeal at the time of *structural* explanations of urban change in which capital was said to dictate outcomes to the state.

Business elites

Another group of political theorists recognise that 'historically founded elites combining limited experience of the diversity of urban life with narrowly defined self-interest' may come to have an undue influence over decision-making in cities (McDermott 1998, 166). When this power extends to influencing, or even exercising a veto over, spending priorities in the city budget, less favoured groups of citizens tend to be excluded and marginalised. Molotch (1976) draws on the theory of business elites to develop the notion of cities as 'urban growth machines':

> For those who count, the city is a growth machine, one that can increase aggregate rents and trap related wealth for those in the right position to benefit. The desire for growth creates a consensus among a wide range of elite groups no matter how split they might be on other issues.
> (Logan and Molotch 1987, 50)

In formulating positions on Californian transportation options in the 1970s, Whitt (1982) found that elites carefully coordinated not only the positions they would take, but also the amounts each was prepared to commit to winning the outcomes they sought.

In a review of the role of business elites in British urban development, Peck (1995, 41) concludes that 'business leadership activities – and business politics in general – must be understood in the context of, first, the political economy of capital, and, secondly, the restructuring of the state'. This moves us some distance towards the theoretical insights provided by the regulationists.

Regulation theory

Regulation theory attempts to show how changes to the form of urban governance and power relationships are conditioned by, and relate to, the different phases of capitalism (Peck and Tickell 1992). Hence regulationists like Lauria (1997) suggest that an important shift has taken place in the approach to managing cities in conjunction with the transition from a Fordist–Keynesian regime of capital accumulation to a regime of 'flexible accumulation' (Chapter 3). For a time in the 1970s and 1980s, when disinvestment, plant closures and redundancies were running at record levels, older European and North American cities appeared powerless to halt urban decline. This phase of economic restructuring was marked by pressure from central government and business to loosen regulatory guidelines at a local level, and by the onset of disillusion with comprehensive-rational planning solutions. In Britain, the decade of conservative government led by Mrs Thatcher (1979–90) ushered in a new mode of social regulation characterised by a stronger appeal to market rationality, deregulation, micro-economic reform and the privatisation of publicly owned enterprises and council housing (Thornley 1991). In the United States similar shifts in the regulatory framework took place under President Reagan between 1980 and 1988. In an influential book called *Reinventing Government*, the authors describe how an entrepreneurial approach replaced direct provision in the US public sector including urban government (Osborne and Gaebler 1992). The same general tendencies can be seen in other OECD countries, including Australia and New Zealand – though in the latter case in a more extreme form.

Yet in the 1990s, with globalisation and deregulation proceeding apace, politicians, business interests, unions and citizens' groups began to join forces within some of these cities to try to staunch the effects of capital flight and revitalise their local economies. In an important journal article written in the late 1980s, the urban geographer David Harvey (1989) drew attention to a shift from an essentially reactive mode of urban governance, or 'managerialism', to what he called urban entrepreneurialism. In noting that 'the greater emphasis on local action … seems to have something to do with the declining powers of the nation-state to control multinational money flows', he went on to suggest that, in urban regions, 'investment increasingly takes the form of a negotiation between international finance capital and local powers doing the best they can to maximise the attractiveness of the local site as a lure for capitalist development' (Harvey 1989, 5).

Urban regimes

Urban regime theory shifts the focus of the power struggle in urban government away from the class-based dynamics of control, domination and resistance, to coalition-building processes to secure agreed outcomes. According to Stone (1989), the emergence of a coalition or alliance of political, business and community interests with the capacity to give effect to a significant social programme or urban development package is the defining feature of urban regimes. Regime theory accepts that the increasingly plural and fragmented nature of big city populations works against one group monopolising power, but shifts the theoretical question from 'Who governs?' to an interest in the *capacity to govern*. 'Regime theory considers governance to be conditioned not by the dominance of a particular group, but by the processes of coalition-building and maintenance' (McDermott 1998, 167).

It starts with the premise that the structure of capitalist society privileges dominant interest groups and stakeholders. But in an era of individualism, disenchantment with formal political organisations and the distancing of political institutions from citizens, governments have to work at gaining a broadly based mandate for their programmes. Some of the hallmarks of an urban regime are an entrepreneurial political culture, effective working relationships between major stakeholders, the active sponsorship of major development projects by local government and resort to public–private partnership or joint ventures.

One of the present ambiguities of local politics is the need for cities and localities within them to become more attractive to international investment. The formation by city boosters of 'pro-growth' business coalitions, together with the hype of 'place marketing' (Chapter 3), is a strategy for regaining lost shares of national investment. Local property developers, retailers, property investors and home-owners, employers and unions are generally united in their desire for the boost that outside investment gives to fiscal health, spending power and jobs. The growing role of business interests in formal and informal political leadership in cities, and in partnering the state as lead agency in urban development activity, is consistent with the emergence of new elites to stimulate local economic growth.

American research on urban regimes confirms that elected officials and business are the key power-sharing stakeholders. Engineers, accountants, lawyers and planners in government departments may have a monopoly on the disciplinary knowledge and technical expertise required to process and manage urban affairs, but ultimately they work under instruction. Stoker (1995, 273) comments that, while some long-serving officials and technical advisers become indispensable to governing regimes in US cities, 'in other Western democracies, especially in Europe, their leading role is difficult to deny'. Beyond this, minorities, neighbourhood associations and even organised labour may be drawn into a governing coalition especially by well-entrenched and secure regimes. Regimes may also practise a politics of exclusion to see to it that troublesome or insurgent interests do not gain access to the decision-making process.

Stone (1989) provides a case study of Atlanta between 1946 and 1988. He describes a resolute commitment to 'full-tilt' development and an activist agenda of economic growth underwritten by extensive public spending, but not without considerable risk. The regime was dominated by the downtown business elite headed up by Coca-Cola, which has its global headquarters in Atlanta, Georgia. The political force formed by black mayors with the help of black clergy and Atlanta's black middle class constitutes the other arm of the urban coalition. 'For the black middle class it was the selective incentives of high-quality housing, employment and small-business opportunities that encouraged a desire to go along' (Stoker 1995, 276). It was the cohesion of this well-established regime that ultimately won Atlanta the right to host the Olympic Games in 1996. Notably, the 2000 census confirms that Atlanta recorded one of the fastest urban growth rates in the United States during the 1990s.

In the early 1980s, British cities like Birmingham began to develop remedial strategies in response to the catastrophic industrial decline of the preceding decade. In the process it was strongly influenced by the experience of some of the cities on the east coast of the United States with strong urban regimes. Many had developed 'flagship' projects and business tourism initiatives. In 1983, the Birmingham City Council voted to develop Britain's first convention centre and a number of related projects in the city centre (the National Indoor Arena, a symphony hall and the Hyatt Hotel). Other key economic players from business include the Chamber of Commerce and the Training and Enterprise Council. The total cost of the package was £276 million, part of which was met by a £40 million grant from the European Commission and part by Birmingham City Council and the National Exhibition Centre Ltd. The Exhibition Centre was a joint venture between the Council and the Chamber of Commerce (Collinge and Hall 1997, 136).

Central government departments also administer a range of locally specific direct measures in Birmingham, like the Civic Action Team and a Task Force, and operate several grant regimes (Regional Assistance, City Challenge and the Single Regeneration Budget). The Single Regeneration Budget brought together 20 programmes in the mid-1980s run through various central government departments. It gives priority to the creation of local partnerships between agencies drawn from the private, public, statutory and voluntary sectors. 'The Government Office for the West Midlands is responsible and *centrally* accountable for the management of the regional European Operational Programme, and the Single Regeneration Budget' (Collinge and Hall 1997, 136). This is consistent with the greater role and stronger guidance provided by centralist institutions in Europe.

Glasgow led the way in the 1980s with its 'Miles Better' campaign, the National Garden Festival in 1988 and its rebranding as 'Cultural Capital of Europe'. 'Glasgow was the place to visit to explore successful place-marketing; cultural-led urban regeneration programmes and public–private agency partnership at its most successful' (Mooney and Danson 1997, 76). But what the urban regeneration strategy, with its inner-city focus and appeal to international tourists, failed to deliver was any 'trickle down' to the poorest council housing estates on the outer periphery of Glasgow (Easterhouse, Drumchapel, Castlemilk and Pollok). By the early 1990s, the Glasgow City Council's Planning Department warned that the regeneration of the entire city could be jeopardised as a consequence of ignoring these marginalised communities.

Identity-based politics

Some of America's leading social theorists, like Marion Young (1990), Cornel West (1990), bell hooks (1990) and Manuel Castells (1997), postulate that new forms of 'identity politics' are emerging that will increasingly foster a politics of difference in place of a politics of equality. Economic and social change during the last generation has removed some of the traditional certainties in people's lives about full employment, family, public institutions (the state, church, schooling), sexuality and security in old age. The elevation of the individual above cooperative citizenship by public choice theorists, along with the steady retrenchment of services by governments, has chipped away at the precious social capital of communities. The active involvement of citizens in public affairs and community life is undone by this narrowing of interests to private and individual spheres. Castells (1997) has remarked upon the sense of powerlessness of ordinary people and communities in the face of globalisation, the fragmenting of social capital, the proliferation of identities, the social tensions created around cultural difference and the disorientation caused by 'information overload' in the networked society.

Yet at the same time, for many women, gays and minorities this destabilising of identity has been liberating. Theorists like Young (1990), West (1990), hooks (1990) and Castells (1997) foresee the possibility of a 'new cultural politics' growing out of this recasting of identity with its basis in multiculturalism, and the women's and gay movements. They envisage an identity politics that offers an alternative to the traditional political parties and institutions. Jessie Jackson's Rainbow Coalition of black churches, minorities, communities, unions and women, and the loose 'rainbow coalition' that elected Ken Livingstone to the mayoralty of London, possibly contain the seeds of the 'new cultural politics'.

The Los Angeles Alliance for a New Economy (LAANE) is another multiracial and cross-class coalition that seeks 'Growth with Justice' in the free-wheeling global city. For example,

In July 1998, LAANE helped to organize a broad-based coalition of more than sixty unions, religious groups, and community organisations (including some local homeowners' associations) that successfully pressured Universal as well as Loew's theatres into agreeing to pay a living wage and to provide adequate health insurance for the 8000 new service workers that would be added by the expansion ... of Universal Studio's popular CityWalk theme park.

(Soja 2000, 410–13)

Another grassroots initiative to strengthen the social economy in uncertain times is the local exchange trading scheme (LETS), which replaces official currency with a form of bartering (Lee 1996; North 1999). Approximately 30,000 Britons participate in some 350 LETS schemes in the UK. In about 25 North American cities like Ithaca, New York, and Salmon Arm, British Columbia, 'Hours' are exchanged. The point of Hours in Ithaca is, first of all, to add to the local money supply and to keep local wealth circulating within the community and immediate region (they're valid within a 25-kilometre radius of central Ithaca). By trading with Hours, 'Ithacans declare their allegiance to their town and add a touch of warmth to microeconomics' (Spayde 1997, 48). Ithaca is the college town of Cornell University and therefore used to alternative and experimental initiatives like the Green Star Co-op (a supermarket-sized organics store), the Alternatives Federal Credit Union, the Farmer's Market ('traditional' farmers, small organic growers and local craftspeople) and an EcoVillage (co-housing and sustainable living).

In the UK, a radical anarchist movement, 'Reclaim the Streets', which started out campaigning against new road projects in the 1980s, has since broadened its protests from street parties against the car's dominance of the public domain to an assault on global capitalism. On May Day 2000, Parliament Square in the centre of London was invaded, occupied and dug up (Fig. 9.1).

Gender is a primary source of identity. The traditional patriarchal power structure means that the needs of women and children have seldom been recognised, let alone incorporated in the political process by the city fathers (Saegert 1980). However, while the numbers of women involved in public life and leadership are growing, this will not be sufficient

FIGURE 9.1 *Reclaim the Streets dug up a corner of London's Parliament Square on May Day 2000 as part of a wilful slight to global capitalism*
Source: Martin Argles (*Guardian*, 10 May 2000, 5)

to reshape cities and their institutions to meet the needs of the constituencies they represent (Clarke *et al.* 1995). Many feminists believe that they are more likely to make their presence truly felt in urban areas, and increase their effectiveness in addressing gender inequalities, by bypassing traditional electoral politics.

The unsettling nature of globalisation and the spreading disillusionment with mainstream politics are leading to popular forms of citizen re-engagement based upon identity. Attachment to place and solidarity with others within a local community, which Castells (1997, 60–4) calls 'territorial identity', are giving birth to popular forms of community action at grassroots level. Environmentalism has taken root at the local level, particularly in middle-class neighbourhoods, and has spread from suburb to suburb to the urbanised countryside. In the United States, toxic waste dumps, nuclear plants, public housing projects, prisons and mobile home parks naturally engender the greatest hostility in comfortable middle-class neighbourhoods. But in many cases these community associations object to development of any kind and insist on the strictest conservation of local amenity. The Council for the Protection of Rural England (CPRE), for example, is a popular watchdog organisation that was formed to coordinate the separate efforts of villages and small towns overwhelmed by urban expansion in south-east England. In Ithaca, in upper state New York, the Citizen's Planning Alliance was founded in the late

1990s to keep a populist eye on city growth and development.

The widespread adoption in the 1980s of prestigious redevelopment schemes as a means of revitalising declining city centres and as a tool of local economic development was strongly opposed by vulnerable low-income neighbourhoods. But the respective experience of threatened communities in San Francisco, Vancouver and Birmingham would seem to suggest that, at best, they were only able to wrestle concessions from the project developers (Beazley *et al.* 1997). In the case of Expo '86 and Pacific Place in Vancouver, about 10,000 people living in some of Canada's poorest neighbourhoods on the Downtown Eastside were placed at risk. Community resistance was spearheaded by the Downtown Eastside Residents' Association (DERA) with the aim of minimising the numbers of tenants evicted from cheap hotels and rooming houses. Since then, DERA has continued to be involved in developing and managing social housing in the area.

From the mid-1960s onwards, the San Francisco Redevelopment Agency's (SFRA) plans for Yerba Buena led to a major confrontation over the site between development interests and the local community, where several thousand (mostly retired people) were forced out of affordable apartments and homes. Estimates suggest that 4000 people were displaced, 700 businesses employing about 7600 workers were lost, and the mostly working-class minorities living south of Market were broken up. The state and federal governments backed the redevelopment of Yerba Buena, which was staged over a 30-year period by a powerful alliance comprising the City of San Francisco, the city's largest corporations, hotel owners (the Marriott opened in 1989), and others in the convention and tourist sectors, building and construction unions, together with San Francisco's media. Activists like Alvin Duskin, the Tenants and Owners in Opposition to Redevelopment (TOOR), San Francisco Tomorrow, the Sierra Club and other groups alleging failure to comply with state and federal environmental acts fought each stage of the redevelopment through the law courts.

On the other hand, Birmingham's urban regeneration strategy, mentioned above, garnered comparatively little public opposition, owing to the fact that such a solid political consensus formed around the need to tackle structural decline in the early 1980s.

Public consultation and debate about the social costs and benefits were desultory and recourse to litigation much more limited by comparison with the North American legal system. But, crucially, there was no direct threat of displacement, and the only potential source of resistance from a deprived council estate (Central Ladywood) adjacent to the redevelopment site was silenced with a £13-million refurbishment (Beazley *et al.* 1997, 190–1).

More enduring community associations have evolved at the neighbourhood level in cities everywhere. In the 1970s an 'expanded community of limited liability' called Organization of the North

East (ONE) was formed on Chicago's Near Northside about 12–15 km uptown from the Loop. The sociologist Gerald Suttles (1972) defines this as an 'expanded community of limited liability' because the residents are not just mobilising to protect their local patch but are joining a coalition that is regional in coverage. ONE brought together Uptown and Edgewater residents from a myriad of smaller neighbourhoods to try and control the spread of arson along the length of the Winthrop–Kenmore Corridor, and to oppose the further development of federally subsidised rental housing. Edgewater's neighbourhood organisations (Fig. 9.2) provide

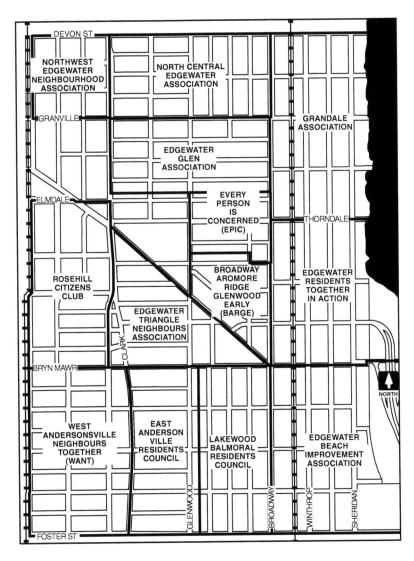

FIGURE 9.2 *Neighbourhood organisations in the Edgewater area on Chicago's Northside*
Source: Extracted from community newspapers

a good approximation to what Suttles (1972) terms a 'defended neighbourhood', or minimal named community.

The Winthrop–Kenmore Corridor is serviced by a good rapid transit connection to downtown Chicago, and traditionally attracted poor elderly and out-of-work tenants, owing to the concentration of single room occupancy (SRO) hotels and day labour, or casual employment agencies, along its track (Rollison 1990). In the late 1970s, Uptown and Edgewater were on the turn with slum landlords looking to quit and getting kids to torch derelict buildings so they could claim the insurance. Between June 1977 and January 1978, over 200 fires were started in the Edgewater/Uptown area, so ONE called a public meeting of residents to stem this threat to life and property (13 people died in two separate fires on 29 December 1977). At the same time ONE campaigned to prevent a proposal from the Chicago Housing Authority to build 119 rental units in two federally subsidised high-rises at 5200 and 5500 North Sheridan. Since many of these units would have been reserved for discharged psychiatric patients, the residents successfully objected to the Chicago City Planning Commission hearings that Edgewater already had its fair share of subsidised housing. A letter from a local citizen published on 25 September 1978 in the *Chicago Sun-Times* claimed that, 'This was true democracy in action and resulted in victory for the community.'

Similar clashes of class and culture are also taking place within the shadow of downtown skyscrapers where revitalisation and gentrification are transforming what were once run-down, 'out-of-the-way' low-rent neighbourhoods (Chapter 6). (This is the subject matter of the Tom Wolfe novel *Bonfire of the Vanities*.) During the 1980s, with young bankers, stockbrokers, media people and the like moving into converted lofts on the Lower Eastside of Manhattan Island in New York, the tenement housing that has always housed New York's newest immigrants came under increasing pressure from apartment renovators (Chapter 6). With rents soaring and the threat of eviction by unscrupulous landlords mounting (Fig. 9.3), tenants organised themselves and called a 'rent strike' in the neighbourhood. This was followed by an even more desperate form of local protest. About 250–300 homeless people evicted from 'squats' on the Lower Eastside, and trying to avoid unhealthy (tuberculosis and AIDS)

and dangerous (drugs and violence) night shelters, pitched themselves a 'tent city' in Tompkins Square Park (Fig. 9.4). But with City Hall and the 'Masters of the Universe' from the financial district and midtown growing more and more alarmed at the menace in their midst, the New York Police Department raided the Square in August 1988, and again just before Christmas in 1990, and literally broke up the tent city.

Lastly, a vast number of poor communities around the world have engaged in grassroots action simply to prolong their survival in the absence of basic welfare (Castells 1997, 62). For example, Rio de Janeiro's

FIGURE 9.3 *Street art depicting police harassment in response to the growing popular resistance to eviction on the Lower Eastside of Manhattan, New York City*
Source: Author's collection

FIGURE 9.4 *The tent city pitched in Tompkins Square Park by 250 to 300 homeless people during the winter of 1990/91*
Source: Author's collection

Institute of Religious Studies reports that the number of volunteer organisations in Brazil grew from 1041 in 1988 to over 4000 in 1998 (reported in *Guardian Weekly*, 8–14 February 2001, 27). Communal kitchens began to spring up in Santiago de Chile and Lima during the 1980s. Churches like the Catholic Church and internationally sponsored NGOs like World Vision are working with local activists in São Paulo and Bogota to develop the skills and capacity of these communities. Where a sense of communal identity takes shape in these circumstances, it is often underpinned by a radical awareness of being the exploited and marginalised. Castells (1997, 64) sums up this worldwide mushrooming of identity-based politics in cities in the following way:

local communities, constructed through collective action and preserved through collective memory, are specific sources of identities. But these identities, in most cases, are defensive reactions against the imposition of global disorder and uncontrollable, fast-paced change. They do build havens, but not heavens.

STRATEGIES FOR SUSTAINABLE AND LIVEABLE CITIES

The remainder of this chapter outlines some of the key strategies being adopted by urban authorities in the drive for enhanced environmental sustainability and improved living conditions in cities. Emphasis is placed upon urban strategy and policies, even though the contribution of other sectors of government and private enterprise to these broad social and environmental objectives is possibly more significant. There is a definite limit to what can be achieved by better physical planning and urban design. The approaches to sustainable urban development and liveable cities singled out for closer attention speak to those issues raised in the previous chapter. Topics to be covered include urban regeneration, strategies for better housing and social integration in cities, energy efficiency and urban transportation innovation, 'smart growth' and compact city principles, and Agenda 21.

The influence of ideology and values on urban policy

Along with all other forms of public policy, urban programmes are subject to the ideology and values cultivated by the political party in power. For example, the ideology of the 'enterprise culture' espoused by the Thatcher and Reagan administrations permeated urban and housing policy during the 1980s. Then, during the course of the 1990s, proponents of a 'Third Way' in politics grew in number and began to reshape public policy, especially in the USA, Europe and Australia (Blair 1998; Giddens 1998; Latham 1998). In a jointly authored paper, the leaders of the UK and Germany argue that, since globalisation is unstoppable, the state would be better to devote political action to complementing and improving markets rather than hampering them (Blair and Schröder 1999). A sticking point for many critics is the insistence on the part of Third Way supporters that 'Public policy has to shift from concentrating on the redistribution of wealth to promoting wealth creation' (Giddens 2000, 3). Critics say there is no Third Way, just different ways of combining state action with the operation of markets (Lloyd 1999). Regardless, Third Way thinking is bound to have a lasting effect on future public policy in those societies where it is pursued.

In trying to articulate a Third Way in politics, Clinton and Blair turned to the writings of Amitai Etzioni (1995), who advocates the restoration of a sense of personal duty and social responsibility in public life. His impulse is 'to push the balance in the public consciousness between entitlement and obligation further in the direction of accepting greater moral responsibility for self, family, neighbours and community' (Badcock 1998, 590). Etzioni calls this 'communitarianism'. The 'end of welfare as we know it' during the second Clinton administration represents one of the political responses to communitarian values.

Other social philosophers, like Cox (1995), Putnam (1995), Saul (1997) and Theobald (1997), are also in search of an ethical antidote to the rampant individualism, community asset-stripping and running down of social capital that was the hallmark of the 1980s. Social capital is the 'glue' that binds communities together. It is formed by the myriad, usually informal, networks for local cooperation and community support mechanisms that are so vital to the development of healthy communities. These include voluntary contributions to community well-being ranging from care-givers, to membership of Meals on Wheels, Neighbourhood Watch, Safe Houses, school committees, resident associations, sporting clubs and so on. According to Robert Putnam (1995) and Eva Cox (1995), neo-liberalism, with its stress on individual rights and self-reliance, has badly eroded social capital and undermined community cohesion in many city neighbourhoods and suburbs.

For their part, more and more citizens in cities are looking for ways to resist globalisation and are demanding to share some of the powers of decision-making with politicians and business interests. The Catholic doctrine of 'subsidiarity' is gaining in popularity at the community level as localised groups of citizens challenge a top-down approach to urban development. In political terms, subsidiarity requires the state to refrain from interfering in, or supplanting, lower-level forms of civil society where they are capable of managing their own affairs. Some of the case studies show how groups of local citizens have become actively involved in the political process to create more sustainable and liveable cities for themselves and future generations.

Regenerating cities and urban communities

Regenerating parts of old cities badly hit by economic restructuring and responding to social exclusion on run-down housing estates represent two of the most formidable challenges facing urban politicians and managers in North America, Europe and Australasia. The context is decidedly different from place to place, as is the amalgam of actors and institutions that get involved in each city. Although the political theories canvassed in the first part of this chapter may not be in agreement about where the power ultimately lies, they all ascribe a role to: citizens exercising their rights within civil society; the political community made up of politicians, bureaucrats and the courts; and the business and financial elites (Sandercock and Friedmann 2000, 529).

For example, although the federal Department of Housing and Urban Development was responsible under President Clinton for initiating overdue changes to urban and housing policy (US Department of Housing and Urban Development 1995), the clearance of project housing and neighbourhood redevelopment currently taking place in the inner core of US cities relies much more heavily on the drive of city politicians, funding from private investors and foundations, and grassroots initiatives at community level. In March 1995, US$17 billion was cut from the federal urban and housing programme as part of Gingrich's Contract with America. Reforms to public housing in the 1990s under Clinton authorised the demolition of 100,000 of the worst public rental units, but combined this with a freeze on Section 8 housing vouchers. At the time, it was concluded that this would leave poor Americans with access to 'fewer units when Bill Clinton leaves office than when he came in' (reported in *The Economist*, 7 June 1997, 34–5). Many of the changes imposed upon the poorest residents of US cities in the 1990s reflected a political will to introduce a market regime into the public housing sector (US Department of Housing and Urban Development 1996). Generally speaking, American progressives like Drier (1997) recognise that much public housing in the biggest cities was in a woeful condition and that management systems needed reforming. But they recoil from the resort to corporatism, and argue that the reliance on market-based solutions and philanthropy as a response to the housing needs of the very poor is quite misguided.

Governments in the UK have been trying to regenerate the inner cities for 30 years. Yet inner parts of some northern cities remain run-down and semi-abandoned. Along with this, the plight of council housing estates at the edge of cities in England, Scotland and Northern Ireland has become more and more pressing in recent years. The protracted efforts of British governments to address the related problems of inner-city decline and urban poverty have their genesis in the 1970s in a series of programmes that collectively came to be known as the Urban Aid Programme (Fig. 9.5). During the Thatcher era, the 'enterprise culture' ethos was applied to deregulating the planning system and privatising urban policy (Thornley 1991). Enterprise Zones and Urban Development Corporations were created in designated areas like the London Docklands and Liverpool's Merseyside. These 'reforms' also reflected a loss of confidence on the part of central government in the ability of local authorities to manage urban regeneration.

By the late 1980s, the limitations of a 'property-led' urban regeneration strategy 'which did little more than hand "those inner cities" to the property development industry on a plate' had become apparent (Geddes 1997, 211). Widespread criticism saw the re-instatement of a role for both local authorities and local communities in the planning

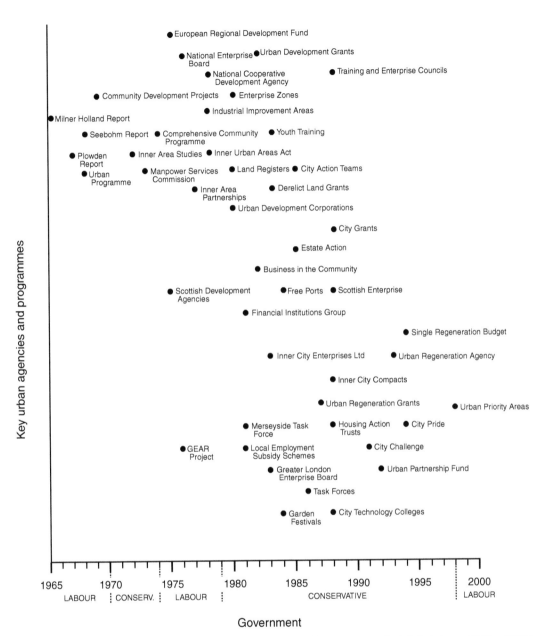

FIGURE 9.5 *Principle agencies and programmes contributing to urban governance in Britain between 1965 and 2000*
Source: Pacione (1997). Reproduced by permission of Routledge

and management of urban regeneration projects. Foremost among these central government programmes were City Challenge, which enabled cities to bid for funds in a competitive budget round, and the Urban Partnership Fund. City Challenge was launched in 1991 and local partnerships were invited to compete for a standard funding package (£37.5 million over five years) in the 57 Urban Programme areas. It was discontinued after two rounds. Like its predecessors under the Conservatives, City Challenge imposed a business-led framework on urban regeneration activity, while drawing local authorities and communities in on a partnership basis. Some City Challenge proposals made provision for 'community forums' or Community Development Trusts as a way of engaging the views of local residents. But research casts doubt on the overall efficacy of these attempts at community empowerment:

In particular it seems that civil servants have embraced notions of greater community involvement but are unaware of, or underestimate the severity of, the practical problems that those working within communities are likely to encounter ... This suggests a need for considerable and sustained investment in community capacity building if the bottom-up approaches currently being espoused by government are to be successful.

(Foley and Martin 2000, 785)

In November 1993, the Major government announced further reforms to the urban regeneration strategy. A Single Regeneration Budget was established in response to criticism that the plethora of government programmes amounted to a 'patchwork quilt'. Twenty regeneration programmes run by five different departments were merged. Secondly, a network of ten integrated Government Offices for the Regions were created in an effort to coordinate programme delivery. However, the potential of these initiatives was never fully realised owing to a shortfall in funding (Mawson and Hall 2000, 68).

Following the election of Labour in 1997 and a Comprehensive Spending Review (CSR), departmental programmes were re-aligned once more to reflect New Labour's renewed commitment to attacking the multiple causes of social and economic decline. The areas that are targeted include former industrial districts with large amounts of derelict, vacant and under-used land and buildings – brownfield sites – and publicly owned housing estates suffering from concentrated social deprivation. The CSR made available £5 billion for housing and regeneration, including £2.3 billion for a 'refocused' Single Regeneration Budget. A Social Exclusion Unit was established in the Prime Minister's Office to address the needs of England's worst housing estates and poorest neighbourhoods. The New Commitment for Regeneration (NCR) pilot programme was launched in 1998 with the aim of formally drawing together the regeneration initiatives of central government and local authorities.

Along with these measures, the Blair government commissioned an Urban Task Force (1999)

headed by Lord Richard Rogers to establish a new vision for urban regeneration in England and recommend practical solutions to attract people back into cities losing population – cities like Liverpool, Manchester, Newcastle, Birmingham and Sheffield. The Task Force concluded that people are leaving these cities because of deteriorating living conditions, and called for an Urban Renaissance. The central thrust of the report comes down to four points.

- A realisation that increased spending on health and welfare, which are priorities of the Blair government, makes little sense without also improving the urban neighbourhood in which the services are delivered. According to the Task Force, this requires better designed, and more compact infill housing mixed in with other uses.
- Financial incentives to entice private capital into depressed areas. In certain Urban Priority Areas, dedicated Urban Regeneration Companies and Housing Regeneration Companies would be given tax breaks and planning powers. By 'harmonising' the VAT (value-added tax) on greenfield housing development, which is presently zero-rated, and home and property improvement, which attracts a 17.5 per cent VAT, the Task Force hoped to encourage urban regeneration.
- Articulation of the virtues of higher urban densities and infill housing on brownfield sites as a way of checking suburban growth in rural England, on the one hand, and creating city environments that support pedestrians, cyclists and public transport users, on the other.
- A call to put local authorities, not government departments, at the heart of the regeneration process. In the opinion of the Task Force, 'They should lead the urban renaissance. They should be strengthened in powers, resources and democratic legitimacy to undertake this [regeneration] role in partnership with the citizens and communities they represent' (reported in the *Guardian*, 10 May 2000, 41).

The White Paper to be based on the Urban Task Force recommendations was repeatedly delayed. The Treasury, fearing loss of control and revenue, opposed not just the tax changes but the whole concept of urban priority areas.

'SMART GROWTH' AND COMPACT CITY PRINCIPLES

In 1975, Richard Register and a few friends in Berkeley, California, founded Urban Ecology as a non-profit organisation to 'rebuild cities in balance with nature'. Since then, Urban Ecology has built a 'Slow Street' in Berkeley, opened up part of a creek bed that had been re-routed through a stormwater culvert 80 years earlier, planted fruiting street trees, designed and built solar greenhouses, passed energy ordinances, established a bus company, and promoted bikeways and walking trails as alternatives to road transport (Register 1987). These are a few of the eco-city principles that have been proselytised by Urban Ecology through its journal the *Urban Ecologist* and the international eco-city conferences held in Berkeley (1990), Adelaide, Australia (1992), and Yoff, Senegal (1996).

Similar principles sit behind the concept of 'smart growth' and the push for more compact urban development in the future. Various terms such as 'eco-city' (Roseland 1997) and 'sustainable cities' (Haughton and Hunter 1994) are used to convey the same broad thrust towards ecologically sustainable urban development. Managing future urban growth to minimise its impact upon the environment and conserve energy has recently pushed its way on to the political agenda in Britain. In particular, a growing body of planners, urban designers and environmentalists envisage the possibility of achieving significant energy savings by raising residential densities in cities and re-zoning for a richer mix of land uses. They are promoting the compact city as a more sustainable urban form (Jencks *et al.* 1996).

However, the claim that energy can be conserved with a growth management strategy rests on a crucial assumption about the likelihood of getting *enough* North American, British and Australasian households who have grown used to suburban living, together with the comfort and convenience of cars, to make the switch to higher-density housing and public transport. Urban planners like Patrick Troy (1996) and Michael Breheny (1997) argue that the environmental benefits of compact cities have been oversold and doubt whether suburban densities can ever be increased to the point where new-generation public transport systems are viable. They also share a concern that, in parts of Australia and

New Zealand, the market is already crowding households with more modest incomes into 'landless' housing on the fringe (Troy 1992).

On the other hand, Britain's Urban Task Force (1999) points out that much recent housing has been built to quite low densities, giving England the lowest residential densities in the European Union. In parts of east Manchester, for example, densities have fallen to eight per hectare. Yet in some continental cities, like Barcelona, the ratio of housing is 10 to 50 times higher (reported in *The Economist*, 5 August 2000, 57). In responding to the natural phobia that people have about high-density living following the disastrous record of many of the council tower blocks built in the post-war era, the Urban Task Force points out that it is wrong automatically to equate urban living with high-rise. Apartments in Paris and Berlin, or loft living in Amsterdam and Sydney, or redecorated Georgian terraces around communal gardens in the West End of London provide very acceptable, and even fashionable, alternatives to the detached bungalow in the suburbs.

While supporters of 're-urbanisation' like Newman and Kenworthy (1999) accept that NIMBY-ism is going to prevent any genuine intensification of car-dependent suburbs in the foreseeable future, they none the less argue that metropolitan strategy plans should be setting aside mixed-use zones for higher-density redevelopment in the future. These 'dense nodes' or high-rise 'urban villages' placed at key nodes on the commuter bus and rail system already exist throughout Europe. The Barbican in London is an early example of a high-density residential precinct with good connections to the London Underground. A few North American cities (Toronto, Boston, Vancouver, Portland, Boulder) and Australian cities (Perth, and North Sydney and Pyrmont-Ultimo in Sydney) have also built their own versions of 'urban villages'.

The master-plan for the redevelopment of a 275-ha site in South Sydney aims to combine all the key ecological and urban design principles that will typify new-generation 'eco-cities'. To be known as Green Square, this urban village will ultimately be home to 25,000 residents and a workplace for 15,000 workers. Lying 3 km south of central Sydney, it will be a stop on the new underground line connecting the CBD with Sydney's main airport. Public transport, cycling and walking will be encouraged, with policies and street hierarchies designed to calm traffic and deter cars. No point in the Green Square precinct will be more than 900 m from the subway station. Block-edge, courtyard housing rising to 8–14 storeys on street frontages will grade inwards and down. The public space will be dominated by a civic square and central park system, pocket parks and an 'irregular necklace' of woods and water criss-crossing the street network. Detention and retention ponds and open stormwater channels will be part of a water management system along the lines of a 'mini-Amsterdam or Bangkok' (reported in the *Australian*, 1 May 1998, 44).

The necessity for a curb on urban sprawl has been accepted at a governmental level in both Australia, where it is called 'urban consolidation' (House of Representatives Standing Committee for Long Term Strategies 1992), and in the UK, where it is referred to as 'urban compaction' (Breheny 1997; Urban Task Force 1999). Projections produced in the UK indicate that England will need 3.8 million new homes between 1996 and 2021, and that the demand will be concentrated in the south-east, where it could cause an overheating of land and house prices. In order to slow the further consumption of rural land, the government originally proposed that 60 per cent of all new housing should be built on recycled brownfield sites. Environmental lobby groups like Friends of the Earth and the CPRE were pressing for a target of 75 per cent of new housing on derelict or abandoned urban land. Since the greatest potential lies in the conurbations in the Midlands and the north – where the demand is relatively low – eventually a compromise had to be struck. In March 2000, the Blair government finally approved plans for 215,000 new houses during the ensuing five years and a further 115,000 in London. It also laid down strict planning guidelines giving councils powers to force builders to redevelop recycled brownfield sites and convert empty buildings to flats before turning to untouched greenfield sites. By reviewing demand every five years, 68 square kilometres of countryside could be saved (reported in *Guardian Weekly*, 16–22 March 2000, 11).

URBAN TRANSPORTATION AND ENERGY CONSERVATION

The adoption of environmentally friendly transportation systems in cities is one of the key planks of a sustainable urban development strategy, yet perhaps the most difficult to gain political acceptance for in car-dependent cities. This applies particularly in the case of reforms that would make the user pay for the external costs – environmental and commercial – imposed by traffic congestion. For example, the British government's white paper on transport policy containing proposals to tax congestion through road tolls and company car park charges met with stiff opposition when it appeared in 1998.

Singapore launched the first licensing scheme in 1975 requiring drivers entering the CBD during rush hours to buy a special pass. The Norwegian cities of Oslo, Bergen and Trondheim also successfully operate road tolls. With much more ubiquitous urban road networks, California, Texas and Florida have introduced some high-occupancy toll lanes where charging relates to the volume of traffic on a daily basis. In July 2000, with the backing of a majority of Londoners – 67 per cent favoured charging as long as the revenue was spent on transport improvements – Mayor Livingstone issued a discussion document canvassing options for tackling congestion in central London. About 200,000 private cars and 50,000 commercial vehicles drive into central London daily. Estimates suggest that a £5 toll would reduce traffic volumes by 10 per cent and raise speeds by 2 m.p.h. to 11 m.p.h. in peak periods. This would raise up to £320 million for public transport improvements and free up space for buses (reported in *The Economist*, 29 July 2000, 54). As well as strong opposition from three affluent boroughs where Conservatives govern – Kensington and Chelsea, Westminster and Wandsworth – congestion pricing runs into the sensitive issue of exemptions. Beside emergency services, buses, disabled drivers and possibly taxis/cabs, should the exemption also extend to the sick travelling to central-London hospitals or parents taking children to school?

Reversing, or at least slowing down, the dispersal of urban populations in North America, Europe and Australasia is also attracting mounting attention as research evidence confirms that European cities (Munich, Stockholm, Vienna, Amsterdam) and Asian cities (Hong Kong, Singapore, Tokyo) with mass transit systems are not only more energy efficient, but devote a lower percentage of their gross regional product (GRP) to transportation (Newman and Kenworthy 1999). The Transport Efficiency Act 1998, which was passed in the United States against the road lobby's protests, now ensures that local communities have a voice in determining how federal funds for transport planning are spent. By 2000, over 250 local authorities in the USA had held plebiscites on initiatives to begin to tackle sprawl and car-dependence. With the circumvention of top-down planning, some communities have switched the application of federal funds from highway-building to transit, cycling and walking paths. Allocations moved from only a few per cent on mass transit in 1990 to 24 per cent by 1998. For example, in April 2000, New York's Metropolitan Transportation Authority approved a US$17 billion version of its five-year budget. 'Big ticket' capital items included money for a full-length Second Avenue subway line on Manhattan and clean-fuel buses (Fig. 9.6).

FIGURE 9.6 *Natural Resources Defense Council drive for clean-fuel buses for New York*
Source: The New York Times (April 2000)

But in most other lower-density North American and Australasian cities where ridership levels could never repay the investment in heavy mass transit, other light rail and bus-only systems are being built. Toronto has led the way with a well-integrated system of rapid-rail corridors and high-speed busways. Adelaide's German-based O-Bahn system collects commuters from bus stops near their homes, and then carries them at high speed on a fixed track through a linear park system into the CBD. But, quite often, the decisions as to where these systems get built, and the communities that are to benefit, spark protracted debate and even community protest. A class action filed against LA's Metropolitan Transit Authority (MTA) on behalf of 350,000 transit-dependent bus riders was able to establish that the policies and investment decisions of the MTA were discriminatory. The Bus Riders' Union and its lead counsel, the NAACP Legal Defense Fund, showed that both the existing bus services and the costly fixed-rail track under construction at the time not only favoured affluent white communities but extended a larger subsidy to these commuters than the poor (Soja 2000, 257).

Unlike British, American and Australian commuters, in cities such as Amsterdam, Copenhagen, Stockholm, Zurich, Vienna and Freiburg, Europeans resisted the substitution of cars for public transport, and cycling and walking in their town centres. Table 9.1 illustrates some of these differences in modal split for Southern California and the Netherlands. Cycling and walking in European – and some Asian – cities approach 18–20 per cent, whereas the comparable statistics for Canada, Australia and the United States are 6.2 per cent, 5.1 per cent and 4.6 per cent respectively (Newman and Kenworthy 1999). Nowadays, where inner-city redevelopment occurs in Europe, the idea of car-free neighbourhoods is catching on, especially in countries like Denmark and the Netherlands with their traditional attachment to cooperative housing and cycling. For example, walkers and cyclists account for almost one in three work-trips in the compact 'cycling cities' of Amsterdam and Copenhagen.

Community agenda-setting and action

The global agenda – Agenda 21 – agreed at Rio in 1992, and subsequently worked out in international agreements like Kyoto and Buenos Aires, is about

TABLE 9.1 *Percentage of work-trips by mode in southern California and Dutch urban centres, 1980 and 1990*

	Southern Californian centres						Dutch centres				
Mode	**Los Angeles**	**Orange**	**Riverside**	**San Bernardino**	**Ventura**		**Amsterdam**	**The Hague**	**Utrecht**	**Rotterdam**	**Green Heart**
1980											
Car	85.5	90.9	90.0	90.5	90.2		52.9	50.4	53.9	51.8	54.8
Public	7.0	2.2	0.9	0.9	1.5		17.4	10.5	11.6	16.3	12.2
Bicycle	1.7	2.4	2.3	2.1	2.6		20.8	27.4	27.4	20.4	22.3
Walk	3.7	2.5	3.9	4.1	3.0		6.7	7.7	4.9	6.8	4.7
Home	1.4	1.5	1.9	2.5	1.9		—	—	—	—	—
Other	0.7	0.6	1.0	0.8	0.8		2.2	3.9	2.2	4.8	6.0
1990											
Car	85.9	90.5	91.2	92.1	91.7		51.8	51.0	49.3	55.6	64.9
Public	6.3	2.5	1.0	0.7	0.7		14.6	10.8	18.2	14.0	7.4
Bicycle	1.1	1.5	1.2	1.2	1.6		27.4	30.2	29.0	23.0	22.6
Walk	3.3	2.2	2.4	3.0	2.3		3.9	6.1	2.0	4.8	2.3
Home	2.7	2.7	3.2	2.3	2.9		—	—	—	—	—
Other	0.7	0.7	0.9	0.7	0.8		2.2	2.0	1.6	2.7	2.7

Source: compiled by Clark and Kuijpers-Linde (1994) from Public Use Microdata samples and the DVG. Reproduced by permission of Taylor & Francis Ltd

'reducing the source of environmental impacts, not playing with how you manage or mitigate the outcomes of development' (Newman 1999, 93). The challenge for local communities in cities around the world, therefore, is to forge a consensus for sustainable development, and then build partnerships between local authorities, business and community groups and citizens simultaneously to protect the environment and to improve the quality of life. By 2000 more than 65 per cent of local councils in the UK had made a formal commitment to Local Agenda 21. The borough of Greenwich in the southeast of London, for example, initiated an Agenda 21 process with round-table discussions involving community representatives in 1994. This was followed by the development of a community planning process setting out priorities for action. Greenwich is one of 37 pilot authorities submitting to a best-value performance review at four-yearly intervals. The draft local authority indicator list for performance reviews includes measures of: domestic energy and water consumption; household waste; number of air pollution days; water quality in rivers and lakes; net change in natural habitat, including given species of wildlife; companies with environmental management systems; noise and crime levels; traffic volumes and quality of access; health and education levels; housing conditions and homelessness; access to key services; benefit recipients; social support and community participation rates, including among minorities; social and community enterprises; business start-ups and closures; job opportunities for young people; and demonstration of cultural interest (see http://www.la21-uk.org.uk/clip/guide.rtf).

After some 'blood-letting' on Honolulu's Council in the mid-1990s, the Mayor's Action Plan annually assigned a base-line US$2 million to each of 19 neighbourhoods for Construction and Improvement Projects that they nominated following a process of community debate. One effect of handing this decision-making to neighbourhood Vision Teams is that there is no longer the same degree of obstruction endemic to a top-down model. Similarly, for over 12 years the radical Brazilian Workers' party has governed the poor southern Brazilian city of Porto Alegre (population 1.3 million). Popular support is obtained for a participatory model that empowers neighbourhood assemblies. They decide priorities and elect delegates who must apply city-wide criteria to draw up a budget that is finally agreed on by the whole municipality (reported in the *Guardian Weekly*, 8–14 February 2001, 26). Corruption has been confronted, wealth transferred and, for its efforts, Porto Alegre has been awarded UN Habitat prize for the world's best-governed city.

The same readiness to experiment with capacity-building and open governance marks the ground-breaking experience of the citizens of another city in southern Brazil, Curitiba (population 1.5 million). Much of the credit is given to Jaime Lerner, who has twice been elected to run the city and served as state governor (reported in *Guardian Weekly*, 16 June 1996, 24). The city has the same problems as any other city in the South including the shanty towns, poverty and crime. The population grew by 200 per cent during the 1980s and 1990s, and 32 per cent of the residents earned less than US$200 a month in the mid-1990s. Lerner contends that the poorer you are, the better the services you should have: 'The difference is in the respect for people, the quality of the service provided. People feel part of the city, they belong to the city, they are proud of it and responsible for what happens.'

In the 1970s, with people flooding in from the countryside, Curitiba's transport system was not coping. All the best features of a subway system were adapted and applied to large, red articulated buses running on dedicated lanes, and passenger-friendly bus stations and ticketing equipment. Without any subsidy, 80 per cent of workers get to work by bus, including 28 per cent of car owners. This has led to a 20 per cent drop in fuel consumption in Curitiba. The concept of waste has been turned on its head by establishing a 'food-for-waste' programme for 35,000 low-income families. Each 4 kg of waste is exchanged for 1 kg of fruit and vegetables purchased with the proceeds from the sale of the recycled waste. The produce is bought from small farmers in the surrounding countryside. In one month, the 54 exchange depots collect an average of 282 tonnes of waste. Benefits include a better diet for these families, less risk of flooding from rubbish in the canals and a modest cash income for at least some small farmers who might otherwise migrate to Curitiba. In May 1996, 700 schoolchildren from some of the city's poorest neighbourhoods each paid 4 kg of recyclable rubbish to see *King Lear* performed by one of Brazil's best theatre companies.

Lerner's rejoinder to those who cannot be convinced that these schemes will also work in bigger cities is to dismiss the constraints of scale and money: 'Every city could do the same. Curitiba is different only because it has made itself different, it has gone against the flow and made itself a human city' (reported in *Guardian Weekly*, 16 June 1996, 24).

CONCLUSION

The forms of governance and regulation that civil societies – the polity – opt for when citizens exercise their right to vote ultimately determine how well cities function as living and working environments. The balance struck between state, market and community (Fig. 0.3) runs the gamut from highly centralised and regulated systems with a strong collective ethos to much more devolved and *laissez-faire* systems that place fewer restrictions on individuals and firms. Thus communist states like the former Soviet and East European economies gave rise to centrally planned and managed cities in which production took precedence over living standards and the urban environment (Andrusz *et al.* 1996). By contrast, the foundations for Hong Kong's prosperity can be traced to the comparative advantage it derived as a 'freeport' with very low taxes and minimal restraint on business and trade. Countries throughout Asia, and beyond, in the case of London's Docklands Enterprise Zone, have set up their own versions of free enterprise zones and special economic zones. But most societies occupy the broad middle ground, veering either more towards state intervention in the case of Europe's social democracies or towards market liberalism in the case of the USA, the UK, and Australia.

Cities as local economies and communities owe much, therefore, to the political and economic orientation of the state as the embodiment of civil society. Yet, even though cities are subject to state intervention,

urban regimes have emerged in places as dissimilar as Atlanta and Curitiba radically to alter outcomes with respect to property development, job creation, social justice and the environment. For example, by actively engaging in urban politics, the citizens of both Vancouver and Toronto have successfully pursued liveable city and environmentally friendly agendas. And midway through 2001, Christchurch, a medium-sized New Zealand city that was scorned as the 'Socialist Republic' through the drier years of the 1990s because it steadfastly refused to privatise Council-run services, was judged the best-run city in an international competition.

FURTHER READING

The edited volume by **Judge** *et al.* (1995) is still one of the best source books on the theory of urban politics. Likewise, despite being a little dated now, Michael **Smith's** (1988) treatment of city, state and markets brings together the essential ingredients of American urban political economy. The volume edited by **Jewson and McGregor** (1997) provides a comparative analysis of recent urban politics and governance in Europe and North America. Jamie **Peck's** review article (1995) on business elites for *Progress in Human Geography* also contains references to useful case studies.

Issues centring on the management of cities and policy debates about urban form are laid out, for example, in the **Urban Task Force Report** (1999) on the plight of Britain's inner cities, in a special issue of *Cities* (**Breheny** 1997), and in a volume edited by **Jencks, Burton and Williams** (1996). *Policies for a Just Society* (**Donnison** 1998) tackles the issues of poverty and social exclusion in urban Britain. Given the part it has played in reshaping politics in the early part of the twenty-first century, see **Giddens** (2000) for an elaboration and defence of the Third Way agenda.

10

OUTLOOK FOR THE URBAN MILLENNIUM

INTRODUCTION

This chapter offers a glimpse of the possible future of cities. Urbanists have no 'over-the-horizon' radar, so, at best, they can only surmise about the likely impact on cities of indicative trends as they are taking shape at present. In any form of prognosis there are some useful analytical devices that can help us keep our feet on the ground. The 'What if?' question is an indispensable safeguard when engaging in speculative activity. Where are the boundaries of the best- and worst-case scenarios? Does casting back 25 years help with projecting forward 25 years in deciding the fate of cities? Only sometimes. The spectacular rebuilding of Berlin as the capital of a unified Germany was unimaginable prior to the collapse of communism in 1989. The post-war record of the People's Republic of China evoked no confidence that, after 1997, Hong Kong would be allowed to continue as a free port and economic gateway to China's interior. Not all cultures and societies will chart the same course in terms of national development. Malaysia spurned the IMF/World Bank prescription for economic recovery following the Asian crisis in 1997 and appears not to have suffered any more than neighbours like Thailand and Indonesia.

A new geopolitics is working to alter the economic status of global cities. Berlin, which sits at the crossroads of continental Europe, Scandinavia and Russia, will ultimately challenge both London and Paris as the economic powerhouse of a unified Europe. As China is progressively integrated into the world economy, Shanghai will begin to compete seriously with other Pacific Rim cities like Tokyo and Los Angeles. The decision was taken under Deng Xiaoping at the start of the 1990s to build China's first global city eventually to rival New York. Shanghai is fast replacing Hong Kong as China's nerve centre within the international finance system. Clusters of high technology are being formed, and role-model engineers, scientists and academics are being recruited from all over China. But the workforce building Shanghai is being sent back to the provinces so that living standards can improve to the point where they match levels in other global cities. As elsewhere throughout the south, the gulf between an 'information rich' and privileged urban elite, and 'information poor' rural regions depleted of talented human capital, is likely to widen steadily.

Stilwell's (2000) conceptualisation of the relationships forging the nature, direction and pace of change in cities provides a useful frame of reference for the discussion. These are the relationships between 'interconnected systems' (economy, ecology and society), 'interlocking institutions' (market, state and communities), and the different levels of 'spatial scale' (global, national and local) set out in Figure 0.3. The ways in which these systems, institutions and levels of spatial scale articulate will continue to evolve as citizens in local communities take stock of the effects of globalisation and the growing subservience of nation-states. The grossly uneven impact of globalisation upon nations, regions and cities has strengthened the case for a new geography of international governance that also includes it own global architecture for regulating fickle financial flows. After supplanting the welfare state with a regulatory regime designed to free up markets and reduce dependency on social services, a number of states are once more seeking new means of accommodation with global corporations, labour unions and environmental lobbyists. With the glimmerings of a post-scarcity society in affluent economies (Giddens 1996), middle-class citizens in British, American and Australasian cities are urging their politicians and managers to break with unsustainable forms of urban development to safeguard their existing quality of life and the future of the 'global commons'. But that many of these same people seem reluctant to do anything about the long-term unemployed, or the one in three children growing up in poverty in their own societies, is hard to come to terms with. 'For in the long run, the just city and the sustainable city have to be the same place. The fair city it might be called' (Donnison 1998, 167).

PRESSURES OF POPULATION GROWTH AND URBANISATION

According to a report released by the Population Division of the United Nations Department of Economic and Social Affairs in March 2001, the world's population could increase from its reported level of 6.1 billion to about 9.3 billion by 2050. The projection takes account of an expected decline in fertility rates as women access educational opportunities, and the ravages of the HIV/AIDS epidemic on the African continent. If current trends continue, in 50 years' time 90 per cent of the world's population will be living in today's developing countries – especially India, China, Pakistan, Nigeria, Bangladesh and Indonesia. Africa could have three times as many people as Europe. With no slowing in sight of the exodus off the land and into Asian, African and Latin American cities, two-thirds of the world's population will be urbanised by 2050.

While the downside of global urbanisation is fairly clearly understood, the social and environmental effects of urban development on this scale are difficult to grapple with, let alone prepare for. For example, in pointing out that urban areas in the East Asia and Pacific Region will soon account for more than half of its population, the World Bank's urban development unit for the region realises that in their role as the 'drivers of economic growth' these cities will also have to make provision for the

burgeoning urban poor. 'The challenge throughout urban Asia is to build up social services and poverty safety nets now, when the wealth is available, before the most urgent need arises' (Gittings 2001, 6).

Chinese cities, with their huge pools of workers contracted from the countryside, will be particularly vulnerable in a world recession, which is bound to occur during the first two or three decades of the twenty first century. Even though the communist government has tried to prevent the influx for fear it could lead to urban unrest, escaping rural poverty by migrating to the more prosperous cities is a major aim of many Chinese farmers who have seen their real income fall in recent years. Yet over 12 million factory workers in the cities lost their jobs prior to China's entry into the World Trade Organization and many more can be expected to follow as the bankrupt state-run industrial sector sheds workers. One strategy devised by the State Development Planning Commission envisages the creation of nearly 10,000 New Towns in the first decade of the twenty-first century to help modernise the countryside and absorb at least 100 million rural residents. Unlike many of its Asian neighbours, because China's urbanisation rate stands at only about 32 per cent, the Commission has concluded that urbanisation should be a key component of the 10th Five-Year Plan, starting in 2001.

The limits to globalisation

Early in the twenty-first century the relationships set out in Figure 0.3 are in a state of flux and it is not yet clear how nation-states and corporations are going to respond in the years ahead to the forces that are beginning to stir locally in cities, and organise globally in opposition to trade liberalisation, unrestrained capital flows and environmental irresponsibility on a global scale. Nor is it clear how national urban systems are going to adapt as the global orientation and economic linkages between cities joining the 'world-city club' set them further and further apart from the regional cities nestling beneath them in the urban hierarchy. What is beginning to crystallise in the minds of some observers, however, is a confidence that, while the global momentum of urbanisation might appear to be unstoppable, *there is nothing inevitable about globalisation.*

How nation-states and cities decide to respond over the next decade or so, as multinational corporations endeavour to extend their markets for goods and services, will be crucial to the living standards and environmental conditions of the two-thirds of the world's population living in cities by 2050. Significantly, the main challenge to globalisation is being mounted on the streets of cities like Seattle, Melbourne, Prague, Geneva, Quebec and Genoa. In the recent past, each of these cities has played host to meetings of corporate CEOs and high-level politicians and bureaucrats trying to push the globalisation agenda. Ironically, the people most opposed to these meetings – city workers, farmers and environmentalists from all over Europe, and from countries forming the global South – are meant to be the beneficiaries of globalisation.

Multinationals have exerted an undue influence in these deliberations by networking together in closed 'think-tanks' like the World Economic Forum, which is based in Davos, Switzerland, and the Transatlantic Business Dialogue (TABD). The TABD is a working group of the West's 100 most powerful chief executives, who strategise to defeat or water down environmental, consumer and worker protection laws in selected nations (reported in *Guardian Weekly*, 26–31 May 2000, 14). By controlling agenda-setting in multilateral negotiations like the General Agreement on Tariffs and Trade (GATT), the North American Free Trade Agreement (NAFTA) and the Asia-Pacific Economic Council (APEC), multinational corporations gained advantage in their drive to globalise because member states have been complicit in keeping their citizens in the dark.

But now there is a growing awareness among NGOs and citizens working at the grassroots level in cities around the world that 'Governments will reassert their control over corporations when people reassert their control over governments' (Monbiot 2000, 13). In March 1998, talks sponsored by the OECD that would have led to a Multilateral Agreement on Investment (MAI) broke down. The MAI was an attempt on the part of these wealthy nations to protect and expand cross-border investment worth more than US$350 billion a year, notably by obliging governments to treat foreign investors on the same terms as domestic counterparts. Multinationals would have gained the power to sue nations that harmed their interests. Environmental and labour groups campaigned against MAI, but ultimately the real sticking point was the exemptions demanded for particular industries such as the French film industry.

Both the IMF and the World Bank have come to realise that 'the steering mechanism for the global economy has become unbalanced' and that 'Europeans have been over-represented at almost every level' (Brummer 1999, 12). What the European members mostly have in common is that they were formerly colonial powers. The proposal is to form a new grouping, the GX, which better reflects the new world economic order. The G7 countries – the USA, Japan, Germany, France, the UK, Italy and Canada – are to be joined by the largest emerging economies; China, India, Brazil, Mexico and South Korea satisfy this criterion. Russia gets in because of its strategic and economic significance, and South Africa as representative of the African continent. Saudi Arabia is to be included because it is the world's largest oil producer.

There is growing support for a regulatory tightening of the international financial system after the fiscal crisis of 1997. Suggestions include a 'Tobin tax' on foreign exchange speculation, the creation of a world financial authority with a lender-of-last-resort function, and reform or replacement of the International Monetary Fund and World Bank. A 'Tobin tax', named after the economist who first suggested it,

would penalise short-term financial speculation. If collected internationally it could fund the United Nations or pay for development and emergency programmes. War on Want calculates that a 0.25 per cent tax would raise an annual US$400 billion.

(Monbiot 2000, 13)

However, so long as the concerns of wealthy nations dominate the debate, and the employment and environmental issues affecting the urban poor tend to be down-played, the will to act remains half-hearted.

While the forces of globalisation have irrevocably transformed the metrics of spatial scale, something of an illusion has been created around the false promises relating to the alleviation of poverty. The benefits of ten years of economic growth in the 1990s trickled down to the urban poor in very few countries, China being one of the exceptions. By 2000, the top 20 per cent of workers on the global income scale earned 86 times more than the fifth in the bottom bracket. In 1997 the ratio was 74, while back in the 1960s it was 30 (reported in *Guardian Weekly*, 8–14 February 2001, 26). Without a concerted effort on the part of nation-states and local authorities to tackle the issue of redistribution at every level of spatial scale in Figure 0.3, the income gaps between the rich and the poor in different countries, regions and cities will continue to widen in the future. It has to be said that the social and environmental prospects for many mega-cities in the less developed countries appear quite alarming should urbanisation continue to move the rural poor into these cities at the present rates. For example, in Asia the average rate of urban growth in the first half of the 1990s ranged from 3.7 per cent in China to 4.6 per cent in Indonesia.

Bill Gates's dream that the Internet and the transfer of electronic technology would enable the benefits of globalisation to be shared worldwide ran into practical difficulties. Cities in the North with about 15 per cent of the world's population have about 85 per cent of Internet connections. Even in the capital cities of countries like Cambodia and Tanzania, several hundred people often have to share access to a single telephone. Four-fifths of the websites globally are in English, a language used by only one-tenth of the world's population. The 1999 United Nations Human Development Report estimates that a 'bit tax' on information sent via the Internet could raise well over US$70 billion annually for use in poor countries to improve electronic communications and promote education; but it would be a very indiscriminate form of taxation.

The boom in world markets caused by the digital revolution began to peter out towards the end of 2000. The much-trumpeted 'new economy' was unable to deliver endless growth to the United States, let alone to the rest of the world. As the pain began to spread from the stock market to the real economy, workers were laid off and unemployment returned to US and British cities. As predicted by social policy analysts, the inadequacies of 'workfare' were revealed when employment levels dropped. After eight years of uninterrupted growth, the cycle of economic growth and decline resumed in the United States, and it was apparent that the Internet revolution, like the telegraph before it, was just another phase in a worldwide industrial revolution that began 200 or 300 years ago:

What we also see if we look back over the past couple of centuries is that it was a time of great social and political upheaval; industrialisation did not go unopposed. Trade unions and social democratic parties were part of a powerful backlash that included less benign movements such as communism and fascism.

(John Gray, Professor of European Theory, reported in *Guardian Weekly*, 8–14 March 2001, 13)

In much the same way, unions, environmental groups and local communities in cities are capitalising upon the opportunity to network globally via the Internet and protest against hitherto unaccountable organisations like the WTO, World Bank and the IMF, or to demonstrate against socially or environmentally irresponsible corporations. Around Australia, the first May Day of the new millennium was marked by rallies in the state capitals against the unbridled power of global corporations (Fig. 10.1). And, although the blockading of the local stock exchange by a motley crowd of assorted anarchists and progressives was dismissed as counterproductive in some quarters (pun intended!), these rallies may well represent the tip of an iceberg that will eventually overwhelm those governments that fail to listen to the concerns of ordinary citizens. Hence, in addition to stimulating growth in new sectors of the economy and enhancing the productive capacity of the 'old economy', the utilisation of the same intelligent technology by citizens now enables them to coordinate their political action. The exploitation of the Internet to create a model for global activism was pioneered by a tiny London-based organisation with just 15 staff, called Jubilee 2000. They turned debt relief for insolvent nations into a *cause célèbre* by getting citizens in northern countries to apply pressure to their own politicians while providing Jubilee movements in less developed countries with information on financial mismanagement by their own governments.

As well as mobilising to disrupt successive meetings of the WTO in Seattle (November 1999) and Melbourne (September 2000), and the gathering of IMF and World Bank officials in Prague (December 2000), on any given day local protesters can picket an offending multinational if they have a plant or office in their city. By enabling citizens to communicate and coordinate their actions against a global corporation, regardless of where the offence against labour or the environment is being committed, the Internet opens up the possibility of acting in unison as part of a worldwide movement. For example, Shell's decision not to sink an offshore oil rig in the North Sea suggests that the threat of consumers withholding their buying power internationally is

Figure 10.1 The spread of community-based opposition to globalisation reaches Adelaide, Australia, with the blockading of the Stock Exchange on 1 May 2001
Source: street poster

to be avoided at all costs. In the United States, fast-growing underground groups like the Earth Liberation Front (ELF) and the Coalition to Save the Preserves (CSP) have begun targeting housing developments on the edge of cities like Louisville, Phoenix, Portland and Long Island in an effort to stop urban sprawl (reported in *Guardian Weekly*, 15–21 March 2001, 7).

Trade unions are also beginning to join with progressive communities like the Brazilian city of Porto Alegre to expose the inconsistencies that apply to the international movement of labour in an era of globalisation. Western governments champion the free movement of capital whilst decrying the free movement of labour. And, as part of this moral dilemma, they reserve the right to 'cherry pick' workers from less developed countries to fill gaps in their labour markets. In February 2001, Porto Alegre convened the first Social Forum in what may turn out to be the forerunner of a global movement for sustainable development, environmental protection and social justice. One of the actions taken will lead to a 'long march' in 2003 from São Paulo in Brazil to the border separating Mexico from the United States. Organised by the radical Brazilian Workers' Party, the march will draw attention to the fact that, under NAFTA, 'We take their goods, but they don't like our people, so we are taking our people to them ourselves' (reported in *Guardian Weekly*, 8–14 February 2001, 26).

Indeed, with the projected growth in the world population, the main gateway cities in the United States are likely to remain a prime target for migrants. With an annual influx of about one million people, the United States population is expected to reach about 400 million by 2050.

Separatist tendencies in multicultural urban societies

The rising numbers of newly arrived migrants, refugees and temporary workers drawn to large North American and European cities from countries around the globe have fuelled the anxieties of urban communities already troubled about unsafe streets, poor schools and deteriorating services. The exodus of largely white, low-middle-class urban families observed in Chapter 2 as part of the broader process of counter-urbanisation has the potential to open up racial and cultural divisions in societies receiving large numbers of migrants in the decades ahead.

If the current migration patterns persist in the United States, by 2020 the country will be divided into distinctive ethno-cultural regions (Kotkin 1996, 15). These trends are building on immigration and trade patterns set in the 1970s. In 12 states – mostly in the Plains, upper New England and the Intermountain West – more than 80 per cent of young people under 17 will be white, while in blocks of states like California, New Mexico and Texas, and the north-east, young whites will be in a distinct minority. Because the new migrants to the conservative heartland tend to be older, less affluent and less well educated, and often closer to retirement age, rather than infusing local politics with their urban sensibilities they are helping to fortify traditional values and strengthen right-wing parties.

According to Kotkin, the political contest in the United States between what he calls the Valhallan and Cosmopolitan visions will more than likely shape American society during this century:

Ultimately it may determine whether this society meets the challenge of becoming a harbinger of a new world culture, or whether it will seek to freeze itself like other declining civilizations, in the comforting outlines of its imagined past.
(Kotkin 1996, 15)

Yet partly because of the influx of immigrants, and the skills and drive they bring with them, the great metropolitan centres will remain the incubators of America's commercial, technological and artistic life. Virtually all of the top 10 graduate departments in the sciences and engineering are located either on the west coast, in the upper mid-west, or in the north-east. Most of the nation's key exporting industries are also located in these urban regions. In terms of global competition, Silicon Valley, Hollywood and Wall Street are not likely to relinquish the lead they have in software and computers, multimedia and financial services.

Although not to be compared with the political landscape in the United States, the ultraconservative One Nation party had its basis in a regional rift between rural and urban Australians which shows no signs of waning as yet. Meanwhile in New Zealand, Auckland is destined to become an increasingly multicultural city given the differential growth rates of the Pakeha (European), Maori, Pacific and Asian populations. Yet the presence of Maori, Pacific Islanders and Asians diminishes in the other urban centres the further south one gets in New Zealand.

THE OUTLOOK FOR LESS UNSUSTAINABLE CITIES

Ecologically sustainable urban development (ESD) calls for the reconciling of the often contrary goals of increased urban productivity and environmental protection. None the less, the essence of sustainability as captured in the ethos of the 'triple bottom line' – social responsibility, profitability, environmental stewardship – is catching on as *the* corporate challenge of the twenty-first century (Dunphy *et al.* 2000). Enlightened shareholders are directing fund managers to invest in those corporations that can demonstrate an ethical approach to employment conditions, community relations and environmental impact. Affluent communities, at least, will demand that future urban development utilises 'clean' technologies and eliminates waste. And, indeed, new-generation production cycles will become progressively cleaner, as will industrial parks in edge city locations. Educated consumers are shifting brand loyalty in increasing numbers to environmentally conscious companies and retailers.

With many nation-states vacating significant areas of public enterprise, regulatory oversight and social support, corporations are being called upon to exercise more social and environmental responsibility. Multinational corporations like Volvo, BMW, Johnson and Johnson, Merck, Shell, Ikea, Electrolux and Interface are prominent among the visionary companies that are providing a lead as global 'good citizens'. These corporations, and over 100 more like them, have adopted a set of sustainable principles devised by the Swedish cancer researcher Dr Karl Henrik Robèrt with the help of some colleagues, and known as the Natural Step. Moreover, a study of visionary companies by Collins and Porras (1994) revealed that these companies were up to 15 times more profitable than the general market, even though profitability was not their top priority. The Nagoya Plant of the Kirin Brewery company in Shinkawa, Japan, reached zero emissions in 1997.

With governments chastened by popular reaction to top-down policy- and decision-making, genuine efforts are being made in many countries and cities to democratise the processes governing change in local communities. Sometimes the appeal to self-reliance and voluntarism is a subterfuge to cut back on funding levels, but in most case resources follow the transfer of power. In the United Kingdom, 'beacon councils' are singled out as models of best practice. Their councillors and staff listen to and consult with local people, involve them in the planning and use of local services and environmental improvement, and are not defensive of vested interests. Community action at the level of street-block organisations is helping to rebuild dilapidated housing and social capital in the inner neighbourhoods of many US cities. Privately funded foundations and trusts play a much more active role in resourcing community rebuilding and neighbourhood renewal in American cities than in Europe or Australia.

Although examples of sustainable urban development could be drawn from many other less developed countries, since the collapse of apartheid in 1994 most South African cities have embarked on ambitious versions of ESD. In historically deprived communities there is a staggering backlog of housing, sanitation, employment, literacy, and social and health service 'deficits' to be addressed. Community-based organisations have assumed responsibility for training local officials in management, educating residents in participatory procedures, promoting development planning and developing environmental maintenance strategies for low-income neighbourhoods. Communities are directly represented on all standing committees of local authorities to ensure they approve all zoning/re-zoning and development applications, and urban renewal schemes affecting their residential environment. Still, these reforms run into daunting problems in the black townships, including mounting unemployment and crime. More than 250 of the 850 municipalities in South Africa are in financial trouble as their tax base has declined since 1994. Besides, many are plagued by a high incidence of fraud and administrative incompetence reflecting a lack of effective governance at the local level in the black townships (Williams 2000, 174–9).

Finally, the global warming caused by emitting and dispersing greenhouse gases into the atmosphere provides a very potent reminder of the eventual cost of failing to take the concept of the ecological footprint seriously. The projected rates of urbanisation will greatly compound the threat over the next 50 years unless cleaner fuels are developed in the meantime, since cities are the primary source of greenhouse gases. Industrial expansion and rising consumption levels in mega-cities like Shanghai, Mexico City, São Paulo and Tokyo will increase the demand for energy. Global warming also reminds us how interdependent humankind and nature are, and that our actions as citizens can jeopardise the life of others literally at the ends of the earth. The difficulty is that the early industrialisers – Europe, North America, Japan, Australia – created their wealth by burning fossil fuels and are now seeking to persuade the late industrialisers that they should do otherwise. However, as Atkinson (1994, 97) points out, 'if it is the lifestyle of the rich minority in Third World cities and the industrialised north that is the main cause of the unsustainability of the present development path, then why should the Third World urban poor be restrained in their demands?'

In part the answer is contained in the Agenda 21 document signed by heads of state at the United Nations Conference on Environment and Development (UNCED) held in Rio de Janeiro in June 1992. Over two-thirds of the actions to be taken in support of sustainable development have to be implemented at the local level. Therefore, as a growing proportion of the global population concentrates in cities, the onus will shift to urban citizens to ensure that economic development is sustainable. China, for example, reduced its greenhouse emissions by 17 per cent in the three years to 2001. Much of that contribution is due to the conversion of coal-fired generators and domestic stoves in the cities. Indeed, the central message of *Cities for a Small Planet* is that, 'Tackling the global environmental crisis from the vantage point of each city brings the task within the grasp of the citizen' (Rogers 1997, 2.32).

CONCLUSION

This concluding chapter of *Making Sense of Cities* was written at a time when the promise and limits of economic and social transformations that we summarily and all too simplistically call 'globalisation' were becoming more and more apparent. The technological gear-shifts that paved the way for the knowledge economy (Fig. 3.1) so reduced the costs of telecommunications, data processing and air travel in real terms (Fig. 1.4) that finance and information can now move between cities anywhere around the globe with frictionless rapidity.

With the premium that exists for knowledge workers, countries playing 'catch-up' now have to compete openly in a global labour market for their human capital requirements. Many of the brightest and most able graduates throughout Latin America, Asia, Africa and Oceania are leaving home to undertake study or to work in leading cities throughout their region. Then the best of those gravitate to global centres of learning and business, invariably in Europe and North America. In the process, the selective drift of talented human capital on a global scale is reinforcing the dominance of 'information-rich' cities and regions like San Francisco/Los Angeles, Boston/New York/Washington DC, Singapore/Johore/Riau, the south-east of England, Tokyo/Nagoya/Osaka, Rio de Janeiro/São Paulo, Shanghai, Hong Kong/Shenzhen/Guangzhou or Sydney/Canberra/Melbourne.

The populations of cities that do successfully capture greater shares of mobile capital and human talent will become increasingly multicultural and energised by the presence of international migrants. Urban regimes that can foster a climate of acceptance and tolerance towards others stand to benefit from the creativity that flows from the mingling of peoples and cross-fertilisation of cultures and ideas (Hall 1998). Cities like Amsterdam, San Francisco, Honolulu, Vancouver, Toronto and Sydney seem more willing to embrace minorities and outsiders. But in other cities – most recently in Minneapolis–St Paul and Oldham – housing markets continue to be used to perpetuate mistrust and divide communities. In the long run, a reputation for racial tension and violence will undermine the prosperity of these cities.

Also, in an era of globalisation and the Internet, cities with strong links to the global economy and an abundance of knowledge workers now contain people who interact professionally and socially via a global network of contacts. Futurists like Mitchell (1999) suggest that this has the potential to create a new social divide within cities between communities that are 'information rich' and have a monopoly over global connectedness, and those that are 'information poor' and languish in a digital backwater. At the heart of big cities everywhere social classes can be found living side by side in contiguous neighbourhoods; but they might as well be on other sides of the world for all the contact that passes between them. While this new divide around command over knowledge and information technology is bound to overlap strongly with class, only time will tell how it might reshape social life in the post-modern city.

Finally, for all the social and environmental pressures that the worldwide concentration of people in cities bring to bear in the twenty-first century, it also has to be said that urbanisation contains the seeds of hope. Throughout history, urban societies have fostered education and learning, and have been the bearers of creativity and material progress (Hall 1998). With people on the move globally and making new lives in multicultural societies, there is hope for the spreading of acceptance of difference. As more people move into big cities, political momentum will build from the grassroots for healthier and Greener urban environments, and less unsustainable development. And, as social equity and justice are gradually embraced by more and more environmentalists as necessary conditions for environmentally sustainable urban development, there is even hope that the redistribution sufficient to make inroads into urban poverty might eventually follow.

FURTHER READING

E-topia: Urban Life, Jim – but not as we know it (**Mitchell** 1999) speculates on how the global digital network will alter both the form of cities and social life within them. The author is Dean of the School of Architecture and Planning at MIT.

Dunphy and his co-writers (2000) spell out the possibilities and chart the progress that corporations are making in the pursuit of sustainability.

In Chapter 9 of *Policies for a Just Society*, **David Donnison** (1998) explores the common ground between egalitarians and Greens in order to show that environmental policy and job creation need not be in conflict.

Two works that share the concluding optimism of this chapter are **David Harvey's** *Spaces of Hope* (2000) and **Leonie Sandercock's** *Towards Cosmopolis: Planning for Multi-cultural Cities* (1998).

REFERENCES

Abrahamson, M. (1996) *Urban Enclaves: Identity and Place in America*. New York: St Martin's Press.

Abu-Lughod, J. (1997) The specificity of the Chicago ghetto: comment on Wacquant's 'Three pernicious premises'. *International Journal of Urban and Regional Research* 21, 357–62.

Adams, J.S. (1969) Directional bias in intraurban migration. *Economic Geography* 45, 303–23.

Agnew, J., Mercer, J. and Sopher, D. (eds) (1984) *The City in Cultural Context*. Boston: Allen & Unwin.

Alexander, I. (1979) *Office Location and Public Policy*. London: Longman Group.

Alexander, J.W. (1954) The basic–nonbasic concept of urban economic functions. *Economic Geography* 30, 246–61.

Alexandersson, G. (1956) *The Industrial Structure of American Cities*. Lincoln: University of Nebraska.

Allen, J.P. and Turner, E. (1989) The most ethnically diverse urban places in the United States. *Urban Geography* 10, 523–39.

Alonso, W. (1964) *Location and Land Use: Toward a General Theory of Land Rent*. Cambridge, MA: Harvard University Press.

Amin, A. and Graham, S. (1997) The ordinary city. *Transactions, Institute of British Geographers* NS 22, 411–29.

Anderson, K. (1998) Sites of difference: Beyond a cultural politics of race polarity. In Fincher, R. and Jacobs, J.M. (eds), *Cities of Difference*. New York and London: The Guilford Press, 201–25.

Andrusz, G., Harloe, M. and Szelenyi, I. (eds) (1996) *Cities after Socialism*. Oxford and Cambridge, MA: Blackwell.

Atkinson, A. (1994) The contribution of cities to sustainability. *Third World Planning Review* 16, 97–101.

AURDR (1995) *Green Cities. Strategy Paper #3*. Australian Urban and Regional Development Review. Canberra: Commonwealth of Australia.

Badcock, B. (1984) *Unfairly Structured Cities*. Oxford: Basil Blackwell.

Badcock, B. (1986) Land and housing policy in Chinese urban development, 1976–86. *Planning Perspectives* 1, 147–70.

Badcock, B. (1994) 'Snakes or ladders?' The housing market and wealth distribution in Australia. *International Journal of Urban and Regional Research* 18, 609–27.

Badcock, B. (1998) Ethical quandaries and the urban domain. *Progress in Human Geography* 22, 586–94.

Badcock, B. (2000) The imprint of the post-Fordist transition on Australian cities. In Marcuse, P. and van Kempen, R. (eds), *Globalizing Cities: A New Spatial Order?* Oxford and Malden, MA: Blackwell, 211–27.

Badcock, B. and Beer, A. (2000) *Home Truths: Property Ownership and Housing Wealth in Australia*. Carlton South, Vic.: Melbourne University Press.

Bairoch, P. (1988) *Cities and Economic Development: From the Dawn of History to the Present*. Chicago: University of Chicago Press.

Ball, M. (1983) *Housing Policy and Economic Power: The Political Economy of Owner-Occupation*. London: Methuen.

Ball, M., Harloe, M. and Martens, M. (1988) *Housing and Social Change in Europe and the USA*. London: Routledge.

Banfield, E.C. (1961) *Political Influence*. New York: Macmillan.

Bartone, C., Bernstein, J., Leitmann, J. and Eigen, J. (1994) *Toward Environmental Strategies for Cities: Policy Considerations for Urban Environmental Management in Developing Countries*. Washington, DC: World Bank.

Bater, J.H. (1980) *The Soviet City: Ideal and Reality*. London: Edward Arnold.

Beauregard, R.A. (ed.) (1989) *Atop the Urban Hierarchy*. Totowa, NJ: Rowman & Littlefield.

Beauregard, R.A. (1995) Edge cities: peripheralizing the center. *Urban Geography* 16, 708–21.

Beaverstock, J.V. (1996) Revisiting high-waged labour market demand in the global cities: British professional and managerial workers in New York City. *International Journal of Urban and Regional Research* 20, 422–45.

Beaverstock, J.V. and Smith, J. (1996) Lending jobs to global cities: skilled international labour migration, investment banking and the City of London. *Urban Studies* 33, 1377–94.

Beazley, M., Loftman, P. and Nevin, B. (1997) Downtown redevelopment and community resistance: an international perspective. In Jewson, N. and MacGregor, S. (eds), *Transforming Cities: Contested Governance and New Spatial Divisions*. London and New York: Routledge, 181–92.

Bell, G. and Tyrwhitt, J. (eds) (1972) *Human Identity in the Urban Environment*. Harmondsworth: Penguin.

Bernstein, N. (2000) Poverty found to be rising in families considered safe. *New York Times* 20 April, B1 + B9.

Berry, B.J.L. (1967) *Geography of Market Centers and Retail Distribution*. Englewood Cliffs, NJ: Prentice-Hall.

Berry, B.J.L. (ed.) (1971) Comparative factorial ecology. *Economic Geography* 47 (supplement).

Berry, B.J.L. and Horton, F.E. (1970) *Geographic Perspectives on Urban Systems*. Englewood Cliffs, NJ: Prentice-Hall.

Berry, B.J.L. and Kasarda, J.D. (1977) *Contemporary Urban Ecology*. New York: Macmillan.

Berry, B.J.L. and Rees, P.H. (1969) The factorial ecology of Calcutta. *American Journal of Sociology* 74, 445–92.

Blair, A. (1998) *The Third Way*. London: Fabian Society.

Blair, A. and Schröder, G. (1999) *Europe: The Third Way – die Neue Mitte*. London: Labour Party and SPD.

Blakely, E.J. and Snyder, M.G. (1997) *Fortress America: Gated Communities in the United States*. Washington, DC: Brookings Institution Press.

Bourassa, S.C., Greig, A.W. and Troy, P.N. (1995) The limits of housing policy: home ownership in Australia. *Housing Studies* 10, 83–104.

Bourne, L.S. (1981) *The Geography of Housing*. London: Edward Arnold.

Boyle, P. and Halfacree, K. (eds) (1998) *Migration into Rural Areas: Theories and Issues*. Chichester: John Wiley & Sons.

Braunerhjelm, P., Faini, R., Norman, V., Ruane, F. and Seabright, P. (2000) *Integration and the Regions of Europe: How the Right Policies can Prevent Polarisation*. London: Centre for Economic Policy Research.

Breheny, M. (1997) Urban compaction: feasible and acceptable? *Cities* 14, 209–17.

Briggs, X.S. (1997) Moving up versus moving out: neighbourhood effects in housing mobility programs. *Housing Policy Debate* 8, 195–234.

Brotchie, J., Batty, M., Hall, P. and Newton, P. (eds) (1991) *Cities of the 21st Century: New Technologies and Spatial Systems*. New York: Longman Cheshire.

Brotchie, J., Batty, M., Hall, P. and Newton, P. (eds) (1995) *Cities in Competition: Productive and Sustainable Cities for the 21st Century*. Melbourne: Longman Australia.

Bruegel, I. (2000) The restructuring of London's labour force: migration and shifting opportunities, 1971–91. *Area* 32, 79–90.

Brummer, A. (1999) The new world economic order. *Guardian Weekly* 23–29 September, 12.

Brush, J.E. and Bracey, H.E. (1955) Rural service centers in southwestern Wisconsin and southern England. *Geographical Review* 45, 559–69.

Bull, P.J. (1978) The spatial components of intra-urban manufacturing change: suburbanization in Clydeside, 1958–68. *Transactions, Institute of British Geographers* NS 3, 91–100.

Burbidge, A. (2000) Capital gains, homeownership and economic inequality. *Housing Studies* 15, 258–80.

Burgess, E.W. (1925) The growth of the city: an introduction to a research project. In Park, R.E., Burgess, E.W. and McKenzie, R.D. (eds), *The City*. Chicago: University of Chicago Press, 47–62.

Burnley, I.H. (1989) Settlement dimensions of the Vietnamese-born population in Sydney. *Australian Geographical Studies* 27, 129–54.

Burnley, I.H. (1996) Relocation of Overseas-born populations in Sydney. In Newton, P.W. and Bell, M. (eds), *Population Shift: Mobility and Change in Australia*. Canberra: Australian Government Publishing Service, 83–102.

Business Victoria (1997) *Advantage Melbourne: The Comparison of Investment Locations in Asia-Pacific*. Melbourne: Business Victoria.

Butler, T. (1997) *Gentrification and the Middle Classes*. Aldershot, Hants: Ashgate.

Butlin, N.G. (1964) *Investment in Australian Economic Development, 1861–1900*. Cambridge: Cambridge University Press.

Cairncross, F. (1997) *The Death of Distance: How the Communications Revolution will Change our Lives*. Boston, MA: Harvard Business School Press.

Case, K.E. (1992) The real estate cycle and the economy: consequences of the Massachusetts boom of 1984–87. *Urban Studies* 29, 175–83.

Castells, M. (1977) *The Urban Question: A Marxist Approach*. London: Edward Arnold.

Castells, M. (1989) *The Informational City: Information Technology, Economic Restructuring, and the Urban–Regional Process*. Oxford: Basil Blackwell.

Castells, M. (1994) European cities, the informational society, and the global economy. *New Left Review* 204, 18–32.

Castells, M. (1997) *The Power of Identity*. Cambridge, MD and Oxford: Blackwell.

Cervero, R., Rood, T. and Appleyard, B. (1999) Tracking accessibility: employment and housing opportunities in the San Francisco Bay Area. *Environment and Planning A* 31, 1259–78.

Champion, A.G. (ed.) (1989) *Counterurbanization: The Changing Pace and Nature of Population Deconcentration.* London: Edward Arnold.

Champion, A.G. (1992) Urban and regional demographic trends in the developed world. *Urban Studies* 29, 461–82.

Cheshire, P. (1990) Explaining the recent performance of the European Community's major urban regions. *Urban Studies* 27, 311–33.

Cheshire, P. (1995) A new phase of urban development in western Europe? The evidence for the 1980s. *Urban Studies* 32, 1045–63.

Chinitz, B. (1961) Contrasts in agglomeration: New York and Pittsburgh. *American Economic Review: Papers and Proceedings* May 279–89.

Chisholm, M. (1979) *Rural Settlement and Land Use. An Essay in Location.* London: Hutchinson.

Christaller, W. (1933) *The Central Places of Southern Germany.* Translated by C. Baskin in 1966. Englewood Cliffs, NJ: Prentice-Hall.

Chu, D.K.Y. (1996) The Hong Kong–Zhujiang Delta and the world city system. In Lo, F. and Yeung, Y. (eds) 1996: *Emerging World Cities in Pacific Asia.* Tokyo: United Nations University Press, 465–97.

Clark, G.L. and Dear, M. (1981) The state in capitalism and the capitalist state. In Dear, M. and Scott, A.J. (eds), *Urbanization and Urban Planning in Capitalism and Society.* London: Methuen, 45–62.

Clark, W.A.V. and Kuijpers-Linde, M. (1994) Commuting in restructuring urban regions. *Urban Studies* 31, 465–81.

Clark, W.A.V. and Schultz, F. (1997) The geographical impacts of welfare reform in the United States. *Environment and Planning A* 29, 762–70.

Clarke, S.E., Stacheli, L.A. and Brunell, L. (1995) Women redefining local politics. In Judge, D., Stoker, G. and Wolman, H. (eds), *Theories of Urban Politics.* Thousand Oaks, CA and London: Sage Publications, 205–27.

Climatic Change Secretariat (1999) *The Kyoto Protocol to the Convention on Climatic Change.* Geneva: UNEP/IUC.

Cloke, P., Crang, P. and Goodwin, M. (eds) (1999) *Introducing Human Geographies.* London: Arnold.

Cohen, M.A. (1999) Impact of international economic change on cities. *Australian Planner* 36, 72–74.

Cohen, P. (1998) A transforming San Francisco industrial landscape. *Pacifica* Fall, 1 and 7–12.

Cohen, R.B., Fleton, N., Nkosi, M. and Van Liere, J. (1979) *The Multinational Corporation: A Radical Approach.*

Papers by Stephen Herbert Hymer. Cambridge: Cambridge University Press.

Collinge, C. and Hall, S. (1997) Hegemony and regime in urban governance: towards a theory of the locally networked state. In Jewson, N. and MacGregor, S. (eds), *Transforming Cities: Contested Governance and New Spatial Divisions.* London and New York: Routledge, 129–40.

Collins, J.C. and Porras, J.I. (1994) *Built to Last: Successful Habits of Visionary Companies.* London: Century.

Costanza R. and Segura, O. (eds) (1996) *Getting Down to Earth.* Washington, DC: Island Press.

Cox, E. (1995) *A Truly Civil Society (1995 Boyer Lectures).* Sydney: Australian Broadcasting Corporation.

Cronon, W. (1991) *Nature's Metropolis: Chicago and the Great West.* New York and London: W.N. Norton & Co.

Dahl, R.A. (1961) *Who Governs? Democracy and Power in an American City.* New Haven, CT: Yale University Press.

Daly, H. and Cobb, J.B. (1990) *For the Common Good.* Boston: Beacon Press.

Danziger, S. and Reed, D. (1999) Winners and losers. The era of inequality continues. *Brookings Review* Fall, 15–17.

Davies, N. (1998) There is nothing natural about poverty. *New Statesman* 6 November, 32–4.

Davis, M. (1990) *City of Quartz. Excavating the Future in Los Angeles.* London and New York: Verso.

Davis, M. (1999a) *Ecology of Fear: Los Angeles and the Imagination of Disaster.* New York: Vintage Books.

Davis, M. (1999b) *Gangland, Cultural Elites and the New Generationalism.* Sydney: Allen & Unwin.

Dear, M. (1988) The postmodern challenge: reconstructing human geography. *Transactions, Institute of British Geographers* NS, 262–74.

Dear, M. and Flusty, S. (1998) Postmodern urbanism. *Annals of the Association of American Geographers* 88, 50–72.

Dear, M. and Wolch, J. (1992) Learning from Los Angeles. *Environment and Planning A* 24, 917–20.

Deskins, D.R. (1996) Economic restructuring, job opportunities and black social dislocation in Detroit. In O'Loughlin, J. and Friedrichs, J. (eds), *Social Polarization in Post-Industrial Metropolises.* Berlin and New York: Walter de Gruyter, 258–82.

DETR (2000) Housing: Key Figures – April 2000. http://www.housing.detr.gov.uk/information/keyfigures/index.ht.

Dick, H.W. and Rimmer, P.J. (1998) Beyond the Third World city: the new urban geography of South-east Asia. *Urban Studies* 35, 2303–22.

Dicken, P. (1998) *Global Shift: Transforming the World Economy*. London: Paul Chapman.

Dieleman, F.M. and Hamnett, C. (1994) Globalisation, regulation and the urban system: Editors' introduction to the special issue. *Urban Studies* 31, 357–64.

Donnison, D. (1998) *Policies for a Just Society*. London: Macmillan Press.

Drier, P. (1997) The new politics of housing. How to build a constituency for a progressive federal housing policy. *Journal of the American Planning Association* 63, 5–27.

Dunn, E.S. (1980) *The Development of the US Urban System. Vol. 1. Concepts, Structures, Regional Shifts*. Baltimore: Resources for the Future and Johns Hopkins University Press.

Dunn, K.M. (1993) The Vietnamese concentration in Cabramatta: site of avoidance and deprivation, or island of adjustment and participation? *Australian Geographical Studies* 31, 228–45.

Dunphy, D., Benveniste, J., Griffiths, A. and Sutton, P. (eds) (2000) *Sustainability. The Corporate Challenge of the 21st Century*. Sydney: Allen & Unwin.

Eckersley, R. (ed.) (1998) *Measuring Progress. Is Life Getting Better?* Collingwood, Vic.: CSIRO Publishing.

Elliott, L. (1999) Unless the World Trade Organization cleans up its act there will be more issues for the protesters to trade on. *Guardian Weekly* 9–15 December, 12.

Elwood, D. (1986) The spatial mismatch hypothesis: are there teenage jobs missing in the ghetto? In Freeman, R. and Holzer, H. (eds) *Black Youth Employment Crisis*. Chicago: University of Chicago Press, 147–90.

England, K.V.L. (1993) Suburban pink collar ghettos: the spatial entrapment of women? *Annals of the Association of American Geographers* 83, 225–42.

Esping-Andersen, G. (1990) *The Three Worlds of Welfare Capitalism*. Cambridge: Polity Press.

Etzioni, A. (1995) *The Spirit of Community: Rights, Responsibilities and the Communitarian Agenda*. London: Fontana Press.

Evans, A.W. (1973) *The Economics of Residential Location*. London: Macmillan.

Expert Group on the Urban Environment (1996) *European Sustainable Cities*. Brussels: European Commission.

Fainstein, S.S. (1994) *The City Builders: Property, Politics, and Planning in London and New York*. Oxford and Cambridge, MA: Blackwell.

Fainstein, S.S., Gordon, I. and Harloe, M. (eds) (1992) *Divided Cities: New York & London in the Contemporary World*. Cambridge, MA and Oxford: Blackwell.

Fales, R.L. and Moses, L.N. (1972) Land-use theory and spatial structure of the nineteenth-century city. *Papers of the Regional Science Association* 28, 49–80.

Feagin, J.R. and Beauregard, R.A. (1989) Houston: hyperdevelopment in the sunbelt. In Beauregard, R.A. (ed.), *Atop the Urban Hierarchy*. Totowa, NJ: Rowman & Littlefield, 153–194.

Fergusson, R. (ed.) (1990) *Out There: Marginalization and Contemporary Culture*. Cambridge, MA: MIT Press.

Fielding, A.J. (1992) Migration and social mobility: south east England as an escalator region. *Regional Studies* 26, 1–27.

Fincher, R. and Jacobs, J.M. (eds) (1998) *Cities of Difference*. New York and London: The Guilford Press.

Foley, P. and Martin, S. (2000) Perceptions of community led regeneration: community and central government viewpoints. *Regional Studies* 34, 783–7.

Folke, C., Larsson, J. and Sweitzer, J. (1996) Renewable resource appropriation. In Costanza R. and Segura, O. (eds), *Getting Down to Earth*. Washington, DC: Island Press.

Forbes, D. (1997) Metropolis and megaurban region in Pacific Asia. *Tijdschrift voor Economische en Sociale Geografie* 88, 457–68.

Ford, J. and Wilcox, S. (1998) Owner-occupation, employment and welfare: the impact of changing relationships on sustainable home ownership. *Housing Studies* 13, 623–39.

Forrest, R. and Kearns, A. (1999) *Joined-up Places? Social Cohesion and Neighbourhood Regeneration*. York: Joseph Rowntree Foundation.

Forrest, R. and Williams, P. (1990) *Home Ownership: Differentiation and Fragmentation*. London: Unwin Hyman.

Forrester, V. (1999) *The Economic Horror*. Oxford and Malden, MA: Polity Press.

Forster, C. (1999) *Australian Cities. Continuity and Change*. Melbourne: Oxford University Press (2nd edn).

Freeman, R.B. (1995) Are your wages set in Beijing? *Journal of Economic Perspectives* 9, 15–32.

Freeman, R. and Holzer, H. (eds) (1986) *Black Youth Employment Crisis*. Chicago: University of Chicago Press.

Freestone, R. and Murphy, P. (1998) Metropolitan restructuring and suburban employment centres. Cross-cultural perspectives on the Australian experience. *American Planners Association Journal* 64, 286–97.

French, R.A. and Hamilton, F.E. (eds) (1979) *The Socialist City: Spatial Structure and Urban Policy*. Chichester: John Wiley & Sons.

Frey, W.H. (1993) The new urban revival in the United States. *Urban Studies* 30, 741–74.

Frey, W.H. (1995) Immigration and internal migration 'flight' from US Metropolitan Areas: towards a new demographic balkanisation. *Urban Studies* 32, 733–57.

Frey, W.H. and Farley, R. (1996) Latino, Asian, and Black segregation in US metropolitan areas: are multiethnic metros different? *Demography* 33, 335–50.

Friedmann, J. (1986) The world city hypothesis. *Development and Change* 17, 69–83.

Friedmann, J. and Wolff, G. (1982) World city formation: an agenda for research and action. *International Journal of Urban and Regional Research* 4, 309–43.

Fröebel, F., Heinrichs, J. and Kreye, O. (1980) *The New International Division of Labour.* Cambridge: Cambridge University Press.

Galbraith, J.K. (1992) *The Culture of Contentment.* London: Penguin.

Gans, H.J. (1962) *The Urban Villagers: Group and Class in the Life of Italian-Americans.* New York: Free Press of Glencoe.

Garreau, J. (1991) *Edge City: Life on the New Frontier.* New York: Doubleday.

Geddes, M. (1997) Poverty, excluded communities and local democracy. In Jewson, N. and MacGregor, S. (eds), *Transforming Cities: Contested Governance and New Spatial Divisions.* London and New York: Routledge, 205–18.

Gertler, M.S. (1988) The limits to flexibility: comments on the post-Fordist vision of production and its geography. *Transactions, Institute of British Geographers* NS 13, 419–32.

Giddens, A. (1996) Affluence, poverty and the idea of a post-scarcity society. *Development and Change* 27, 365–77.

Giddens, A. (1998) *The Third Way: The Renewal of Social Democracy.* Cambridge: Polity Press.

Giddens, A. (2000) *The Third Way and its Critics.* Cambridge: Polity Press.

Gilbert, M. (1997) Feminism and difference in urban geography. *Urban Geography* 18, 166–79.

Giradet, H. (1992) *Cities: New Directions for Sustainable Urban Living.* London: Gaia Books.

Gittings, J. (2001) Social safety nets must be a priority as urban populations leap ahead. *Guardian Weekly* 4–10 January, 6.

Glover, J., Harris, K. and Tennant, S. (1999) *A Social Health Atlas of Australia. Vol. 1: Australia.* Adelaide: Public Health Information Development Unit (2nd edn).

Glynn, S. (1970) *Urbanisation in Australian History, 1788–1900.* Sydney: Nelson.

Godfrey, B.J. and Zhou, Y. (1999) Ranking world cities: multinational corporations and the global urban hierarchy. *Urban Geography* 20, 268–81.

Gottmann, J. (1961) *Megalopolis: The Urbanized Northeastern Seaboard of the United States.* New York: Twentieth Century Fund.

Green, H.L. (1955) Hinterland boundaries of New York City and Boston in Southern New England. *Economic Geography* 31, 283–300.

Gregory, R.G. (1993) Aspects of Australian living standards: the disappointing decades 1970–90. *Economic Record* 69, 61–76.

Gregory, R.G. and Hunter, B. (1995) The macro-economy and the growth of ghettos and urban poverty in Australia. Discussion Paper No. 325, Centre for Economic Policy Research. Australian National University, Canberra.

Grigsby, W.G. (1963) *Housing Markets and Public Policy.* Philadelphia: University of Pennsylvania Press.

Haila, A. (1995) Real estate markets in Los Angeles and Helsinki. Paper presented at the International Congress of Real Estate, 24–26 April, Singapore.

Haila, A. (1999) Why is Shanghai building a giant speculative property bubble? *International Journal of Urban and Regional Research* 23, 583–8.

Hall, P. (1966) *The World Cities.* New York: McGraw-Hill.

Hall, P. (1998) *Cities in Civilization.* London: Weidenfeld and Nicolson.

Hall, P. (1999) How cities can be expected to change. *Australian Planner* 36, 66–71.

Hall, P. (2000) Creative cities and economic development. *Urban Studies* 37, 639–49.

Hamnett, C. (1991) The blind men and the elephant: the explanation of gentrification. *Transactions, Institute of British Geographers* NS 16, 173–89.

Hamnett, C. (1994) Social polarisation in global cities: theory and evidence. *Urban Studies* 31, 401–24.

Hamnett, C. (1999a) The city. In Cloke, P., Crang, P. and Goodwin, M. (eds), *Introducing Human Geographies.* London: Arnold, 246–55.

Hamnett, C. (1999b) Time to reflect on house price booms. *Independent*, 30 August, Business/15.

Hamnett, C. (1999c) *Winners and Losers: Home Ownership in Modern Britain.* London: UCL Press.

Hamnett, C. and Randolph, B. (1988) *Cities, Housing and Profits: Flat Break-up and the Decline of Private Renting.* London: Hutchinson Education.

Hannigan, J. (1999) *Fantasy City: Pleasure and Profit in the Postmodern Metropolis.* London: Routledge.

Hanson, S. and Pratt, G. (1988) Spatial dimensions of the gender division of labor in a local labor market. *Urban Geography* 9, 367–78.

Hanson, S. and Pratt, G. (1994) On suburban pink collar ghettos: the spatial entrapment of women? *Annals of the Association of American Geographers* 84, 500–4.

Hanson, S. and Pratt, G. (1995) *Gender, Work and Space.* London: Routledge.

Harloe, M. (1977) *Captive Cities: Studies in the Political Economy of Cities and Regions*. London: Wiley.

Harrison, B., Kelley, M.R. and Gant, J. (1996) Innovative firm behavior and local milieu: exploring the intersection of agglomeration, firm effects, and technological change. *Economic Geography* 72, 233–58.

Harvey, D. (1973) *Social Justice and the City*. London: Arnold.

Harvey, D. (1978) The urban process under capitalism: a framework for analysis. *International Journal of Urban and Regional Research* 2, 101–31.

Harvey, D. (1985) *The Urbanization of Capital*. Oxford: Blackwell.

Harvey, D. (1989) From managerialism to entrepreneurialism: the transformation in urban governance in late capitalism. *Geografiska Annäler* 71B, 3–17

Harvey, D. (2000) *Spaces of Hope*. Edinburgh: Edinburgh University Press.

Haughton, G. (1997) Developing sustainable urban development models. *Cities* 14, 189–95.

Haughton, G. and Hunter, C. (1994) *Sustainable Cities*. Regional Policy and Development Series 7. London: Regional Studies Association.

Hayden, D. (1995) *The Power of Place. Urban Landscapes as Public History*. Cambridge, MA and London: MIT Press.

Hillson, J. (1977) *The Battle of Boston: Bussing and the Struggle for School Desegregation*. New York: Pathfinder Press.

Holloway, S.R. (1996) Job accessibility and male teenage employment, 1980–1990: the declining significance of space? *Professional Geographer* 48, 445–57.

Holzer, H.J. (1991) The spatial mismatch hypothesis: what has the evidence shown? *Urban Studies* 28, 105–22.

hooks, b. (1990) *Yearning: Race, Gender, and Cultural Politics*. Boston: South End Press.

House of Representatives Standing Committee for Long Term Strategies (1992) *Patterns of Urban Settlement: Consolidating the Future?* Canberra: Australian Government Publishing Service.

Hoyt, H. (1939) *The Structure and Growth of Residential Neighbourhoods in American Cities*. Washington, DC: Federal Housing Administration.

Hugo, G. (1996) Counterurbanisation. In Newton, P.W. and Bell, M. (eds), *Population Shift: Mobility and Change in Australia*. Canberra: Australian Government Publishing Service, 126–45.

Hugo, G. (1997) Asia and the Pacific on the move: workers and refugees, a challenge to nation states. *Asia Pacific Viewpoint* 38, 267–86.

Hutton, T. (1992) *Sustainable cities: a socio-economic perspective*. Vancouver, BC: Centre for Human Settlement, University of British Columbia.

Hutton, W. (1995) High-risk strategy is not paying off. *Guardian Weekly* 12 November, 13.

ICLEI (1993) *Local Action Plans of the Municipalities in the Urban CO_2 Reduction Project*. Toronto: International Council for Local Environmental Initiatives.

Ihlanfeldt, K.R. and Sjoquist, D.L. (1990) Job accessibility and racial differences in youth unemployment rates. *American Economic Review* 80, 267–76.

IMF (1997) *World Economic Outlook: A Survey by the Staff of IMF*. Washington, DC: International Monetary Fund.

Immergluck, D. (1998) Job proximity and the urban unemployment problem: do suitable nearby jobs improve neighbourhood employment rates? *Urban Studies* 35, 7–23.

Jacobs, J. (1961) *The Death and Life of Great American Cities*. New York: Vintage Books.

Jacobs, J. (1969) *The Economy of Cities*. New York: Random House.

Jacobs, J. (1984) *Cities and the Wealth of Nations: Principles of Economic Life*. New York: Random House.

Jacobs, J.M. (1993) The city unbound: qualitative approaches to the city. *Urban Studies* 30, 827–48.

Jacobs, J.M. (1996) *Edge of Empire: Post-colonialism and the City*. London and New York: Routledge.

Jencks, C. and Peterson, P.E. (eds) (1991) *The Urban Underclass*. Washington, DC: Brookings Institution Press.

Jencks, M., Burton, E. and Williams, K. (eds) (1996) *The Compact City: A Sustainable Form?* London: E&FN Spon.

Jewson, N. and MacGregor, S. (eds) (1997) *Transforming Cities: Contested Governance and New Spatial Divisions*. London and New York: Routledge.

Johnston, R.J., Gregory, D., Pratt, G. and Watts, M. (eds) (1999) *The Dictionary of Human Geography*. Oxford: Blackwell (4th edn).

Judge, D., Stoker, G. and Wolman, H. (eds) (1995) *Theories of Urban Politics*. Thousand Oaks, CA and London: Sage Publications.

Kain, J.F. (1968) Housing segregation, negro employment and metropolitan decentralisation. *Quarterly Journal of Economics* 82, 175–97.

Keeble, D. and Tyler, P. (1995) Enterprising behaviour and the urban–rural shift. *Urban Studies* 32, 975–97.

Keil, R. and Ronneberger, K. (1994) Going up the country: internationalization and urbanization on Frankfurt's northern fringe. *Environment and Planning D: Society and Space* 12, 137–66.

Keil, R. and Ronneberger, K. (2000) The globalization of Frankfurt am Main: core, periphery and social conflict. In Marcuse, P. and van Kempen, R. (eds), *Globalizing Cities: A New Spatial Order?* Oxford and Malden, MA: Blackwell, 228–48.

Kellett, J.R. (1969) *The Impact of the Railways on Victorian Cities.* London: Routledge & Kegan Paul.

King, A.D. (1990a) *Urbanism, Colonialism, and the World Economy.* London and New York: Routledge.

King, A.D. (1990b) *Global Cities: Post-Imperialism and the Internationalisation of London.* London and New York: Routledge.

Kirwan, R. (1980) *The Inner City in the United States.* London: Social Science Research Council.

Knopp, L. (1998) Sexuality and urban space: Gay male identity politics in the United States, the United Kingdom, and Australia. In Fincher, R. and Jacobs, J.M. (eds), *Cities of Difference.* New York and London: The Guilford Press, 148–76.

Knox, P.L. (1982) *Urban Social Geography: An Introduction.* Harlow, Essex: Longman Group.

Knox, P.L. (1992) Suburbia by stealth. *Geographical Magazine* August, 26–9.

Knox, P.L. (1993) *The Restless Urban Landscape.* Englewood Cliffs, NJ: Prentice-Hall.

Knox, P.L. (1994) *Urbanization: An Introduction to Urban Geography.* Englewood Cliffs, NJ: Prentice-Hall.

Knox, P. and Pinch, S. (2000) *Urban Social Geography: An Introduction.* Harlow: Pearson Education (4th edn).

Korteweg, P. and Lie, R. (1992) Prime office locations in the Netherlands. *Tijdschrift voor Economische en Sociale Geografie* 83, 250–62.

Kotkin, J. (1996) Flight to Valhalla. *Guardian Weekly* 24 March, 15.

Kotkin, J. (2000) A city with few pillars looks to its bricks. *New York Times* 23 April, 12.

Krongkaew, M. (1996) The changing urban system in a fast-growing city and economy: the case of Bangkok and Thailand. In Lo, F. and Yeung, Y. (eds) *Emerging World Cities in Pacific Asia.* Tokyo: United Nations University Press, 286–34.

Latham, M. (1996) Making welfare work. *Policy* 12, 18–24.

Latham, M. (1998) *Civilising Global Capital.* St Leonards, NSW: Allen & Unwin.

Lauria, M. (ed.) (1997) *Reconstructing Urban Regime Theory: Regulating Urban Politics in a Global Economy.* Thousand Oaks, CA: Sage Publications.

Lawless, P. (1995) Inner-city and suburban labour markets in a major English conurbation: processes and policy implications. *Urban Studies* 32, 1097–125.

Lawton, R. (1989) Introduction: aspects of the development and role of great cities in the western world.

In Lawton, R. (ed.), *The Rise and Fall of Great Cities.* London and New York: Bellhaven Press, 1–19.

Le Corbusier (1933) *The Radiant City: Elements of a Doctrine of Urbanism to be Used as the Basis of our Machine-Age Civilization.* London: Faber & Faber.

Lee, R. (1996) Moral money? LETS and the social construction of local economic geographies in southeast England. *Environment and Planning A* 28, 1377–94.

Le Gates, R.T. and Hartman, C. (1986) The anatomy of displacement in the United States. In Smith, N. and Williams, P. (eds), *Gentrification of the City.* Boston: Allen & Unwin, 178–203.

Leitmann, J. (1994) The World Bank and the Brown Agenda. *Third World Planning Review* 16, 117–27.

Lemon, A. (ed.) (1991) *Homes Apart: South Africa's Segregated Cities.* Cape Town: David Philip.

Lever, W.F. and Turok, I. (1999) Competitive cities: introduction to the review. *Urban Studies* 36, 791–93.

Lewine, E. (1997) Is downtown dead? *New York Times* 25 May, Section 13.

Ley, D. (1983) *A Social Geography of the City.* New York: Harper & Row.

Ley, D. (1996) *The New Middle Class and the Remaking of the Central City.* Oxford: Oxford University Press.

Leyshon, A. and Thrift, N. (1997) *Money/Space. Geographies of Monetary Transformation.* London and New York: Routledge.

Lloyd, J. (1999) Falling out. *Prospect* 45, 22–7.

Lo, F. and Yeung, Y. (eds) (1996) *Emerging World Cities in Pacific Asia.* Tokyo: United Nations University Press.

Logan, J.R. and Molotch, H.L. (1987) *Urban Fortunes: The Political Economy of Place.* Berkeley: University of California Press.

Lösch, A. (1941) *The Economics of Location.* Translated by Woglom, W.H. and Stolper, W.F. in 1954. New Haven: Yale University Press.

Lovering, J. (1990) Fordism's unknown successor: a comment on Scott's theory of flexible accumulation and the re-emergence of regional economies. *International Journal of Urban and Regional Research* 14, 159–74.

Lyons, M. (1996) Employment, feminisation, and gentrification in London, 1981–93. *Environment and Planning A* 28, 341–56.

McDermott, P. (1998) Review of 'Theories of urban politics'. *Urban Policy and Research* 16, 166–9.

McDowell, L. (1997) *Capital Culture: Gender of Work in the City.* Oxford and Malden, MA: Blackwell.

McGranahan, G. and Songsore, J. (1994) Wealth, health and the urban household: weighing environmental burdens in Accra, Jakarta and São Paulo. *Environment* 36, 4–11 and 40–5.

McKenzie, E. (1994) *Privatopia: Homeowner Associations and the Rise of Residential Private Government*. New Haven, CT: Yale University Press.

McLafferty, S. and Preston, V. (1996) Spatial mismatch and employment in a decade of restructuring. *Professional Geographer* 48, 420–30.

Maclennan, D. and Pryce, G. (1996) Global economic change, labour market adjustment and the challenges for European housing policies. *Urban Studies* 33, 1849–65.

Macleod, S. and McGee. T.G. (1996) The Singapore–Jahore–Riau growth triangle: an emerging extended metropolitan region. In Lo, F. and Yeung, Y. (eds), *Emerging World Cities in Pacific Asia*. Tokyo: United Nations Press, 417–64.

Maher, C.A. and Burke, T. (1991) *Informed Decision-making: The Use of Secondary Data Sources in Policy Studies*. Melbourne: Longman Cheshire.

Marcuse, P. (1996) *Is Australia Different? Globalisation and the New Urban Poverty*. Melbourne: Australian Housing and Urban Research Institute.

Marcuse, P. and van Kempen, R. (2000) *Globalizing Cities. A New Spatial Order?* Oxford and Malden, MA: Blackwell Publishers.

Markusen, A. (1996) Sticky places in slippery space: a typology of industrial districts. *Economic Geography* 72, 293–313.

Markusen, A. (1999) Fuzzy concepts, scanty evidence, policy distance: the case for rigour and policy relevance in critical regional studies. *Regional Studies* 33, 869–84.

Markusen, A. and Gwiasda, V. (1994) Multipolarity and the layering of functions in world cities: New York City's struggle to stay on top. *International Journal of Urban and Regional Research* 18, 168–93.

Markusen, A., Lee, Y.S. and Digiovanna, S. (eds) (1999) *Second Tier Cities: Explaining Rapid Growth beyond the Metropole*. Minneapolis: University of Minnesota Press.

Marsh, I. and Levy, S. (1998) *Sydney Gay and Lesbian Mardi Gras: Economic Impact Statement*. Sydney: SGLMG Ltd.

Marshall, A. (1920) *Industry and Trade: A Study of Industrial Technique and Business Organization: Their Influences on the Condition of Various Classes and Nations*, 3rd edn. London: Macmillan.

Massey, D. (1984) *Spatial Division of Labour*. London: Macmillan.

Massey, D. (1997) Space/power, identity/difference: tensions in the city. In Merrifield, A. and Swyngedouw, E. (eds), *The Urbanization of Injustice*. New York: New York University Press, 100–16.

Massey, D., Allen, J. and Pile, S (eds) (1999) *Understanding Cities*. London: Routledge.

Massey, D. and Jess, P. (1995) *A Place in the World? Places, Cultures and Globalization*. Oxford: Oxford University Press.

Massey, D. and Meegan, R. (1982) *The Anatomy of Job Loss: The How, Why and Where of Employment Decline*. London: Methuen.

Mawson, J. and Hall, S. (2000) Joining it up locally? Area regeneration and holistic government in England. *Regional Studies* 34, 67–79.

Meijer, M. (1993) Growth and decline of European cities: changing positions in Europe. *Urban Studies* 30, 981–90.

Mera, K. (1998) Global forces and the city region. *Australian Planner* 35, 180–4.

Merrifield, A. and Swyngedouw, E. (eds) (1997) *The Urbanization of Injustice*. New York: New York University Press.

Miliband, R. (1969) *The State in Capitalist Society*. London: Weidenfeld and Nicolson.

Mills, E.S. (1972) *Studies in the Structure of the Urban Economy*. Baltimore: Resources for the Future Inc and Johns Hopkins Press.

Mitchell, W.J. (1999) *E-topia: Urban Life, Jim – but not as we know it*. Cambridge, MA: MIT Press.

Molotch, H. (1976) The city as growth machine. *American Journal of Sociology* 82, 309–55.

Molotch, H. and Logan, J. (1987) *Urban Fortunes. The Political Economy of Place*. Berkeley, CA: University of California Press.

Monbiot, G. (2000) Devil take the hindmost. *Guardian Weekly* 30 March–5 April, 13.

Mooney, G. and Danson, M. (1997) Beyond 'Culture City': Glasgow as a 'dual city'. In Jewson, N. and MacGregor, S. (eds), *Transforming Cities: Contested Governance and New Spatial Divisions*. London and New York: Routledge, 73–86.

Morris, A.E.J. (1974) *History of Urban Form: Before the Industrial Revolution*. New York: John Wiley & Sons.

Morrison, P.S. (2001) Unemployment and urban labour markets. *Urban Studies* 38.

Moses, L.N. and Williamson, H.F. (1967) The location of economic activity in cities. *American Economic Review, Papers and Proceedings* May, 211–22.

Moss, M. (1991) The information city in the global economy. In Brotchie, J., Batty, M., Hall, P. and Newton, P. (eds), *Cities of the 21st Century: New Technologies and Spatial Systems*. New York: Longman Cheshire, 181–92.

Moulaert, F. and Scott, A.J. (eds) (1997) *Cities, Enterprises and Society on the Eve of the 21st Century*. London and Washington: Pinter.

Mounfield, P.R. (1984) The deindustrialisation and reindustrialisation of the UK. *Geography* 69, 141–6.

Muller, P.O. (1997) The suburban transformation of the globalizing American city. *Annals of the American Academy of Political and Social Science* 551, 44–58.

Mumford, L. (1961) *The City in History.* London: Penguin.

Murphey, R. (1984) City as a mirror of society: China, traditional and transformation. In Agnew, J., Mercer, J. and Sopher, D. (eds), *The City in Cultural Context.* Boston: Allen & Unwin, 188–204.

Murphy, P. and Watson, S. (1997) *Surface City. Sydney at the Millennium.* Annandale, NSW: Pluto Press.

Murray, C.A. (1984) *Losing Ground: American Social Policy 1950–1980.* New York: Basic Books.

National Housing Strategy (1991) *The Findings of the Housing and Locational Choice Survey: An Overview.* Background Paper 11. Canberra: Australian Government Publishing Service.

National Housing Strategy (1992) *Housing Location and Access to Services.* Issues Paper No. 5. Canberra: Australian Government Publishing Service.

Neutze, M. (1965) *Economic Policy and the Size of Cities.* Canberra: Australian National University Press.

Newman, P.W.G. (1999) Sustainability and Australian cities. *Australian Planner* 36, 93–100.

Newman, P.W.G. and Kenworthy, J.R. (1999) *Sustainability and Cities: Overcoming Automobile Dependence.* Washington, DC: Island Press.

Newton, P.W. and Bell, M. (eds) (1996) *Population Shift: Mobility and Change in Australia.* Canberra: Australian Government Publishing Service.

Newton, P.W., Brotchie, J.F. and Gibbs, P.G. (1997) *Cities in Transition: Changing Economic and Technological Processes and Australia's Settlement System.* Highett, Vic.: CSIRO.

Nijman, J. (1996) Ethnicity, class, and the economic internationalisation of Miami. In O'Loughlin, J. and Friedrichs, J. (eds), *Social Polarization in Post-Industrial Metropolises.* Berlin and New York: Walter de Gruyter, 283–303.

North, P. (1999) Explorations in heterotopia: LETS and the micropolitics of money and livelihood. *Environment and Planning D: Society and Space* 17, 69–86.

Norton, R.D. (1992) Agglomeration and competitiveness: from Marshall to Chinitz. *Urban Studies* 29, 155–70.

Nourse, H.O. (1968) *Regional Economics: A Study of the Economic Structure, Stability and Growth of Regions.* New York: McGraw-Hill.

O'Brien, R. (1992) *Global Financial Integration: The End of Geography.* London: Royal Institute of International Affairs.

O'Connor, K. and Stimson, R.J. (1997) Population and business in metropolitan and non-metropolitan Australia: the experience of the last decade. *People and Place* 5, 40–9.

O'Donoghue, D. (1999) The relationship between diversification and growth: some evidence from the British urban system 1978 to 1991. *International Journal of Urban and Regional Research* 23, 549–66.

OECD (1996) *Report on Urban Regeneration.* Paris: Organisation for Economic Cooperation and Development.

Office for National Statistics (1998) *Social Trends 28. 1998 Edition.* London: The Stationery Office.

Ohmae, K. (1995) *The End of the Nation State.* New York: Harper Collins.

O'Loughlin, J. and Friedrichs, J. (eds) (1996) *Social Polarization in Post-Industrial Metropolises.* Berlin and New York: Walter de Gruyter.

Olds, K. (1997) Globalizing Shanghai: the 'global intelligence corps' and the building of Pudong. *Cities* 14, 1–15.

Ong, P. and Blumenberg, E. (1996) Income and racial inequality in Los Angeles. In Scott, A.J. and Soja, E.W. (eds), *The City: Los Angeles and Urban Theory at the End of the Twentieth Century.* Berkeley, CA: University of California Press, 311–35.

O'Rourke, K.H. and Williamson, J.G. (1999) *Globalisation and History: The Evolution of a Nineteenth-Century Atlantic Economy.* Cambridge, MA: MIT Press.

Osborne, D. and Gaebler, T. (1992) *Reinventing Government: How the Entrepreneurial Spirit is Transforming the Public Sector.* Reading, MA: Addison-Wesley.

Özüekren, S. and van Kempen, R. (eds) (1997) *Turks in European Cities: Housing and Urban Segregation.* Utrecht: European Research Centre on Migration and Ethnic Relations.

Pacione, M. (ed.) (1997) *Britain's Cities: Geographies of Division in Urban Britain.* London: Routledge.

Pahl, R.E. (1975) *Whose City? And Further Essays on Urban Society.* Harmondsworth: Penguin Books.

Park, R.E. (1925) The city: suggestions for the investigation of human behavior in the urban environment. In Park, R.E., Burgess, E.W. and McKenzie, R.D. (eds), *The City.* Chicago: Chicago University Press, 1–46.

Park, R.E., Burgess, E.W. and McKenzie, R.D. (eds) (1925) *The City.* Chicago: Chicago University Press.

Parker, J. (1995) Turn up the lights: a survey of cities. *The Economist* 29 July.

Parliamentary Commissioner for the Environment (2000) *Ageing Pipes and Murky Waters: Urban Water System Issues for the 21st Century.* Wellington: Government Printer.

Peach, C. (1975) *Urban Social Segregation.* London and New York: Longman.

Peach, C. (1996) Good segregation, bad segregation. *Planning Perspectives* 11, 379–98.

Peck, J. (1995) Moving and shaking: business elites, state localism and urban privatism. *Progress in Human Geography* 19, 16–46.

Peck, J. and Tickell, A. (1992) Local modes of social regulation? Regulation theory, Thatcherism and uneven development. *Geoforum* 23, 347–64.

Peterson, P. (1981) *City Limits*. Chicago: University of Chicago Press.

Piore, M.J. and Sabel, C.F. (1984) *The Second Industrial Divide: Possibilities for Prosperity*. New York: Basic Books.

Pirenne, H. (1948) *Medieval Cities: Their Origins and the Revival of Trade*. Princeton, NJ: Princeton University Press (revised English edn).

Pivo, G. (1990) The net of mixed beads. Suburban office development in six metropolitan regions. *American Planners Association Journal* 56, 457–69.

Policy Studies Institute (1997) *Ethnic Minorities in Britain: Diversity and Disadvantage*. London: PSI.

Pollard, J. and Storper, M. (1996) A tale of twelve cities: metropolitan employment change in dynamic industries in the 1980s. *Economic Geography* 72, 1–22.

Porter, M.E. (1990) *The Competitive Advantage of Nations*. New York: Free Press.

Powell, D. (1993) *Out West: Perceptions of Sydney's Western Suburbs*. Sydney: Allen & Unwin.

Power, A. and Mumford, K. (1999) *The Slow Death of Great Cities? Urban Abandonment or Urban Renaissance*. York: Joseph Rowntree Foundation.

Pred, A.R. (1966) *The Spatial Dynamics of US Urban-industrial Growth, 1800–1914: Interpretive and Theoretical Essays*. Cambridge, MA: MIT Press.

Price, K.A. (1986) *The Global Financial Village*. London: Banking World.

Putnam, R. (1995) Tuning in, tuning out: the strange disappearance of social capital in America. *Political Science and Politics* 28, 664–83.

Raban, J. (1974) *Soft City*. London: Hamish Hamilton.

Rees, P.H. (1971) Factorial ecology: an extended definition survey, and criteria of the field. *Economic Geography* 47, 220–30.

Rees, W. (1992) Ecological footprints and appropriated carrying capacity: what urban economics leaves out. *Environment and Urbanization* 4, 121–30.

Register, R. (1987) *Eco-city Berkeley: Building Cities for a Healthy Future*. Berkeley: North Atlantic Books.

Reich, R.B. (1993) *The Work of Nations: Preparing ourselves for 21st Century Capitalism*. New York: Simon & Schuster.

Reich, R.B. (1998) *Locked in the Cabinet*. New York: Random House.

Reid, H. (1993) Demand for housing types in the Sydney region. *Urban Futures Journal* 3, 33–7.

Reilly, W.J. (1931) *The Law of Retail Gravitation*. New York: Reilly.

Richards, L. (1990) *Nobody's Home: Dreams and Realities in a New Suburb*. Melbourne: Oxford University Press.

Rifkin, J. (2000) *The End of Work: The Decline of the Global Workforce and the Dawn of the Post-Market Era*. Harmondsworth: Penguin (2nd edn).

Robinson, V. (1996) *Transients, Settlers and Refugees: Asians in Britain*. Oxford: Clarendon Press.

Robson, B.T. (1969) *Urban Analysis: A Study of City Structure with Special Reference to Sunderland*. Cambridge: Cambridge University Press.

Robson, B.T. (1975) *Urban Social Areas*. London: Oxford University Press.

Rogers, R. (1997) *Cities for a Small Planet*. London: Faber & Faber.

Rohe, W.M. and Kleit, R.G. (1997) From dependency to self-sufficiency: an appraisal of the Gateway transitional families program. *Housing Policy Debate* 8, 75–108.

Rollison, P. (1990) The everyday geography of poor elderly hotel tenants in Chicago. *Geografiska Annäler* HG 72B, 47–57.

Rose, D. (1988) A feminist perspective on employment restructuring and gentrification: the case of Montreal. In Wolch, J. and Dear, M. (eds), *The Power of Geography: How Territory Shapes Social Life*. Winchester, MA: Unwin Hyman, 118–38.

Roseland, M. (1997) Dimensions of the eco-city. *Cities* 14, 197–202.

Rosenbaum, J.E. (1995) Changing the geography of opportunity by expanding residential choice lessons from the Gautreaux program. *Housing Policy Debate* 6, 231–68.

Ross, A. (1999) *The Celebration Chronicles – Life, Liberty and the Pursuit of Property Value in Disney's New Town*. New York: Ballantine Books.

Saegert, S. (1980) Masculine cities and feminine suburbs: polarised ideas, contradictory realities. *Signs. Journal of Women in Culture and Society* 5, 96–111.

Sandercock, L. (1998) *Towards Cosmopolis: Planning for Multi-cultural Cities*. Chichester: John Wiley & Sons.

Sandercock, L. and Friedmann, J. (2000) Strategising the metropolis in a global era. *Urban Policy and Research* 18, 529–34.

Sassen, S. (1988) *The Mobility of Labour and Capital: A Study in International Investment and Labour Flow*. Cambridge and New York: Cambridge University Press.

Sassen, S. (1991) *The Global City: New York, London, Tokyo*. Princeton: Princeton University Press.

Sassen, S. (1997) New employment regimes in cities. In Moulaert, F. and Scott, A.J. (eds), *Cities, Enterprises and Society on the Eve of the 21st Century*. London and Washington: Pinter, 129–50.

Saul, J.R. (1997) *The Unconscious Civilization*. Ringwood, Vic.: Penguin.

Saunders, P. (1990) *A Nation of Home Owners*. London: Unwin Hyman.

Schneider, W. (1992) The suburban century begins. *The Atlantic Monthly* July, 33–44.

Scott, A.J. (1988) Flexible production systems and regional development: the rise of new industrial spaces in North America and western Europe. *International Journal of Urban and Regional Research* 12, 171–86.

Scott, A.J. and Soja, E.W. (eds) (1996) *The City: Los Angeles and Urban Theory at the End of the Twentieth Century*. Berkeley, CA: University of California Press.

Seabrook, J. (1996) *In the Cities of the South: Scenes from a Developing World*. London and New York: Verso.

Sennett, R. (ed.) (1969) *Classic Essays on the Culture of Cities*. New York: Appleton-Century Crofts.

Setchell, C. (1995) The growing environmental crisis in the world's mega-cities: the case of Bangkok. *Third World Planning Review* 17, 1–18.

Sforzi, F. (1987) Riflessionisul distretto industriale: un'ipotesi di identificazione spaziale. In Becattini, G. (ed.), *Mercato e forze locali: il distretto industriale*. Bologna: Il Mulino.

Shen, Q. (1998) Location characteristics of inner-city neighbourhoods and employment accessibility of low-wage workers. *Environment and Planning D: Society and Space* 16, 345–65.

Shevky, E. and Bell, W. (1955) *Social Area Analysis: Theory, Illustrative Application and Computational Procedures*. San Francisco: Stanford University Press.

Sjoberg, G. (1960) *The Preindustrial City: Past and Present*. Glencoe, IL: Free Press.

Skinner, G.W. (1964) Marketing and social structure in rural China. *Journal of Asian Studies* 34.

Smailes, A.E. (1947) The analysis and delimitation of urban fields. *Geography* 32, 151–64.

Smith, M.P. (1988) *City, State and Market: The Political Economy of Urban Society*. Cambridge, MA and Oxford: Blackwell.

Smith, N. (1979) Towards a theory of gentrification: a back to the city movement by capital not people. *American Planning Association Journal* 45, 538–48.

Smith, N. (1996) *The New Urban Frontier: Gentrification and the Revanchist City*. London and New York: Routledge.

Smith, R.H.T. (1965) Method and purpose in functional town classification. *Annals of the Association of American Geographers* 55, 539–48.

Smith, S.J. (1999) Society-space. In Cloke, P., Crang, P. and Goodwin, M. (eds), *Introducing Human Geographies*. London: Arnold, 12–23.

Smith, N. and Williams, P. (eds) (1986) *Gentrification of the City*. Boston: Allen & Unwin.

Soja, E.W. (2000) *Postmetropolis: Critical Studies of Cities and Regions*. Malden, MA and Oxford: Blackwell.

Sommer, J. and Hicks, D.A. (eds) (1993) *Rediscovering Urban America: Perspectives on the 1980s*. Washington, DC: Department of Housing and Urban Development.

Spayde, J. (1997) Ithaca, NY: Our kind of town. *UTNE Reader* May–June, 45–8.

Stacey, J. (1990) *Brave New Families: Stories of Domestic Upheaval in Late Twentieth Century America*. New York: Basic Books.

Stalker, P. (2000) *Workers without Frontiers – the Impact of Globalisation on International Migration*. Geneva: International Labour Organisation.

Statistics New Zealand (1998a) *New Zealand Now: Incomes*. Wellington: Statistics NZ.

Statistics New Zealand (1998b) *New Zealand Now: Housing*. Wellington: Statistics NZ.

Stedman Jones, G. (1971) *Outcast London: A Study in the Relationship between Classes in Victorian Society*. Oxford: Clarendon Press.

Stilwell, F. (1992) *Understanding Cities and Regions*. Sydney: Pluto Press.

Stilwell, F. (1997) *Globalisation and Cities: An Australian Political-Economic Perspective*. Urban Research Program, Working Paper No. 49, ANU, Canberra.

Stilwell, F. (2000) Ways of seeing: Competing perspectives on urban problems and policies. In Troy, P. (ed.), *Equity, Environment and Efficiency: Ethics and Economics in Urban Australia*. Carlton, Vic.: Melbourne University Press, 13–37.

Stoker, G. (1995) Regime theory and urban politics. In Judge, D., Stoker, G. and Wolman, H. (eds), *Theories of Urban Politics*. Thousand Oaks, CA and London: Sage Publications, 54–71.

Stone, C. (1989) *Regime Politics: Governing Atlanta, 1946–1988*. Lawrence: University of Kansas Press.

Stopford, J. and Strange, S. (1991) *Rival States, Rival Firms: Competition for World Market Share*. Cambridge: Cambridge University Press.

Storper, M. and Walker, R. (1989) *The Capitalist Imperative: Territory, Technology, and Industrial Growth*. New York: Basil Blackwell.

Stretton, H. (1978) *Urban Planning in Rich and Poor Countries*. Oxford: Oxford University Press.

Suttles, G. (1972) *The Social Construction of Communities*. Chicago: University of Chicago Press.

Szelenyi, I. (1978) Ecological change and residential mix in Adelaide. In Urlich-Cloher, D.U. and Badcock, B. (eds), *Proceedings of RGS Symposium in Residential Mix in Adelaide*. Adelaide: Royal Geographical Society of Australasia, 15–18.

Taylor, P. (1999) Globalization and World Cities (GaWC). www.lboro.ac.uk/departments/gy/research/gawc.htm

Theobald, R. (1997) *Reworking Success: New Communities at the Millennium*. Gariola Island, BC: New Society Publishers.

Thomas, B. (1972) *Migration and Urban Development: A Reappraisal of British and American Long Cycles*. London: Methuen.

Thompson, W.R. (1965) *A Preface to Urban Economics*. Baltimore: Johns Hopkins University Press for Resources for the Future Inc.

Thornley, A. (1991) *Urban Planning under Thatcherism: The Challenge of the Market*. London: Routledge.

Thrift, N. (1994) Globalisation, regulation, urbanisation: the case of The Netherlands. *Urban Studies* 31, 365–80.

Thrift, N. and Olds, K. (1996) Refiguring the economic in economic geography. *Progress in Human Geography* 20, 311–37.

Thurow, L.C. (1999) Building wealth. *The Atlantic Monthly* 283, 57–69.

Tiebout, C.M. (1956) The urban economic base reconsidered. *Land Economics* February, 95–9.

Timms, D.W.G. (1971) *The Urban Mosaic: Towards a Theory of Residential Differentiation*. Cambridge: Cambridge University Press.

Tönnies, F. (1887) *Community and Association (Gemeinschaft und Gesellschaft)*. Translated by C.P. Lomis in 1955. London: Routledge & Kegan Paul.

Toynbee, A. (1970) *Cities on the Move*. Oxford: Oxford University Press.

Troy, P. (1992) The new feudalism. *Urban Futures Journal* 2, 36–44.

Troy, P. (1996) *The Perils of Urban Consolidation*. Annandale, NSW: Federation Press.

Troy, P. (2000) The direction of urban planning. Paper presented to the New Zealand Planning Institute Conference, Christchurch 20 May 2000.

Turok, I. (1999) Urban labour markets: the causes and consequence of change. *Urban Studies* 36, 893–915.

Turok, I. and Edge, N. (1999) The jobs gap in Britain's cities. *Network* 2, 1.

United Nations (1996) *An Urbanizing World: Global Report on Human Settlements, 1996*. Oxford: Oxford University Press for United Nations Centre for Human Settlements (HABITAT).

Urban Task Force (1999) *Urban Renaissance: Sharing the Vision: Summary of Responses to the Urban Task Force Prospectus*. London: Urban Task Force.

US Bureau of Labor Statistics (1996) *Labor Market Review*. Washington, DC: US Government Printing Office.

US Department of Housing and Urban Development (1995) *HUD's Reinvention: From Blueprint to Action. Summary*. Washington, DC: Office of Policy Development and Research.

US Department of Housing and Urban Development (1996) *Public Housing in a Competitive Market: An Example of How it Would Fare*. Washington, DC: Office of Policy Development and Research.

US Office of Technology Assessment (1995) *The Technological Reshaping of Metropolitan America*. Washington, DC: US Government Printing Office.

Vance, J.E. (1960) Labour shed, employment field and dynamic analysis in urban geography. *Economic Geography* 36, 189–220.

Vance, J.E. (1971) Land assignment in the precapitalist, capitalist and postcapitalist city. *Economic Geography* 47, 101–20.

van Grunsven, L. (2000) Singapore: the changing residential landscape in a winner city. In Marcuse, P. and van Kempen, R. (eds), *Globalizing Cities: A New Spatial Order?* Oxford and Malden, MA: Blackwell, 95–126.

van Kempen, R. and Özüekren, A.S. (1998) Ethnic segregation in cities: new forms and explanations in a dynamic world. *Urban Studies* 35, 1631–56.

van Weesep, J. and Musterd, S. (1991) *Urban Housing for the Better-off: Gentrification in Europe*. Utrecht: Stedelijke Netwerken.

Wackernagel, M. and Rees, W. (1996) *Our Ecological Footprint: Reducing Human Impact on the Earth*. Gabriola Island, BC: New Society Publishers.

Wacquant, L.J.D. (1997) Three pernicious premises in the study of the American ghetto. *International Journal of Urban and Regional Research* 21, 341–62.

Wallerstein, I. (1974) *The Modern World System*. New York: Academic Press.

Walmsley, D.J. (1988) *Urban Living: The Individual in the City*. Essex: Longman Group.

Warde, A. (1991) Gentrification as consumption: issues of class and gender. *Environment and Planning D: Society and Space* 9, 223–32.

Warf, B. (1990) The reconstruction of social ecology and neighbourhood change in Brooklyn. *Environment and Planning D: Society and Space* 8, 97–121.

Waring, M. (1988) *Counting for Nothing: What Men Value and what Women are Worth*. Auckland: Allen & Unwin.

Warner, S.B. (1962) *Street-car Suburbs: the Process of Growth in Boston, 1870–1900*. Cambridge, MA: MIT Press.

Watson, S. and Gibson, K. (1995) *Postmodern Cities and Spaces*. Oxford and Cambridge, MA: Blackwell.

Wekerle, G.R. (1984) A woman's place is in the city. *Antipode* 16, 11–19.

West, C. (1990) The new cultural politics of difference. In Fergusson, R. (ed.), *Out There: Marginalization and Contemporary Culture*. Cambridge, MA: MIT Press.

Wheatley, P. (1971) *The Pivot of the Four Quarters*. Chicago: Aldine.

Whitt, J.A. (1982) *Urban Elites and Mass Transportation: The Dialectics of Power*. Princeton, NJ: Princeton University Press.

Williams, J.J. (2000) South Africa: Urban transformation. *Cities* 17, 167–83.

Wilson, J. (1999) Jobs to prompt exodus from north to south. *Guardian Weekly* 19–25 August, 7.

Wilson, W. (1990) Residential relocation and settlement adjustment of Vietnamese refugees in Sydney. *Australian Geographical Studies* 28, 155–77.

Wilson, W.J. (1987) *The Truly Disadvantaged: The Inner City, the Underclass, and Public Policy*. Chicago: Chicago University Press.

Wilson, W.J. (1996) *When Work Disappears*. New York: Alfred A. Knopf.

Wirth, L. (1938) Urbanism as a way of life. *American Journal of Sociology* 44, 1–24.

Wohl, A.S. (1977) *The Eternal Slum*. London: Edward Arnold.

Wolch, J. (1996) From global to local: The rise of homelessness in Los Angeles during the 1980s. In Scott, A.J. and Soja, E.W. (eds), *The City. Los Angeles and Urban Theory at the End of the Twentieth Century*. Berkeley, CA: University of California Press, 390–425.

Wolch, J. and Dear, M. (eds) (1988) *The Power of Geography: How Territory Shapes Social Life*. Winchester, MA: Unwin Hyman.

Woodley, B. (1998) Homie truths. *The Australian Magazine* 17–18 January, 11–15.

World Commission on Environment and Development (1987) *Our Common Future* (the Brundtland Report). Oxford: Oxford University Press.

Wyly, E.K. (1998) Containment and mismatch: gender differences in commuting in metropolitan labour markets. *Urban Geography* 19, 395–430.

Yates, J. (2000) Is Australia's home ownership rate really stable? An examination of change between 1975 and 1994. *Urban Studies* 37, 319–42.

Young, M. (1990) *Justice and the Politics of Difference*. Princeton, NJ: Princeton University Press.

Young, M. and Wilmott, P. (1957) *Family and Kinship in East London*. London: Routledge & Kegan Paul.

Zipf, G.K. (1949) *Human Behavior and the Principle of Least Effort*. Reading, MA: Addison-Wesley.

Zorbaugh, H. (1929) *The Gold Coast and the Slum*. Chicago: University of Chicago Press.

Zukin, S. (1982) *Loft Living: Culture and Capital in Urban Change*. Baltimore, MD: Johns Hopkins University Press.

Zukin, S. (1995) *The Culture of Cities*. Cambridge, MA: Blackwell.

Zukin, S. (1997) Cultural strategies of economic development and the hegemony of vision. In Merrifield, A. and Swyngedouw, E. (eds), *The Urbanization of Injustice*. New York: New York University Press, 100–16.

Zukin, S. (1998) Urban lifestyles: diversity and standardisation in spaces of consumption. *Urban Studies* 35, 825–40.

INDEX

Milton Keynes UK
Ingram Content Group UK Ltd.
UKHW051856071024
449327UK00025B/1981